Bruce Eure

Management Systems
Conceptual Considerations

D1365287

Management Systems
Conceptual Considerations

PETER P. SCHODERBEK, Ph.D.
The University of Iowa

ASTERIOS G. KEFALAS, Ph.D.
University of Georgia

CHARLES G. SCHODERBEK, Ph.D.
Northern Illinois University

1975

BUSINESS PUBLICATIONS, INC. *Dallas, Texas 75243*
Irwin-Dorsey Limited   Georgetown, Ontario   L7G 4B3

First Printing, April 1975
Second Printing, November 1976

ISBN 0-256-01575-9
Library of Congress Catalog Card No. 74–24461
Printed in the United States of America

All the time the Guard was looking for her, first through a telescope, then through a microscope, and then through an opera-glass. At last he said, "You are travelling the wrong way . . ."

Through the Looking-Glass
Lewis Carroll

Systems are components in hierarchies. In another sense, hierarchies are systems and each system is itself a hierarchy.

ERVIN LASZLO

Foreword

A NOTE ON SYSTEMS SCIENCE*

World War II marked the end of an era of Western culture that began with the Renaissance, the Machine Age, and the beginning of a new era, the Systems Age.

In the Machine Age man sought to take the world apart, to analyze its contents and our experiences of them down to ultimate indivisible parts: atoms, chemical elements, cells, instincts, elementary perceptions, and so on. These elements were taken to be related by causal laws, laws which made the world behave like a machine. This mechanistic concept of the world left no place in science for the study of free will, goal seeking, and purposes. Such concepts were either taken to be meaningless or were relegated to the realm of pure speculation, metaphysics.

It was natural for men who believed (1) the world to be a machine that God had created to serve his purposes, and (2) that man was created in His image, to seek to develop machines that would do man's work. Man succeeded and brought about mechanization, the replacement of man by machine as a source of physical work.

Work itself was broken down into its smallest elements. These were assigned to machines and men, and assembled into the modern production line. Productivity increased and work was dehumanized. The process which replaced man by machine reduced man to behaving like a machine—to performing simple, dull, repetitive tasks.

* By Russell L. Ackoff who is Professor of Systems Sciences in The Wharton School of Finance and Commerce, University of Pennsylvania. He was Editor of *Management Science* from 1965 to 1970, and is now on the Advisory Board of the *Mathematical Spectrum,* on the Editorial Board of *Management Decision* and is Advisory Editor in Management Sciences for John Wiley & Sons. He is coauthor of more than 10 books and is the author of more than 100 articles in a variety of journals and books. Copyright © 1972, The Institute of Management Sciences.

With World War II we began to shift into the Systems Age. A system is a whole that cannot be taken apart without loss of its essential characteristics, and hence it must be studied as a whole. Now, instead of explaining a whole in terms of its parts, parts began to be explained in terms of the whole. Therefore, things to be explained are viewed as parts of larger wholes rather than as wholes to be taken apart. Furthermore, nonmechanistic ways of viewing the world were developed which were compatible with the older mechanistic view and which made it possible to deal with free will, goal seeking, and purposes within the framework of science. Instead of thinking of men in machine-like terms we began to think of machines in man-like terms.

The Systems Age brought with it the Post Industrial Revolution. This very young revolution is based on machines that can observe (generate data), communicate it, and manipulate it logically. Such machines make it possible to mechanize mental work, to automate.

In the Machine Age science not only took the world apart, but it took itself apart, dividing itself into narrower and narrower disciplines. Each discipline represented a different way of looking at the same world. Shortly before World War II science began to put itself back together again so that it could study phenomena as a whole, from all points of view. As a result, a host of new interdisciplines emerged such as Operations Research, Cybernetics, Systems Engineering, Communications Sciences, and Environmental Sciences. Unlike earlier scientific disciplines which sought to separate themselves from each other and to subdivide; the new interdisciplines seek to enlarge themselves, to combine to take into account more and more aspects of reality. Systems Science is the limit of this process, an amalgamation of all the parts of science into an integrated whole. Thus, Systems Science is not a science, but is science taken as a whole and applied to the study of wholes.

Systems Science goes even one step further; it denies the value of the separation of science and the humanities. It views these as two sides of the same coin; they can be viewed and discussed separately, but cannot be separated. Science is conceived as the search for similarities among things that appear to be different; the humanities as the search for differences among things that appear to be the same. Both are necessary. For example, to solve a problem we need to know both (1) in what respects it is similar to problems already solved so that we can use what we have already learned; and (2) in what respects it differs from any problem yet solved so that we can determine what we must yet learn. Thus the humanities have the function of identifying problems to be solved, and science has the function of solving them.

The emergence of Systems Science does not constitute a rejection of traditional scientific and humanistic disciplines. It supplements them with a new way of thinking that is better suited than they to deal with

large-scale societal problems. It offers us some hope of dealing successfully with such problems as poverty, racial and other types of discrimination, crime, environmental deterioration, and underdevelop-
ment of countries. Systems Science may not only be able to assure man
of a future, but it may also enable him to gain control of it.

Preface

While probing the historical antecedents of systems thinking, future scholars will, no doubt, uncover its roots in the fertile soil of the present. Only the latter half of this century, however, will they characterize as *the* Age of Systems. Probably most often to be cited in their studies will be the treatment of organizations and similar social phenomena not as detached parts but as integral wholes. Underlying this shift in emphasis would be the belief that only in such a holistic framework could a system's essential elements be realistically understood. This shift from part-time systems or cause-effect investigation of a mechanistic type to the Gestalt-like technique of systems thinking will have revealed to its proponents startlingly new organizational properties that the atomistic mechanistic approach had previously failed to disclose.

Unlike other intellectual movements, sprung from a specific discipline and nurtured within restrictive and narrow confines, systems thinking was born free of particularized scientific fetters and reared in an interdisciplinary environment. Because it deals with wholes in general and not with specific parts, it transcends the usual strictly defined disciplinary boundaries of the traditional sciences. It has indeed become an interdisciplinary movement.

Perhaps systems thinking has had its greatest impact in the area of human organization. Most modern writers of organizational theory seem to prefer the systems approach to other fragmented approaches. While firmly convinced of the relevancy of systems thinking for organizational management, the present authors see their real task as one of alerting and exposing management, present and future, to the profound intellectual changes of the last few years in the managerial climate. Familiarity with the systems approach has become a *conditio sine qua non* for understanding modern management thinking.

This book is organized around five logically related conceptual areas.

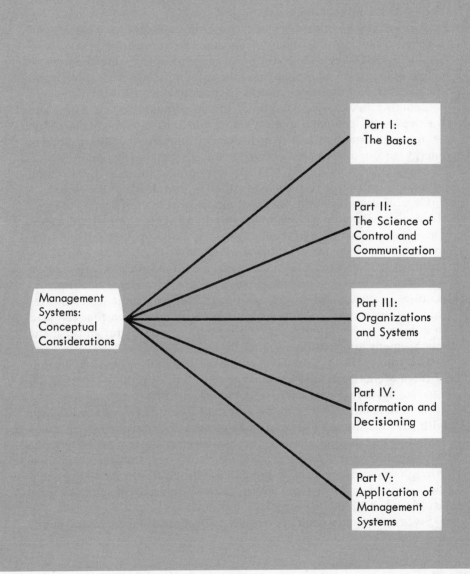

In Part One, "The Basics," the reader is given an initially brief overview of systems thinking, its origins and its development from the seminal ideas of biologists to the developed and developing concepts of its present-day general practitioners. The inquisitive student, while not encumbered with an extensively detailed account of the evolutionary history of systems thinking, will find here enough information to begin his own diggings in this new and exciting field of investigation.

To appreciate what he is digging for, the reader will find useful the presentation of the ABCs of systems, the various concepts needed to understand what a system really is—its inputs, outputs, environment, and its behavioral characteristics.

Part Two treats in greater detail one of the several aspects of systems touched upon in the opening section—self-regulation. Since self-regulation involves *both* control *and* communication, the chapters on cybernetics and on communication theory complement one another and highlight one of the core concepts of modern systems thinking.

Zooming in as it does on human organizations, Part Three provides the needed conceptual framework from which to view the organization as an open system, one in which there is a continuous interchange of energy and/or information between the system and its environment. The organization is linked with its environment through the scanning process. On its effectiveness hangs the life and death of the firm.

Organizations scan the environment for information needed to make the decisions enabling them to survive in a highly competitive world. The main thrust of Part Four, therefore, is on managerial decision making, based on information scanned from the firm's external environment. The logic, use, and impact of information technology on managerial decisions are discussed, as are formalized attempts to acquire, evaluate, and disseminate data. The persistent searcher after information is warned of possible sources of confusion that can lead to nowhere in his attempt to come up with a meaningful and realistic management information system.

The final portion, Part Five, is devoted entirely to the application of the systems approach to management. Besides pulling together the various systemic characteristics treated in previous chapters, the authors here attempt to show how system concepts can be used to deal with the real-life problems of management. Not only are procedural rules and the various phases of the proposed paradigm explained, but the planning and control of large-scale systems, typically accomplished through network analysis, is presented with sufficient detail to enable the reader to grasp the practicality of a systems approach, here particularized in PERT.

Since cybernetics deals specifically with control and communication, the final two chapters of the text are taken up with both cybernetic principles and applications. The principles are both philosophical and operational; the applications sufficiently diverse to astound the tyro and to stimulate the systems user.

We were led to investigate the application of system concepts to management after several years of teaching systems to undergraduate and graduate students and to adult groups. Most of the available literature on management systems seem to emphasize the technical aspects of such application. Writings, however, of men like C. W. Churchman, R. Ackhoff, S. Beer, Sir G. Vickers, K. Boulding, L. Thayer, and of E. Laszlo, have stressed the lack of underlying theory in recent applications of operations research, systems analysis, and systems engineering. We have attempted to fill this need through an examination and refinement of the basic concepts, propositions and laws of general systems theory and of cybernetics and by relating these to the planning and control of today's complex organizations.

Efforts to introduce the subject of systems into the business school's curriculum have generally resulted in fragmenting the field into numerous courses with titles such as Systems Analysis, Management Information Systems, Electronic Data Processing, Computerized Business Systems, Accounting Information Systems. While such subject areas are, undoubtedly, substantive in themselves, they lack a common unifying framework.

This text is designed for an introductory course that could well be entitled Management Systems, Management Information Systems, or Introduction to Systems Concepts. Since it is introductory, it spans many diverse areas, all of which have something to contribute to an understanding of system fundamentals. The overriding objective has been to provide the wherewithal for a clear understanding of system postulates that underlie all applications. While it is indeed introductory, this in no way eliminates the need for the sustained serious concentration that characterizes the student of the sciences or of the humanities. Grappling with concepts, like engaging one's fellow in a game of intellectual wizardry, can be a highly satisfying but demanding pastime. The authors are convinced that the efforts exerted to master the systems concepts will be richly rewarding.

Because of its diversity and scope, this text can serve as an excellent point of departure in a systems curriculum. The material presented in these chapters should provide the springboard in advanced courses for class discussion on loftier, more discerning and recondite levels.

The authors are indeed indebted to John Ivancevich whose insightful comments have enhanced the presentation of this material. They gratefully acknowledge the contribution of their numerous undergraduate and graduate students at the University of Iowa and the University of

Georgia who by their comments and criticisms risked the unknown. To these and to all who by their reviews and evaluation of the manuscript have prodded the authors to ever greater efforts at clarifying and illustrating concepts, the authors tender their sincerest thanks.

March 1975   PETER P. SCHODERBEK
                                       ASTERIOS G. KEFALAS
                                       CHARLES G. SCHODERBEK

Contents

xv

Part One

The Basics

Here and elsewhere we shall not obtain the best insight
into things until we actually see them growing from
the beginning.

Aristotle

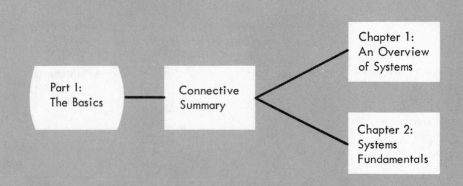

Part 1:
The Basics

Connective
Summary

Chapter 1:
An Overview
of Systems

Chapter 2:
Systems
Fundamentals

CONNECTIVE SUMMARY

The purpose of this part is not to present a complete account of the history of systems thinking in the so-called hard or exact sciences, nor even to display for the general reader the most important attempts hazarded by social scientists in "transplanting" systems thinking into their own arenas of activity or thought. Rather, the intention is to explore certain aspects of systems thinking considered essential for an understanding of its nature, and to elucidate the role of systems thinking in the development of systems theories and disciplines within the social sciences. Unfortunately, the social scientist's eagerness to "get right to the point," a technique widely recommended and practiced in academics, has almost universally forced writers to "skip" this important and indispensable task.

For this reason, Chapter 1 examines the origins of systems thinking (or the systems era) in the social sciences by identifying and clarifying certain common elements which characterize the attempts to utilize theories, principles, postulates, etc., developed in dealing with physical phenomena in the study of man-made systems.

Since the term "system" appears numerous times in the first chapter and will appear many times in subsequent chapters, it was felt necessary to develop a "language" that could serve as the medium over which the messages of the rest of the book could be transmitted. To this end, Chapter 2 presents a basic vocabulary of system's definitions, parameters, properties, and classifications. The diagrammatical presentations are designed to enhance the student's perception by reinforcing the mental images created through the verbal explanations. It is the feeling of the present authors that the material presented in the two chapters of the first part of the book is the minimum amount of "homework" that a student of modern management must do before embarking into "Systems Theory and Management."

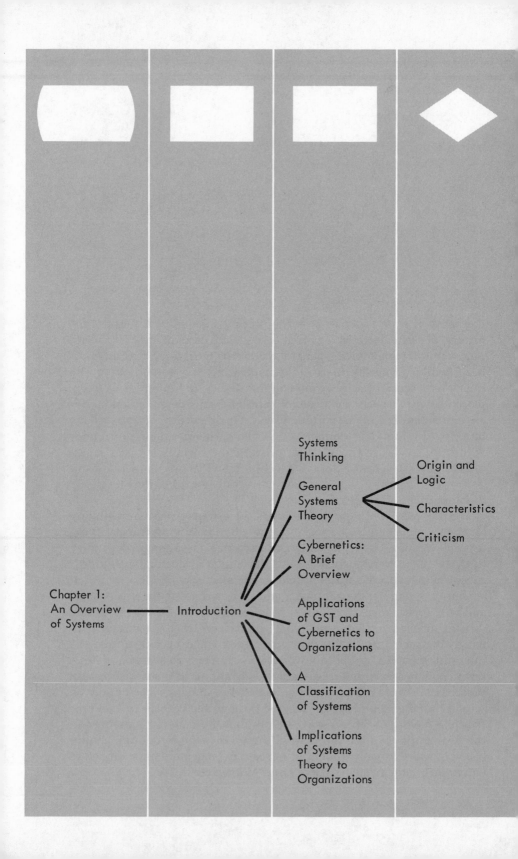

Systems
Thinking

General
Systems
Theory

Origin and
Logic

Characteristics

Criticism

Cybernetics:
A Brief
Overview

Chapter 1:
An Overview
of Systems

Introduction

Applications
of GST and
Cybernetics to
Organizations

A
Classification
of Systems

Implications
of Systems
Theory to
Organizations

Chapter One

An Overview
of Systems

Can we see the order of the parts of the universe, the
subordination among them, and notice how so many different
things compose such a permanent whole, and remain
convinced that the cause of the universe is a principle
without any knowledge of its effects, which without purpose,
without intelligence, relates each being to particular ends,
subordinated to a general end?

Condillac

INTRODUCTION

Man is by nature inquisitive. Since the beginning of time he has been asking questions about the world around him as well as about his own role in that world. The search for knowledge (love of wisdom) was defined by the ancient Greeks as philosophy. As man progressively perceived the world's complexity, he began distinguishing more formally between the external, objective, or physical portion of the world and his own internal world. Knowledge was then subsumed under philosophy in its narrower sense and physical philosophy or science in the broader sense.

Science, exclusively concerned with the material or external portion of man's world, reached its zenith with the publication of Newton's *Principia*, which laid the foundations for today's *classical physics*. Newton's Principia presented the first comprehensive and empirically oriented viewpoint of the physical universe. Classical physics em-

5

bodied a unified scheme of explanatory and predictive laws and principles applicable to a wide range of physical phenomena. At the *macro* level, classical physics dominated scientific thinking for a fairly long time.

Meanwhile, biology was struggling for recognition in the scientific commonwealth dominated by physics. As it developed, classical physics adopted the viewpoint that physical problems could best be solved by breaking them down into their component parts. Thus, the laws and principles of physics were now applied to ever smaller classes of physical phenomena, resulting in the creation of multiple related subdisciplines. Biology and chemistry later followed suit. Thus arose the science of microbiology within biology, organic and inorganic chemistry within chemistry, and so on — and all of these dealt with chemical or biological phenomena at the *micro* level.

Although each discipline and subdiscipline attempted to employ the universal language of mathematics as perfected within physics, there was nevertheless a horrendous communication gap among scientists of differing disciplines, even among scientists of related subdisciplines. With communication impeded, knowledge simply did not diffuse readily.

The 20th century found science fragmented into a vast array of disciplines and subdisciplines, each of which pursued, in piecemeal fashion, its own explanation and prediction of isolated phenomena. As the late Wiener, "Father of Cybernetics," noted back in 1948:

> Since Leibnitz there has perhaps been no man who has had a full command of all the intellectual activity of his day. Since that time, science has been increasingly the task of specialists, in fields which show a tendency to grow progressively narrower. A century ago there may have been no Leibnitz, but there was a Gauss, a Faraday, and a Darwin. Today there are few scholars who can call themselves mathematicians or physicists or biologists without restriction. A man may be a topologist or an acoustician or a coleopterist. He will be filled with the jargon of his field, and will know all the literature and all its ramifications, but more frequently than not, he will regard the next subject as something belonging to his colleague three doors down the corridor, and will consider any interest in it on his own part as an unwarrantable breach of privacy.[1]

The plethora of disciplines and subdisciplines within the natural sciences resulted from the scientists' view that everything within their realm of concern could best be handled by subdividing the problems; they assumed all the time that real world phenomena lent themselves to these more or less arbitrary divisions. Still it occurred to some astute observers that the problems occupying the attention of scientists work-

---

[1] Norbert Wiener, *Cybernetics* (Cambridge, Mass.: MIT Press, 1961), p. 2.

ing in widely different fields were quite similar in nature. Attempts to explicate the common grounds of interest resulted in the formation of certain hybrid disciplines such as biophysics and biochemistry.

While the development of hybrid disciplines constituted the first step toward interdisciplinary research, the real interdisciplinary movement did not begin until scientists realized that most phenomena with which they were concerned shared certain systemic characteristics. Awareness of the impossibility of solving many of the problems of society from a single disciplinary approach also grew. The current problems of smog, pollution, and crime can illustrate this point. Take, for example, the problem of smog in urban areas. A complete understanding of the problem involves knowledge of climate, the molecular behavior of gases, the chemistry of the automobile engine whose exhaust helps create smog, the number of automobiles, the geographic layout of the city, including the location of the homes, work places, and highway arteries, the availability of alternate means of transportation, the speed of traffic and the timing of traffic lights, the incomes of the population, the chemistry and biology of the bloodstream, problems of microbes and virus growth under the chemical, light, and temperature conditions produced by smog.[2]

This way of conceptualizing the world is now commonly referred to as systems thinking, systems approach, systems concept, systems viewpoint, or simply as systems.

SYSTEMS THINKING

The main objective of systems thinking is to reverse the subdivision of the sciences into smaller and more highly specialized disciplines through an interdisciplinary synthesis of existing scientific knowledge. By shaping instead a theoretical framework with relatively general applicability, systems thinkers have effectively changed the intellectual climate and have challenged the validity and general applicability of analytic thinking as utilized and perfected by physicists.

The analytic method rests uneasily upon the following four pillars: (1) preoccupation with the external or physical portion of the universe; (2) emphasis on division and the subsequent composition of phenomena; (3) quantification of causal relations; and, finally, (4) precision as the ultimate ideal of every researcher.

The world of the systems thinker is somewhat different. His way of viewing the universe is also based upon four major pillars: (1) *Organicism,* i.e., the philosophy of putting the organism at the center of

---

[2] Alfred Kuhn, *The Study of Society: A Unified Approach* (Homewood, Ill.: Dorsey Press, 1963), p. 4f.

one's conceptual scheme; (2) *Holism*, in viewing phenomena as organisms that exhibit order, openness, self-regulation, and teleology (goal-directiveness), one focuses on the whole rather than on the parts; (3) *Modeling*, instead of breaking the whole into arbitrary parts, one attempts to map his conception of the real phenomena onto the real phenomena. This is done by abstracting from the real phenomena those characteristics that are relevant and by disregarding those features of the real phenomena that are not needed for the explanation or prediction of the system's behavior; (4) *Understanding*, i.e., realizing (*a*) that life in an organismic system is an ongoing process, (*b*) that one gains knowledge of the whole not by observing the parts but by observing the processes taking place within the whole, and (*c*) that what is observed is not reality itself but rather the observer's conception of reality. In view of these constraints, the systems-oriented researcher strives for an *adequate* knowledge of the whole rather than for an *accurate* knowledge of it. The latter is an ideal he can never hope to achieve.

Thus, there is much more to the difference between analytic and systems thinking than a mere shift in emphasis. Systems thinking is a more meaningful way of looking at and approaching the study of complex phenomena. However, it is not in stark contradiction to analytic thinking. Systems thinking supplements rather than replaces the latter. Systems thinkers, therefore, find it more meaningful to study the processes linking the "parts" together rather than to "micro-analyze" them.

The systems viewpoint is characterized by the development of two independent movements aimed at roughly the same goal: general systems theory (GST) and cybernetics.

GENERAL SYSTEMS THEORY

Origin and Logic

At the 1954 annual meeting of the American Association for the Advancement of Science (AAAS), a society was founded under the leadership of biologist Ludwig von Bertalanffy, economist Kenneth Boulding, biomathematician Anatol Rapoport, and physiologist Ralph Gerard. This society was called the Society for General Systems Theory, later renamed the Society for General Systems Research. Its original purpose and functions were as follows:

> The Society for General Systems Research was organized in 1954 to further the development of theoretical systems which are applicable to more than one of the traditional departments of knowledge. Major functions are to: (1) investigate the isomorphy of concepts, laws, and models in various fields, and to help in useful transfers from one field to another;

(2) encourage the development of adequate theoretical models in the fields which lack them; (3) minimize the duplication of theoretical effort in different fields; (4) promote the unity of science through improving communication among specialists.[3]

GST has its roots in the organismic conception of biology developed by Bertalanffy. This organismic conception (usually called organismic revolution) is summarized by Bertalanffy in the following words: "In contrast to physical phenomena like gravity and electricity, the phenomena of life are found only in individual entities called organisms. Any organism is a system, that is, a dynamic order of parts and processes standing in mutual interaction."[4] Like any other general theory, GST can employ either one of the two approaches suggested by K. Boulding: "The first approach is to look over the empirical universe and pick out certain phenomena which are found in many different disciplines, and seek to build up general theoretical models relevant to these phenomena. The second approach is to arrange the empirical fields in a hierarchy of complexity of organization of their basic "individual" or unit of behavior, and to try to develop a level of abstraction appropriate to each."[5] The second approach will be dealt with in this chapter under the heading "A Classification of Systems." In the next few pages, however, the first approach to the development of a GST viewpoint will be outlined.

As can be seen from Figure 1, General Systems Theory draws heavily upon biology, mathematics, physiology, and economics. Its main area of study, its "domain," so to speak, is the phenomena of growth and evolution. Its main assumption is that the process of growth and its intermediate and final stages (evolution) follow the same pattern, whether the growth is of a single organism, of a group of organisms, or of society itself.

A General Systems Theory–oriented economist, desiring to explain and predict, for example, the growth of a firm or of the economy at large, could "look over the empirical universe" and see that biologists have been studying growth phenomena of organisms and groups of organisms for a long time. He could then "borrow" the knowledge accumulated by the biologist to explain and predict phenomena of economic growth. Is he justified in doing so, or is he merely attempting to capitalize on the prestige of a well-established science? The answer, of course, is that he is fully justified, and for the following reason: both

[3] Ludwig von Bertalanffy, *General Systems Theory* (New York: George Braziller, 1968), p. 15.

[4] Ibid., p. 208.

[5] Kenneth Boulding, "General Systems Theory—The Skeleton of Science," in Peter P. Schoderbek (ed.), *Management Systems*, 2nd ed. (New York: Wiley & Sons, 1971), p. 22.

FIGURE 1–1
The Origins and Development of the Interdisciplinary Movement in Sciences

Specialization through Fragmentation of Sciences ◄————————

Synthesis through Integration of Sciences ————————►

A. Physical or Natural Sciences

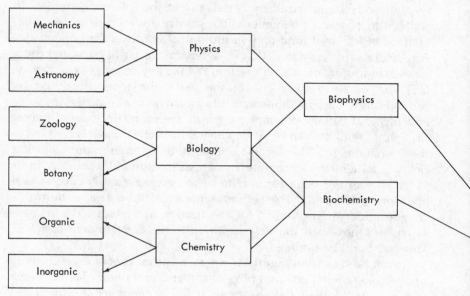

B. Social Sciences

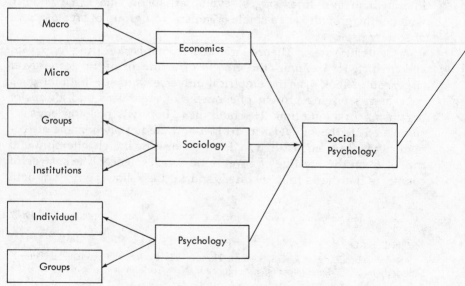

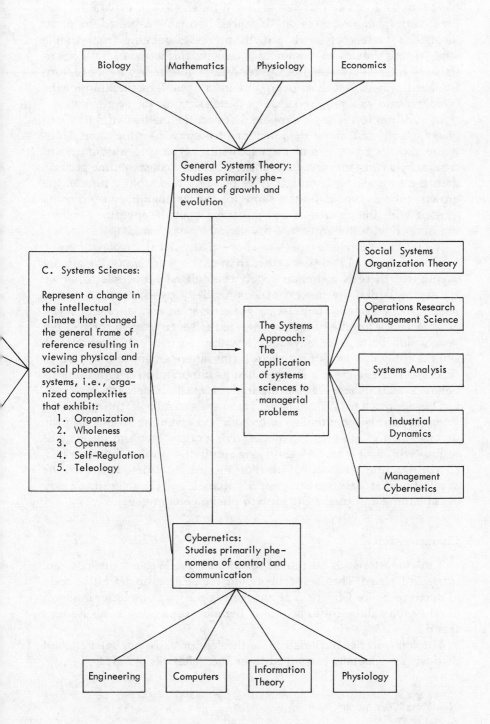

biologists and economists are seriously interested in explaining and predicting phenomena of a "balanced growth" assumed to occur through a variety of small growth processes ranging from simple straight-line growth to exponential growth; furthermore, both scientists are interested in setting the conditions for the avoidance of "unbalanced growth" (such as overpopulation or underpopulation for the biologist, and galloping inflation or depression for the economist).

In addition to the logic explained above, the mathematics (the language used to talk about the phenomena of growth) is the same. Thus, if the growth process is of the exponential type (i.e., a quantity increases by a constant *percentage* of the whole in a constant time period), then the formula for computing, let us say, the doubling time of the growth process would be the same for the demographer concerning himself with the growth of a population as for the investor concerning himself with the growth of his savings in the bank.

It should be obvious from the above that what is involved here is an isomorphism of *processes* rather than of *things or objects*. We are not saying that there is a similarity between dollars and people; what we are saying is that the process of exponentially growing dollars and of people is the same. A population growing at an annual rate of seven percent and a bundle of money invested at seven percent interest per year would both practically double within ten years. In general, one should recognize that the *structure* (the arrangement and nature of processes) of any system is often just as important in determining its behavior as is its *morphology* (the individual components themselves).

This approach to GST can be said to operate as follows: the researcher interested in the phenomena of growth and evolution would scan the diverse sciences of nonliving matter, living organisms, and social institutions with a view to: (1) identifying similarities in these phenomena, (2) constructing a model based upon these similarities which will be applicable to at least two different disciplines, and (3) suggesting a way of explaining and predicting growth phenomena in general.

Characteristics

The characteristics attributed to GST by the systems theorists are many and varied. This is not surprising, for GST has no definitive body of doctrine. Since GST aims at uncovering the laws and order inherent in all systems, it ought to be and is the most contentless of all systems theories.

The following characteristics, neither all-inclusive nor separate and distinct, are generally conceived to be the hallmarks of GST.[6]

---

[6] Joseph A. Litterer, *Organizations: Systems, Control and Adaptation*, vol. 2, 2nd ed. (New York: Wiley & Sons, 1969), pp. 3–6.

1. Interrelationship and interdependence of objects, attributes, events, and the like. Every systems theory must take cognizance of the elements in the system, of the interrelationship existing between the various elements, and of the interdependence of the system components. Unrelated and independent elements can never constitute a system.

2. Holism. The systems approach is not an analytical one, where the whole is broken down into its constituent parts and then each of the decomposed elements is studied in isolation; rather, it is a *Gestalt* type of approach, attempting to view the whole with all its interrelated and interdependent parts in interaction. The system is not a reconstituted one; it is an undivided one.

3. Goal-seeking. Systems embody interacting components. Interaction results in some final state or goal or an equilibrium position where the activities are conducive to goal attainment.

4. Inputs and outputs. All systems are dependent on some inputs for generating the activities that will ultimately result in goal attainment. All systems produce some outputs needed by other systems. In closed systems the inputs are determined once and for all; in open systems, additional inputs are admitted from the environment.

5. Transformation. All systems are transformers of inputs into outputs. That which is received into the system is modified by the system so that the form of the output differs from that which was originally put in.

6. Entropy. Its origin steeped in the field of thermodynamics, entropy designates the state of a closed system where all the elements are in maximum disorder; the system is run down. For living systems, maximum entropy means death; for formal organizations, maximum entropy could mean a lack of all necessary information for running the system, or a maximum condition of disorganization.

7. Regulation. If systems are sets of interrelated and interdependent components in interaction, then the interacting components must be regulated in some fashion so that the systems objectives will ultimately be realized. In human organizations this implies the setting up of objectives and the determining of the activities that will result in goal fulfillment. This constitutes planning. Control implies that the original design for action will be adhered to and that untoward deviations from the plan will be noted and corrected. Feedback is a requisite of all effective control.

8. Hierarchy. Systems are generally complex wholes made up of smaller subsystems. The nesting of systems within other systems is what is implied by hierarchy.

9. Differentiation. In complex systems, specialized units perform specialized systemic functions. This differentiation of functions by components is characteristic of all systems.

10. Equifinality. In open systems the same final state can be reached from several starting points; one result can have different causes.

One can easily envision a firm, a hospital, or a university, as a system and apply the above tenets to that entity. Organizations, for example, obviously have many components which *interact* — production, marketing, accounting, research and development, all of which are dependent upon each other. This interaction can be easily seen by looking at the effects of a decision to increase production of a particular unit. Increased production can affect inventory levels, working capital, purchasing, number of employees required in production, quality control, maintenance, scheduling, transportation, sales levels, equipment utilization, overtime costs, delivery schedules, and a host of other factors. It is precisely this aspect of interrelatedness which is often neglected in the study of organizational problems which produces suboptimal solutions. The importance of interrelatedness will be further noted in the study of industrial dynamics in a later chapter.

In attempting to understand the organization one must view it in its entire complexity rather than simply through one functional area or component. A study of the production system would not yield satisfactory analysis if one ignored the marketing system or the personnel system. While students in their learning process do indeed study functional areas, which is necessary to determine the interactions, it is also necessary to study the entire organization as a system. It is for this precise reason that many schools offer a capstone course in business policies which attempts to view the organization as a *whole* and to integrate its composite dimensions. The use of management games also serves as a vehicle both to see the system in its entirety and to stress the interdependencies of the variables.

All living systems are *goal oriented* and, indeed, this particular element is widely noted and discussed in the literature. In business these goals commonly take the form of profit, share of the market, sales volume, labor productivity, and so on.

The organization is obviously dependent upon *inputs* which are then transformed into *outputs*. This *transformation* function may be production oriented, service oriented, or task oriented. In all cases however, inputs generate the activities to be performed.

In regard to *entropy*, open systems (nature, the individual, organizations) have a continuous flow of inputs which move against the tendency toward entropy. If a living system is to survive, it must have a greater incoming flow of imported energy than an outgoing flow. Indeed, firms attempt to build a reserve both to improve their chances of survival and to provide for growth. Since the concept of entropy is mainly concerned with closed systems (one in which no energy is received from an outside source and one in which no energy is released to its environment) little attention will be devoted to this concept. Organizations, of course, are not closed systems and, indeed, in actuality, no system is ever totally closed; but only relatively so. One employs artificial closure in order to study the system at a particular time, treating the system as if it were shut off from its environment. The reason for doing so is to examine the internal operations of the system and the interdependencies of the components of the system.

Regulation is synonymous with control, and it is quite obvious that organizations employ a spectrum of control methods each designed to measure actual performance with a desired goal. Quality control, cost control, budgetary control, production control, and worker control are all illustrative of vehicles of regulation.

The organization is a *hierarchy* of subsystems and the typical organization chart often depicts this hierarchy. The major goals of the organization are typically segmented into divisional goals which in turn are fragmented further into plant goals, department goals, and individual goals.

Differentiation and specialization are evident in the structuring of tasks, the layout of plants, and the assignment of specialists.

Equifinality simply means that there are alternative ways to reach goals. In some instances it may be achieved through the introduction of new products; in others, it may be through acquisition or more market penetration.

Developments. It would be easier and more meaningful to identify GST-like tendencies among post—World War II research than to pursue the identification of GST disciplines. The reason for this is that most GST research has not been explicitly labeled as such and, more pertinently, most research labeled GST has not been a true application of GST principles but rather has involved unfortunate attempts to transplant terms and concepts from one well-developed discipline to another without much regard for the existence of isomorphism.

GST has had perhaps its greatest impact in its explanation and subsequent popularization of open-systems theory with its main pillars of *organization, wholeness, self-regulation (homeostasis), and teleology.* The applications of open systems to the so-called hard sciences (biology, physics, chemistry, and so on) are too numerous to permit even a brief mention here.[7]

Beyond the individual organism's level, open-systems principles are also used to explain problems of population dynamics, ecological

---

[7] Ludwig von Bertalanffy, "General Systems Theory—A Critical Review," in *Modern Systems Research for the Behavioral Scientist,* W. Buckley (ed.) (Chicago, Ill.: Aldine Publishing Co., 1968), p. 11.

theory, earth science, and so on. In the science of human behavior, GST-like thinking and research can be detected in the earlier Gestalt psychology and personality theorists whose formulations were known long before GST became popular.[8] The study of human behavior on the organizational and societal level as well as of the sociopsychological behavior of human organizations along GST lines is evident in the text of Katz and Kahn, *The Social Psychology of Organizations.*[9]

Criticism

The criticism launched against GST is largely directed against systems thinking in general. If one considers what systems thinking (GST and cybernetics) represents, then one should not be surprised at the criticism aimed at it. Some of the criticism stems from the attempts of hard core status quo defendants to save the rundown mechanistic view of the world. Philosophers of science have struggled with such criticism for a long time. One could dismiss antisystems criticism coming from analytic thinkers as either biased or irrelevant. Some criticism comes from those scientists who see flaws in the so-called new way of thinking. These critics, although accepting the basic philosophy of GST, argue that GST researchers have not been as sophisticated as they should have been. It is this kind of criticism which we will explore here.

Let us, at the risk of duplication, reiterate the basic postulates of systems thinking. Systems thinking represents a move away from:

1. the traditional mechanistic view of the world advocated by the classical analytic thinkers and a move toward an organismic point of view;
2. the strategy of division of complex problems into parts and toward a synthetic view of the whole;
3. the consideration of strictly quantitative relations built into a modeled situation; and finally
4. the pursuit of a precise solution of a limited aspect of a specific problem and toward an understanding of the whole relevant system.

To accomplish all or some of these ambitious objectives GST makes use of the process of analogy. Analogizing is a process of reasoning from parallel causes; in common parlance it means that two situations or events are similar in some respects but not in all. Without under-

---

[8] Wolfgang Koehler, *Gestalt Psychology*, rev. ed. (New York: H. Liveright Publishing Co., 1947); A. Angyal, *Foundations for a Science of Personality* (Cambridge, Mass.: Harvard University Press, 1941).

[9] Daniel Katz and R. L. Kahn, *The Social Psychology of Organization* (New York: Wiley & Sons, 1966).

estimating the widespread use of analogies in science and the meaning-ful achievements resulting from their use, one must ever keep in mind that analogizing is a very tempting but potentially dangerous enter-prise. The early anthropomorphisms of the ancient Greeks, the mach-inomorphisms of the 16th-century mechanists and the ratomorphisms of the earlier behavioralists all point to the dangers of "quick" analo-gies. As Churchman and Ackoff pointed out: "The analogical method, if carried to the extreme, would reduce all sciences to one: mechanics."[10]

GST has been criticized for forming meaningless analogies. There is considerable justification for this criticism, not because there is any-thing inherent in GST that tolerates or calls for weak analogies, but rather because some so-called GST researchers, carried away with apparent similarities among phenomena, extended their generaliza-tions a bit too far.

The same criticism against GST's use of analogies, known also as the "So What" argument, was launched against the writings of J. G. Miller by Buck. Briefly the argument can be stated as follows: Suppose we find an analogy or a formal identity in two systems, what does this mean? It means nothing! So what? Bertalanffy's reply to the "So What" criti-cism is presented below:

> Buck has simply missed the issue of a general theory of systems. Its aim is not more or less hazy analogies; it is to establish principles ap-plicable to entities not covered in conventional science. Buck's criticism is, in principle, the same as if one would criticize Newton's law because it draws a loose "analogy" between apples, planets, ebb and tide and many other entities; or if one would declare the theory of probability meaningless because it is concerned with the "analogy" of games of dice, mortality statistics, molecules in a gas, the distribution of hereditary characteristics, and a host of other phenomena.[11]

A more recent and, to some extent, a more sophisticated criticism of GST and Systems Theory in general comes from D. C. Phillips in his article entitled "Systems Theory — A Discredited Philosophy."[12] Phillips attacked Systems Theory in general and GST in particular for:

1. the failure of systems theorists to appreciate the history of their theory;
2. the failure to specify precisely what is meant by a "system";
3. the vagueness over what is to be included within systems theory;

---

[10] C. West Churchman and Russell Ackoff, *Methods of Inquiry* (St. Louis: Educational Publishers, 1950), p. 489.

[11] L. von Bertalanffy, "General Systems Theory — A Critical Review," pp. 19–20.

[12] Peter P. Schoderbek (ed.), *Management Systems,* 2nd ed. (New York: Wiley & Sons, 1971), p. 55; also, D. C. Phillips, "Organicism in the Late Nineteenth and Early Twentieth Century," *Journal of the History of Ideas,* vol. 3, no. 3, July–September, 1970, pp. 413–32.

4. the weakness of the charges brought against the analytic or mechanistic method;
5. the failure of GST as a scientific theory.

Phillips' objections to GST are quite constructive and deserve to be better known and understood by systems theorists and researchers. To this end, each of the five points of criticism will be briefly recounted.

1. Systems theorists, Phillips asserts, do not sufficiently appreciate the history of their theory. Systems theory really harkens back to Hegelian philosophy, and integral in this organicist conception of the world was the theory of internal relations. This theory asserted that entities are altered by the various relationships into which they enter. However, modern philosophy, with its critical distinction between *defining* and *accompanying* characteristics, rejects the Hegelian theory of internal relations, since it would make all characteristics of an object *defining* characteristics. This is untenable, since many of the characteristics of an object are but *accompanying* characteristics. Furthermore, it cannot be proven that every characteristic *must be a defining* characteristic. Even when one prescinds from the truth or falsity of the theory of internal relations, the corollary to be drawn from it would force one to admit that all knowledge is impossible, for to understand an object, one would necessarily need to know all of its defining characteristics, since these make it precisely what it is. But since, according to the theory, every object is related to everything else in the world and all relationships are necessarily defining, one would need to know all the relationships with which the object enters into in the rest of the universe. This is not only ridiculous; it is impossible. Consequently, the theory of internal relations from which the corollary was logically inferred is untenable.

Other corollaries have been derived from the theory of internal relations: the whole is greater than the sum of its parts, the whole determines the nature of the parts, the parts cannot be understood in isolation from the whole, and so on, but some of these can perhaps be supported on other grounds, as Phillips himself concedes.

2. Systems theorists do not specify precisely what is meant by a *system*. Phillips certainly touches on a sensitive point here. It is evident that many advocates of systems do not pay sufficient attention to their definitions and so propagate definitions that are extremely vague, obtuse, or tautological. Furthermore, the logical implications of these definitions are often never considered. The crucial problems of all systems center around how to conceive of systems in a clear and logically consistent way, how to define the system boundaries unambiguously, and how to segregate effectively the system under study from other systems without violating the basic premise of all systems thinking.

3. Phillips scores the vagueness of what is to be included in systems

theory. The literature he cites does evidence much confusion on the part of systems theorists. It is perhaps an embarrassment for most systems analysts to reiterate Bertalanffy's identification of seven bodies of systems theory (cybernetics, information theory, game theory, decision theory, topology, factor analysis, and general systems theory). Not all of these are on the same level of abstraction or of the same general type. It is indeed difficult to discover any single criterion or group of criteria that could have been used for including these seven and for excluding all others.

4. Modern systems theory apparently developed out of the dissatisfaction experienced, especially by biologists, with the traditional mechanistic or analytical method of studying complex systems. At certain levels of complexity it was found that emergent properties appear that cannot be predicted from the study of the less complex levels. However, Phillips contends that this problem of emergence is not as insoluble as the advocates of general systems imagine. Every mechanistic explanation stipulates two conditions: a knowledge of the law or laws applicable to the system, and a knowledge of the initial systemic conditions. For the more complex systems encountered in biology, it may be extremely difficult to satisfy or realize these basic conditions. However, it has yet to be demonstrated that in principle both these tasks are impossible.

5. One of the hallmarks of any scientific theory is its predictive ability. Phillips asserts that general systems theorists explain post factum what has happened; the predictions made before the event are too vague even to be refuted. This state of affairs was admitted by Bertalanffy himself in 1962.

That some, even perhaps many, systems enthusiasts have not done their homework is a valid criticism. In the past at least, the majority of systems writers, when justifying their assertions, have not really gone beyond Bertalanffy's writings. The reason for this, of course, is that such an endeavor would have taken them far into the realm of the philosophy of a science—a domain in which many of these scientists would have felt ill at ease. The failure to critically analyze concepts and to clearly define terms has done little to further the advancement of GST. The definition of a system as a set of components or objects interacting with each other would make a system out of virtually everything. Such a system would hardly be worth studying, since it would contribute little or nothing to a better understanding of systems theory.

As for the failure of GST as a scientific theory to predict, perhaps one should here make a necessary distinction. If by prediction one means some kind of systematic probability statement about a future event and not a prognostication of the future with total certainty, then one may perhaps confidently hold that GST, together with cybernetics, does possess a measure of predictive power.

CYBERNETICS: A BRIEF OVERVIEW

While the subject of cybernetics will be dealt with later in this book, we are at this time interested in illustrating its interdisciplinary nature. Cybernetics, the science of control and communication in the animal and in the machine, developed somewhat earlier than GST. By drawing heavily from engineering (and especially servomechanics and feedback control theory), computer sciences, mathematics, telecommunications, and physiology, cybernetics attempted to devise general principles and laws by which one can study the phenomenon of control and communication whether in the living or in the nonliving system.

APPLICATIONS OF GST AND CYBERNETICS TO ORGANIZATIONS

It would be convenient to refer to the systems approach as the application of both GST and cybernetics to the study of human—that is, man-made—organizations. It was more or less inevitable that both GST and cybernetics found wide application by researchers and theorists who concerned themselves with organizations. Indeed, here our concern is with the problem of growth (and of course, genesis) and evolution of organizations. GST is an approach to the growth and evolution of systems, but students of organizations are also very much interested in the control aspects of organizations. The researcher is concerned with explaining and predicting the organization's (systems) ability to maintain its actual performance (its output) within certain predetermined limits. Cybernetics is the approach to the control of systems and communication within systems.

On the highest macro level, that of a society as an organization, Buckley in his text *Sociology and Modern Systems Theory* and his thesaurus *Modern Systems Research for the Behavioral Scientist* has developed his comprehensive view of society as a complex, adaptive system.[13]

Still on the macro level, but this time in a smaller subsystem, that of an organization, the so-called Modern Organization Theory (OT) can be considered to lie on the periphery of GST.[14]

On the micro level, operations research represents the application of the systems approach to the firm. Its pioneers in the United States—Churchman, Ackoff, and Arnoff—strove to be as explicit as possible

---

[13] Walter Buckley, *Sociology and Modern Systems Theory* (Englewood Cliffs, N.J.: Prentice-Hall, 1967), and *Modern Systems Research for the Behavioral Scientist* (Chicago, Ill.: Aldine Publishing Co., 1968), pp. 490–513.

[14] William G. Scott, "Organization Theory: An Overview and Appraisal," in *Management Systems*, 2nd ed., Peter P. Schoderbek (ed.) (New York: Wiley & Sons, 1971), p. 40.

about the systemic aspects of operations research. In their *Introduction to Operations Research,* they write: "Central to this discussion is the notion that the *aim* of Operations Research is to obtain a systems or overall approach to problems."[15]

Operations research (OR) emphasizes the operations performed by an organization for accomplishing certain preset goals. Although this approach had a proper start, in its subsequent development it was reduced to an in-depth analysis of certain observable and measurable organizational activities. Recent developments, however, have seen further quantifications, mathematizations and computerizations of the operations rather than a broadening of OR's conceptual foundations. Thus, after two decades of considerable intellectual efforts, today's typical OR problems are still centered around inventory, allocation, replacement, and competition models, the only difference being that the new developments in mathematics, computer languages, and simulation enable the OR man to make his problems even more complicated than before.

As noted above, the GST researcher builds a model as a result of his scanning of the empirical world and of his uncovering of similarities among the phenomena. He engages in an extensive search for relevant information. So, too, the manager who wants to construct a GST model of the total enterprise. So, too, the physician who wants to specialize in, let's say, biomechanics. What all of this implies is that as a result of the interdisciplinary movement among scientists and the development of, to use Boulding's term, "generalized ears" that go along with it, there has arisen a tremendous demand for a technology for the acquisition, evaluation, storage, and transmission of information. This information technology, sometimes referred to as the Second Industrial Revolution, will be discussed in a later chapter. Here we are concerned only with the role of information technology in the application of the systems approach.

A Management Information System (MIS) is an information network designed to provide the right information to the right person at the right time at a minimum cost. Such a system has three basic components: (1) a GST model of the flow of information from certain points of origin to certain points of destination, (2) a computer, i.e., a machine designed to manipulate numbers, and (3) some software or metatechnology for translating the GST model into a language understandable by the machine.

While system analysis is the application of the systems approach to the design and implementation of an MIS, most such treatments of or-

---

[15] C. West Churchman, R. Ackoff, and E. L. Arnoff, *An Introduction to Operations Research* (New York: Wiley & Sons, 1957), p. 1.

ganizations confine themselves to the pure computerization of an existing conceptual model. However, the application of cybernetics to industrial problems, known as Industrial Dynamics (ID) and first developed by MIT's Jay Forrester, is more of a philosophy which asserts that organizations are most effectively viewed (and managed) from the ID's perspective.[16]

In the past several years, Forrester and a team of young scientists have expanded Industrial Dynamics to study the greatest system of all, the world's ecosystem. This study, conducted under the aegis of the Club of Rome (an international informal organization studying the problems of the future world from the global [holistic] viewpoint), constitutes the latest and most ambitious application of the systems approach.[17]

Finally, Beer's management cybernetics is another application of the systems approach to industrial problems. It is a continuous and automatic comparison of some behavioral characteristic or variable of the system (firm) with a standard coupled with continuous and automatic feedback activity. More about this later.[18]

These then are the relationships between (1) systems thinking, i.e., a new way of viewing empirical phenomena as systems; (2) GST, i.e., a formal materialization of systems thinking with respect to problems of growth and evolution; (3) cybernetics, i.e., a formal materialization of systems thinking with respect to problems of control and communication; and (4) systems approach, i.e., the application of both GST and cybernetics to social and industrial problems.

A CLASSIFICATION OF SYSTEMS

For thousands of years man has been occupied with classifying phenomena. While early man must have classified plants and animals as either harmful or not harmful, useful or not useful, one of the early formal attempts at the classification of the thousand or so then known plants and animals was undertaken by Aristotle. His simplified scheme for animals into those with red blood (animals with backbones) and those with no red blood (animals without backbones) and his division of plants (by size and appearance) into herbs, shrubs, and trees served man until the 18th century, when the idea of structure (arrangement of parts) was adopted by Linnaeus.

Every classification scheme, though arbitrary in design, is drawn up

---

[16] Edward B. Roberts, "Industrial Dynamics and the Design of Management Control Systems," in *Management Systems,* 2nd ed., Peter P. Schoderbek (ed.) (New York: Wiley & Sons, 1971), p. 415.

[17] Dennis Meadows, et al., *The Limits to Growth* (New York: Universe Books, 1972).

[18] Stafford Beer, *Management Science: The Business Use of Operations Research,* Doubleday Science Series (Garden City, N.Y.: Doubleday & Co., 1968).

with some particular purpose in mind. Students can be classified by level or by proven ability; climate can be classified on the basis of temperature and rainfall; books can be categorized as either fiction or nonfiction (fact). One classification of planets is by size, another by relative position with respect to the sun. Football players can be classified into offense and defense, into linemen and backs. In all of the above there is some order in the classifying scheme. The scheme adopted evidently presupposes some knowledge of the objects being classified and aids in the study of all other such objects.

As for systems, their classification is necessary if a methodology for their study is to be developed. The first classification of systems in which we are interested is that which utilizes the criterion of *complexity* as its distinctive feature. It is within a hierarchy of complexity that Boulding[19] arranges his theoretical "system of systems" (Figure 1–2). As one progresses from level 1 to level 9, he encounters an increase in system complexity.

FIGURE 1–2
An Ordering of Systems by Complexity

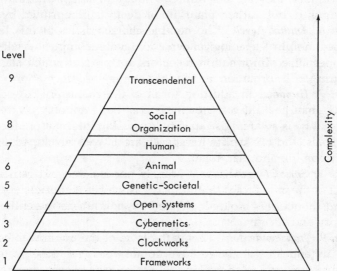

Level	
9	Transcendental
8	Social Organization
7	Human
6	Animal
5	Genetic–Societal
4	Open Systems
3	Cybernetics
2	Clockworks
1	Frameworks

Level 1. Frameworks. This is the level of static structures. Before one can deal with the dynamic behavior of a system, one must first be able to describe accurately their static relationships. These can be described

[19] Kenneth Boulding, "General Systems Theory—The Skeleton of Science," in *Management Systems,* 2nd ed., Peter P. Schoderbek (ed.) (New York: Wiley & Sons, 1971), pp. 20–28.

by function, position, structure, or relationship; e.g., anatomy of an individual, the location of the stars in the solar system.

Level 2. Clockworks.   This is the level of simple dynamic systems with predetermined motions. The movements of the solar system, the theories of physics and chemistry fall into this category. Practically all systems which tend toward equilibrium, including machines, are included.

Level 3. Cybernetics.   This is the level of maintenance of a given equilibrium within certain limits. This level refers both to the engineering type control (thermostat) as well as the physiological (the maintenance of body temperature). Here there is teleological behavior (goal seeking) but no automatic goal changing.

Level 4. Open System.   This level is concerned with self-maintenance of structure and therefore relies on throughput of material and energy. Closely connected with self-maintenance of structure is the property of self-reproduction. Self-maintaining and self-producing systems are definitively living systems. Hence this is the level of the cell.

Level 5. Genetic-Societal.   This level is typified by the plant and is characterized by a division of labor. Although there are many information receptors, they are not refined enough to accept information and to act upon it, but, rather, plant life or death is blueprinted by stages.

Level 6. Animal Level.   The notable characteristics of this level are increased mobility, teleological behavior (goal seeking), and self-awareness. Specialized information receptors are present which allow for a structuring of information and for the storage of information.

Level 7. Human.   In addition to all of the characteristics of animal systems, man has self-consciousness, i.e., he is not only aware, but is aware that he is aware. His capacity for storing information, for formulating goals, and his facility for speech are all well developed. Man can reflect upon life and plan for it.

Level 8. Social Organization.   Man is not isolated but rather is the product of the many roles that he plays as well as that of society in general and therefore is molded, affected by, and affects the entire gamut of history and society.

Level 9. Transcendental.   This is the level of the unknowables which escape us and for which we have no answers. Yet these systems exhibit structure and relationships. These indeed are the most complex of all since they are indescribable.

The first three levels are made up of physical and mechanical systems and have been of particular interest to the physical scientists. The next three levels all deal with biological systems and are of concern to the biologist, botanist, and zoologist. The remaining three levels, that of human, social organizations, and transcendental systems are primarily of interest to the social scientists.

As stated above, systems are classified with certain purposes in mind, and to simply list Boulding's classification scheme without examining the purpose would be unpardonable. Realizing that all theoretical knowledge must have empirical referents, he proposed his categorizations with a view to assessing the gap between theoretical models and empirical knowledge. In this regard he states that, while our theoretical knowledge may extend adequately up to the fourth level, empirical knowledge is deficient at all levels. There is agreement that, even at the first level of static structures, inadequate descriptions of many complex phenomena exist. While adequate models exist at the level of clockworks, and to some extent at the third and fourth levels, these are only a modest beginning. While one could hardly doubt the achievements made in medicine and the systematic knowledge acquired in this area, these too are the mere rudiments of theoretical systems.

A second purpose of the above scheme, according to Boulding, is to "prevent us from accepting as final, a level of theoretical analysis which is below the level of the empirical world which we are investigating."

And thirdly, the scheme serves as a mild warning to the management scientist who, despite the fact that new and powerful tools have led to more sophisticated theoretical formulations, must never forget that he is little beyond the third or fourth level of analysis and therefore cannot expect that the simpler system will hold true for the more complex ones.

Thus, for Boulding's purposes, his classification scheme is quite logical and consistent although one may disagree with his "perception of developments" or even his ordering of complexity. For even here Boulding's concept may indeed have different connotations for different students.

IMPLICATION OF SYSTEMS THEORY TO ORGANIZATIONS

The tenets of general systems theory were simplistically related to an example of the organization earlier in this chapter. Even from such an unpretentious effort one should have become aware of the potential usefulness of this approach. In the process of conceptualizing goals, structure of tasks, regulatory mechanisms, environment, interdependencies of components, boundaries, subsystems, inputs, and their transformation into outputs; all begin to take on more significant meaning.

Indeed, it is only through such conscious recognition of the organization as a system that one can begin to realize the full complexity that must be managed. And, while each of the tenets of general systems theory and organizational relatedness will be examined in detail in the forthcoming chapters, it is useful to briefly mention some of the potential benefits of systems thinking to the managers of organizations:

1. It frees the manager from viewing his task from a narrow functional viewpoint and indeed coerces him to identify other subsystems which are either inputs or outputs to his system. Many corporations utilize products or services of other organizations such as auditors, bankers, brokers, consultants, suppliers, and so on which are external to the organization, but which nevertheless affect the performance of the organization. This identification of systems and subsystems is imperative, since cooperation is required from many segments far beyond the boundaries of the internal organization.

2. It permits the manager to view his goals as being related to a larger set of goals of the organization. It is the manager's task to understand not only his goals but how they are integrated with broader goals which make the organization a system. The manager must realize that the summation of the goals at this level of the organization should be equal to or greater than the goals at the next level. Viewing goals in such a manner focuses attention on the interrelatedness of tasks that must be carried out by the different members of the organization.

3. It permits the organization to structure the subsystems in a manner consistent with subsystems goals. More specifically, it can take advantage of specialization and differentiation within the system and subsystems. Viewing the organization as a system emphasizes the fact that in order to meet the varied requirements of the system the subunits of which the organization is composed must be designed toward the end of goal attainment of the subunits.

4. The system viewpoint with its goal attainment model allows for evaluation of organizational and subsystems effectiveness. Measurement in this type of model is against specific objectives. While certain implicit assumptions are made in the goal-attainment model—e.g., that organizations have goals and that such goals can be identified and progress toward them measured—these assumptions are not formidable in most situations. A detailed discussion on goals and attendant problems will be presented in a subsequent chapter.

SUMMARY

Systems thinking, a logical step in the development of man's approach to the study of complex phenomena, has developed over the years from a shift in emphasis from a macro, to a micro, and back again to a macro viewpoint. The original macro level overlooked the many relevant details of the later micro studies, while these in turn became too divorced from one another to adequately define the working of the whole. Systems thinking with a present macro approach attempts to place components in the proper perspective to one another, to study their mutual interactions and the effect of these interactions on the whole, as well as on the way the whole affects and is affected by its environment.

In this chapter we have reviewed the pillars of systems thinking, noted their role in supplementing rather than replacing analytic processes, examined briefly a classification scheme of systems, and some benefits of this approach.

From this introduction into the "what and why" of systems we go next to a more detailed and definitive presentation of the components and properties of systems.

REVIEW QUESTIONS FOR CHAPTER ONE

1. Briefly describe the "interdisciplinary movement among sciences," and assess its applicability and/or utility to the study of complex organizations.
2. What factors accounted for the development of the systems approach?
3. Trace the historical developments of General Systems Theory and highlight its main characteristics.
4. What are the basic postulates of systems thinking?
5. What are the criticisms presented against General Systems Theory and how would you counteract such criticisms?
6. It has been 20 years since Boulding first set out his classification scheme of systems. What effect has recent research in the behavioral sciences as well as advances in technology had on our understanding of the various levels?
7. List and briefly explain the nine levels of systems presented in Boulding's hierarchy of systems.
8. Choose an example of a functional area in the business school and apply the characteristics of systems to this area.
9. What are some of the implications of systems theory for the management of organizations?
10. Why is it that firms which have never heard of the systems approach or the tenets of General Systems Theory have been successful in spite of this lack of knowledge?

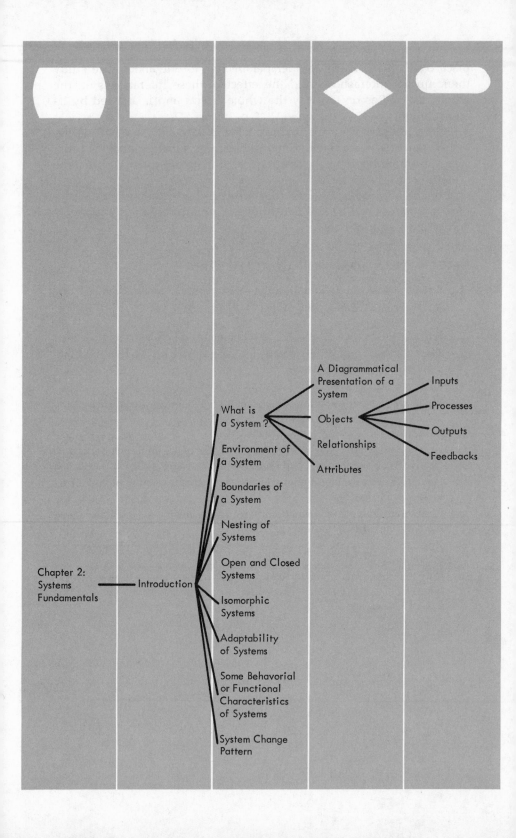

Chapter 2:
Systems
Fundamentals

Introduction

What is
a System?

Environment of
a System

Boundaries of
a System

Nesting of
Systems

Open and Closed
Systems

Isomorphic
Systems

Adaptability
of Systems

Some Behavorial
or Functional
Characteristics
of Systems

System Change
Pattern

A Diagrammatical
Presentation of a
System

Objects

Relationships

Attributes

Inputs

Processes

Outputs

Feedbacks

Chapter Two

Systems
Fundamentals

*Most of the major advances in science and technology have
been in the invisible realm—synergy—the behavior of
whole systems in ways unpredictable by the individual
behavior of their subsystems.*

Fuller

INTRODUCTION

Ackoff in 1971 expressed dissatisfaction with the current state of development and progress in the systems field. He stated his case very colorfully and succinctly:

> Despite the importance of systems concepts and the attention they have received and are receiving, we do not yet have a unified or integrated set (i.e., system) of such concepts. Different terms are used to refer to the same thing and the same term is used to refer to different things. This state is aggravated by the fact that the literature of systems research is widely dispersed and is therefore difficult to track.[1]

In the preceding chapter, some reasons for this lack of a "system" in systems concepts were considered. It may be, as some writers seem to think, that systems thinking is still in its embryonic stage, implying that eventually a rigorous, unified, or integrated set of systems concepts will emerge. However, it appears that the field has already developed far enough, albeit in an abnormal and nonteleological manner.

---

[1] Russell L. Ackoff, "Towards a System of Systems Concepts," *Management Science,* vol. 17, no. 11, July 1971, pp. 661–71.

Perhaps after numerous so-called forward steps toward the "scientifi-cation" of the field, a step backward may be needed. Of course, if this step backward represents a step in the right direction, then it will be a step forward.

We propose to take this step backward in order to provide a more rigorous vocabulary of systems concepts for the reader. The framework will be developed in nonmathematical language. While it does not in-clude all concepts relevant to systems theories and applications, still, enough of the major concepts are included so that the reader will be able to understand the literature in the field.

WHAT IS A SYSTEM?

The "system" concept has been borrowed by the social scientist from the exact sciences, specifically from physics, which deals with matter, energy, motion, and force. All of these concepts lend themselves to exact measurement and obey certain laws. There a system is defined in very precise terms and in a mathematical equation that describes certain relationships among the variables. This kind of definition, however, is of little use to the social scientist, whose variables are very complex and often multidimensional.

The definition given here is a verbal, operational one which, though nonmathematical, is quite precise and as inclusive as that of the exact scientist. A system is here defined as "a set of *objects* together with *relationships* between the objects and between their *attributes* connected or related to each other and to *their environment* in such a manner as to form an *entirety* or *whole*."[2] This definition has a dual property: it is ex-tensive enough to allow for wide applicability and at the same time it is intensive enough to include all the elements necessary for the detec-tion and identification of a system. To further reduce the vagueness in-herent in the terms, the key concepts—namely, objects, relationships, attributes, environment, and whole—will be explained.[3]

A Diagrammatical Presentation of a System

A very detailed explanation of the schematic method of presenting systems can be found in most introductory engineering books. Here

---

[2] This is a commonly accepted definition. See, for example, A. D. Hall and R. E. Fagen, "Definition of System," in W. Buckley, *Modern Systems Research for the Behavioral Scien-tist* (Chicago, Ill.: Aldine Publishing Co., 1968), pp. 81 ff; S. Optner, *Systems Analysis for Industrial and Business Problem Solving* (Englewood Cliffs, N.J.: Prentice-Hall, Inc., 1965); J. J. DiStefano, III, *et al., Schaum's Outline of Theory and Problems of Feedback and Control Systems* (New York: McGraw-Hill Book Co., 1967).

[3] The following analysis and elaboration draw heavily upon Optner, *Systems Analysis,* ch. 2.

FIGURE 2–1

A Diagrammatical Presentation of a System's Parameters, Boundary, and Environment

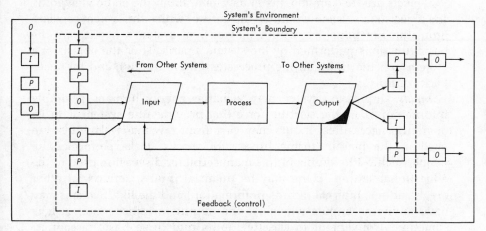

and in the following chapter a detailed but still incomplete explanation of the major symbols used in diagramming a system will be discussed.

The first thing that one should notice when looking at Figure 2–1 is that the input to a system is the output of another system, and that the output of the system becomes the input to another system.

Secondly, one should notice that the line demarcating the system from its environment (i.e., the systems boundary) is not solid. There are two reasons for this: First, such a line indicates that there is a continuous interchange of energy and/or information between the *open* system and its environment. This kind of a boundary serves the same purpose as the cell membrane: it connects the exterior to the interior. Secondly, the broken line indicates that the boundary's actual position is more or less arbitrarily determined by the designer, investigator, or observer of the system's structure. He tentatively assigns a boundary, examines what is happening inside the system, and accordingly readjusts the boundary.

Thirdly, in this diagrammatical presentation of a system the component control positioned over the output or process box in conventional diagrams has been deleted. Instead, the control function has been incorporated into the feedback component for reasons that will become clear when the science of control and communication (i.e., cybernetics) is scrutinized. Finally, it should be noticed that the lines connecting the system's parameters to each other as well as the system to its environment represent the system relationships.

We now turn to an elaboration of the major terms of the definition given above.

Objects

Objects are the components of a system. From the static viewpoint, the objects of a system would be the parts of which the system consists. From the functional viewpoint, however, a system's objects are the basic functions performed by the system's parts. Thus the objects of a system are: the input(s), the process(es), the output(s), and the feed-back control.

Inputs. Inputs to a system may be matter, energy, humans, or simply information. It is the start-up force that provides the system with its operating necessities. Inputs may vary from raw materials which are used in the manufacturing process to specific tasks performed by people such as the typing of this manuscript, or discussion used in the educational setting. There may be financial inputs, services of other organizations, internal records (information) and the like. Systems may have numerous inputs which are the outputs of other systems. It is sometimes convenient to classify inputs into three basic categories: serial inputs, random inputs, and feedback inputs.

A *serial* input is the result of a previous system with which the focal system (system in question) is serially or directly related. These kinds of inputs are easy to identify and study. They present little problem to the researcher because their absence would be felt immediately as the lack of "movement" in the system. Serial or in-line inputs are usually referred to as "direct-coupling" or "hooked-in" inputs.

Let us identify some of the most common serial inputs of a manu-facturing firm. A manufacturing organization can be conceptualized as a transformation system which converts the three basic factors of pro-duction (i.e., men, material, and money) into marketable products. This transformation process is accomplished through the interaction of a vast number of subsystems each performing its own transformation process. The output of each of these subsystems becomes the input to other subsystems.

In Figure 2–2 the interaction of two such subsystems is shown. The production subsystem is concerned with the actual physical conversion or transformation process. In order to perform this task, however, this subsystem needs several inputs, one of which is the volume of produc-tion, that is to say, the number of units (a quantitative attribute of this system's output) as well as certain qualitative characteristics of the output. These inputs are supplied to the production subsystem by the sales subsystem. The output of the sales subsystems which becomes the input to the production subsystem is a serial input because the two subsystems are directly related. In other words, the sales subsystem's output is produced for the specific purpose of providing the energizing or start-up function to the production subsystem without which it cannot function.

Serial inputs may come to the manufacturing firm from the outside environment as well. For example, most of the energy resources needed for the production process (e.g., electricity, water) will be supplied by local subsystems to which the firm is "hooked up." Most of the laborers will also be supplied by the labor force of the community.

One can find analogous examples for a nonmanufacturing firm such as, for instance, a bank or a hospital. Energy inputs to the bank or hospital system, for example, will be provided by the municipal power plant and water reservoir. In all these instances the focal system (i.e., the system whose behavior is under study) is linked directly to a specific system upon which it depends for one or more inputs.

FIGURE 2–2
Serial or In-Line Input

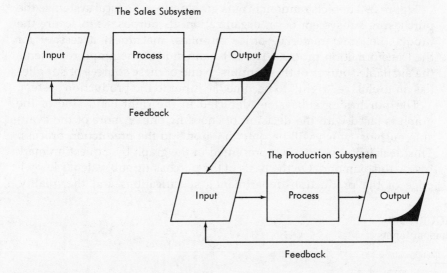

The second form of input is the *random* input. The term "random" is used not in its colloquial sense (meaning *haphazard*) but in its statistical sense. Random inputs represent potential inputs to a system. The focal system must determine which of the available outputs of alternative systems or subsystems will become its inputs. To put it differently, each available output of other systems has a probability of being chosen as an input to the focal system. This probability, which is, of course, less than one for each individual potential input, is determined by the degree of correspondence between the input needs of the focal system and the attributes of the available inputs. The actual selection of the focal system is then based upon this probability distribution and the decision criterion of the system.

Random inputs are the most interesting kind of inputs for any researcher or observer to study. The reason for this is that their presence or absence is not as conspicuous as in the case of serial inputs: They usually affect the *degree* of operation of a system (i.e., its efficiency) rather than the operation itself.

Random inputs range from the limiting case of what is called in genetics "the fertilization sweepstakes," where only one out of 300 million sperm cells penetrates and fertilizes the egg, to the limiting case of the decision maker who must choose between two alternatives. Obviously, the study of random inputs is the study of the decision-making process in a system. Random inputs can also be called *coded* inputs. The code is a systematic arrangement of the attributes or characteristics of the potential input as these relate to the needs of the focal system.

Figure 2–3 depicts random inputs graphically. The focal system is the purchasing subsystem of an organization. Its purpose is to secure the inputs (i.e., raw material, office supplies, machines) necessary for the transformation process. The left-hand side of the graph represents the available sources of these inputs. None of these sources of supplies has an exclusive "right" to become *the* input to the production system.

The purchasing subsystem depicted in the right-hand side of the graph is faced with the decision of choosing one or more of the available outputs which will become the inputs to the production process. This decision situation is represented in the graph by a question mark inside the diamond. On the basis of the purchasing subsystem's knowledge of the production department's specifications and the quality,

FIGURE 2–3
Random Inputs

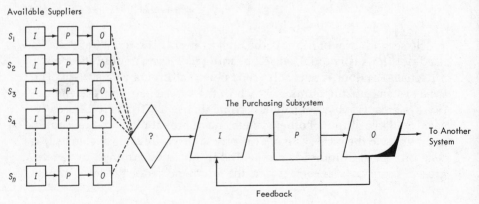

timeliness, and general past experience with the potential suppliers, the purchasing department will design a list of preferences. These preferences will reflect the purchasing department's satisfaction with each one of the suppliers in terms of the likelihood of choosing one or more of them.

A hospital's purchasing department or service is faced with the same situation as is a manufacturing firm's procurement department. There are literally hundreds of suppliers ready and available to supply the hospital with the drugs and other material necessary for the operation of the hospital. Every single drug of a similar nature could perform approximately the same function. Nevertheless, the manager of that department must choose one or several from among the scores of available drugs.

Most of the well-known techniques developed within the field of management science or operations research deal primarily with quantitative methods of assessing the probability of certain outputs which may become the inputs to a certain focal system.

The third kind of input represents a reintroduction of a *portion* of the output of a system as an input to the *same* system. This kind of input has the very descriptive name *feedback.* The use of this kind of input will depend upon its size, as well as upon its sign. Feedback input represents only a very small portion of the system's output. This portion is identified as the difference between a desired state of affairs (i.e., a goal) and the actual performance (A_p). Thus, Goal $-A_p = \pm d$. The researcher who desires to learn something about the behavior of a system and who comes across a feedback input would want to learn several things about it. For example, he would want to know the reason for its existence, its magnitude, its sign, its potential impact on the system when it becomes an input, and a score of other considerations to be discussed later.

Process. The process is that which transforms the input to an output. As such it may be a machine, an individual, a computer, a chemical or equipment, tasks performed by members of the organization, and so on. In the transformation of inputs into outputs we must always know how this transformation takes place, for indeed the processor is designed by the manager. When this is the case this process is termed a "white box." However, in most situations, the process by which inputs are transformed into output is not known in detail because this transformation is too complex. The combination of inputs, or their sequential order, may result in different output states. In this case, the process function is termed a "black box." A process may represent an assembly whereby an array of inputs is transformed into one output (e.g., a car assembly line) or it may be disassembled (e.g., a meatpacking plant where one input is converted into many outputs).

Many managers in large organizations cannot determine the interrelationships of the many components of the systems and therefore cannot understand what factors contribute to the attainment of an objective. For example, if the system objective is profit and indeed it is attained, one should be able to determine the constituents of that result. However, whether it was due to the packaging of the product, the quality, the channels of distribution, service, reputation, advertising, price, design and styling, or some other factor, executives often simply cannot tell you what factor accounted for its success. Since most managerial activities involve transformation of inputs into outputs which cannot be identified in detail and therefore constitute black processes, the next chapter treats the "black box technique" in greater detail.

Outputs. Outputs, like inputs, may take the form of products, services, information, such as a computer printout, or energy, such as the output of a hydroelectric plant. Outputs are the results of the operation of the process, or alternatively, the purpose for which the system exists.

As mentioned repeatedly, the output of one system becomes the input to another system, which, in turn, is processed to become another output, and the cycle repeats itself indefinitely. This is true for all living systems from what biologists call the "food chain" to contemporary product and service enterprises.

All transformation processes lead to more than one type of output. It would be convenient to classify the output of a system into three main categories. One category includes outputs which are directly consumed by other systems. The main output of a business manufacturing firm, for instance, is sold to the customers for either consumption or further processing. A hospital or an educational institution renders services directly to the clients. The system's objective is to maximize this type of output. The percentage ratio of this output to the overall output is usually termed efficiency.

A second category of outputs is the portion of the output which is consumed by the same system in the next production cycle. Defective products of a manufacturing process, for example, are usually reintroduced into the same production process. The output of the accounting subsystem of a bank or a hospital, in addition to being used for satisfying stockholder or taxpayer demand, is used to improve the performance of the system itself.

Finally, a third category of outputs consists of the portion of the total output which is consumed neither by other systems nor by the system itself but rather is disposed of as waste which enters the ecological system as an input. The focal system's objective or goal is to attempt to minimize that kind of output. Recently, this has become a challenging task for the manager.

Feedback. This parameter has been dealt with under "inputs" and will be further examined under the heading "Feedback Control" in the next chapter. The reason that it is mentioned here is to impress upon the reader the necessity of conceiving of this parameter as an integral part of every system which must be considered simultaneously with the other three parameters, namely, inputs, processes, and outputs.

Relationships

Relationships are the bonds that link the objects together. In complex systems in which each object or parameter is a subsystem, relationships are the bonds that link these subsystems together. Although each relationship is unique and should therefore be considered in the context of a given set of objects, still the relationships most likely to be found in the empirical world belong to one of the three following categories: symbiotic, synergistic, and redundant.

A *symbiotic* relationship is one in which the connected systems cannot continue to function alone. Examples of this kind of relationship abound. In certain cases the symbiotic relationship is unipolar, running in one direction; in other situations the relationship is bipolar. For example, the symbiotic relationship between a parasite and a plant is unipolar to the extent that the parasite cannot live without the plant while the latter can — parasitic symbiosis. However, the symbiotic relationship between the production and sales subsystems of a manufacturing system is bipolar: no production — no sales, no sales — no production — mutualistic symbiosis. Despite the tremendous importance of symbiotic relationships, they are the least interesting from the researcher's point of view because they are relatively easy to identify and explain.

A *synergistic* relationship, though not functionally necessary, is nevertheless useful because its presence adds substantially to the system's performance. Synergy means "combined action." In systems nomenclature, however, the term means more than just cooperative effort. Synergistic relationships are those in which the cooperative action of semi-independent subsystems taken together produces a total output greater than the sums of their outputs taken independently. A colloquial and convenient expression of synergy is to say that $2 + 2 = 5$ or $1 + 1 > 2$.[4]

Numerous examples of synergistic relationships can be found in nature as well as in the sciences, especially in chemistry. Fuller states, "Synergy is the essence of chemistry. The tensile strength of chrome-

---

[4] For a more detailed explanation of the concept of synergism as it applies to business enterprises see, "Business Synergism: When $1 + 1 > 2$," *Innovation,* no. 31, May 1972.

nickel steel, which is approximately 350,000 pounds per square inch, is 100,000 P.S.I. greater than the sum of the tensile strengths of each of all its component, metallic elements. Here is a 'chain' that is 50 percent stronger than the sum of the strengths of all its links."[5]

A simple example of a synergistic relationship from the business world would be the following: Suppose a firm aspires to increase its sales by, let us say, ten percent. The firm has two strategies available: (1) a $100,000 expenditure for advertisement which, according to the advertising agency, is supposed to increase sales by five percent; (2) a $100,000 expenditure for increasing the sales force by 20 percent, which is supposed to increase sales by another 5 percent. The two strategies are scheduled to be put into effect sequentially: first advertise, then hit the market with salesmen. Suppose that both strategies were effective; i.e., total increase in sales equals 10 percent.

A synergistically oriented sales promotion manager would have launched both strategies at the same time. Let us say that the increase in sales was 12 percent; the 2 percent difference in increase would be the synergistic effect.

Redundant relationships are those that duplicate other relationships. The reason for having redundancy is reliability. Redundant relationships increase the probability that a system will operate all of the time and not just some of the time. The greater the redundancy, the greater the systems reliability and the greater the expense. Redundant or backup relationships are abundant in the man-made world, and spaceships, satellites, and airplanes have systems with redundant relationships designed to secure operation of the system under virtually any condition.

Attributes

Attributes are properties of objects and of relationships. They manifest the way something is known, observed, or introduced in a process. A machine, for instance, has as its attributes the following characteristics: a machine number, a machine capacity (output per time), a required electrical current, ten years of technical life, six years of economic life, and so on.

Attributes are of two general kinds: defining or accompanying. *Defining* characteristics are those without which an entity would not be designated or defined as it is. *Accompanying* characteristics or attributes are those whose presence or absence would not make any difference with respect to the use of the term describing it.

---

[5] R. Buckminister Fuller, *Operating Manual for Spaceship Earth* (New York: Pocket Books, 1971).

This division of the attributes of a system's objects into defining and accompanying characteristics has some very useful implications for the manager who desires either to design or to use a system. Consider, for example, a company which transports perishable items. The products are transported via refrigerated trucks to various destinations. The company is considering the acquisition of five new refrigerator trucks to replace some old trucks, as well as to increase the size of the fleet. The manager in charge of this acquisition would be interested in certain characteristics of each truck. For instance, he would want to know the maximum load capacity of each truck, its speed, frequency of mainte- · nance, fuel consumption, and several other technical and economic characteristics, all of which are necessary for an accurate description of the equipment. These are the defining characteristics of a truck.

A truck, however, is characterized by certain other features which in a particular timespan do not appear to be necessary for its definition, but nevertheless are attributes of that system. One of these characteristics is, for instance, the amount of pollution created by the engine of the vehicle. If the decision approving the acquisition of the trucks were made ten years ago this attribute of the truck's engine would have been of no significance. That is to say, it would have been an accompanying characteristic. Today, however, this attribute of pollution creation is one of the most significant characteristics for the description of the truck; it is a defining characteristic which must be taken into consideration along with the other defining characteristics of capacity, speed, fuel consumption, and so on.

The reverse situation is also conceivable. Certain characteristics of a system's objects which were considered to be defining may at some other time, or in different circumstances, turn out to be accompanying characteristics. For example, the sex or race of an individual applying for a position within an organization ten years ago may have been considered as a defining characteristic. Today however, with the creation of equal opportunity employment, sex and race are of no real significance in terms of evaluating the applicant's suitability for the particular duty. It is merely an accompanying characteristic. However, for many firms attempting to comply with affirmative action programs, both sex and race are once again defining characteristics.

ENVIRONMENT OF A SYSTEM

Each system has something internal and external to it. What is external to the system can pertain but to its environment and not to the system itself. However, the environment of a system includes not only that which lies outside the system's complete control but that which at the same time also determines in some way the system's performance.

Because the environment lies outside the system, there is little if anything that the system can do to directly control its behavior. Because of this, the environment can be considered to be fixed or a "given" to be incorporated into any system's problems. The environment, besides being external, must also exert considerable or significant influence on the system's behavior. Otherwise, everything in the universe external to the system would constitute the system's environment, something to be programmed into the system's problem-solving framework. Both features must be present together: the environment must be beyond the system's control and must also exert significant determination on the system's performance.

There can be little doubt that the environment affects the performance of a system. Firm X's profits are obviously affected by the number and aggressiveness of its competitors, the number and price of their products, the purchasing power of the dollar, current federal, state, and local tax structures, pending congressional legislation and law suits, the political climate, and a host of other uncontrollable factors.

While the environment is external to the system's control, it is not impervious to its behavior. It is perhaps for this reason that some systems analysts also include in their definition the notion that the environment embraces also those objects whose attributes are changed by the behavior of the system.[6] This makes even more explicit the concept of interaction between systems and environment: environment affects systems and systems in turn affect the environment. Thus Company X and Company Y who are competing with one another must each include the other in its environment. Company Y is in Company X's environment and Company X is in Company Y's environment.

Perhaps one way to reduce the apparent arbitrariness of what constitutes the environment is to pose certain questions proposed by Churchman.[7] First, is the factor in question related to the objective of the system? Secondly, can I do anything about it? If the answer to the first question is "yes" and the answer to the second is "no," then the factor is in the environment. If the answer to both questions is "yes," then the factor is in the system itself. If the answer to the first question is in the negative, then the factor is neither in the system nor in the environment (Figure 2–4).

From Figure 2–4 it should be apparent that relatedness is linked with relevance. What we are concerned with in any system is *relevant* relatedness. When dealing with systems, one must be careful to acknowledge relatedness only when one is ready to declare relevancy. One

---

[6] A. D. Hall and R. E. Fagen, "Definition of System," *General Systems Yearbook*, vol. 1 (1956), pp. 18–28; reprinted in W. Buckley, *Modern Systems Research for the Behavioral Scientist*, p. 83.

[7] C. West Churchman, *The Systems Approach* (New York: Delacorte Press, 1968), ch. 3.

FIGURE 2–4
Environmental Determination

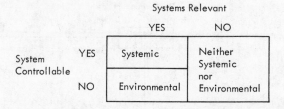

can easily relate something in this world to almost anything else by reason of color, size, shape, density, distance, and so on. Many of these relationships may be spurious; they lack relevance. Perhaps this is why Beer states that there seem to be three stages in the recognition of a system. First, "we acknowledge particular relationships which are obtrusive: this turns a mere collection into something that may be called assemblage. Secondly, we detect a pattern in the set of relationships concerned: this turns an assemblage into a systematically arranged assemblage. Thirdly, we perceive a purpose served by this arrangement: and there is a system."[8]

Figure 2–5 attempts to further clarify the relationship between a system and its environment by using as a criterion of differentiation the relative degree of control which can be exercised by the organization over the factors surrounding it. Ten external factors have been chosen as indicative of the multiplicity of factors usually referred to as "the environment." External factors over which the organization has a high degree of control can be considered the resources of the organization. On the other hand, external factors over which the organization has a relatively low degree of control can be defined as the environment of the organization. The relative degree of control has been depicted in Figure 2–5 as shaded.

As can be seen from this figure, the four major inputs of the organization, that is, the so-called major factors of production (labor, material/equipment, capital, and land) are relatively highly controllable by the organization. These are, therefore, the organization's major resources. On the other hand, the degree of control of the four major external factors depicted in the right-hand side of Figure 2–5 (ecology, government, general public, and competitors) is very low. These are, therefore, the organization's major environmental factors. Between these two extremes of the largely controllable factors (resources) and the largely uncontrollable variables (environment) lie two additional sets of factors

---

[8] Stafford Beer, *Decision and Control* (London: John Wiley and Sons, 1966), p. 242.

FIGURE 2-5
The Organization, Its Resources and Its Environment

The Environment

```
Indicates degree of control or, alternatively, resource

Indicates degree of independence or, alternatively, environment

Indicates the line demarking the system from its environment, i.e., boundary
```

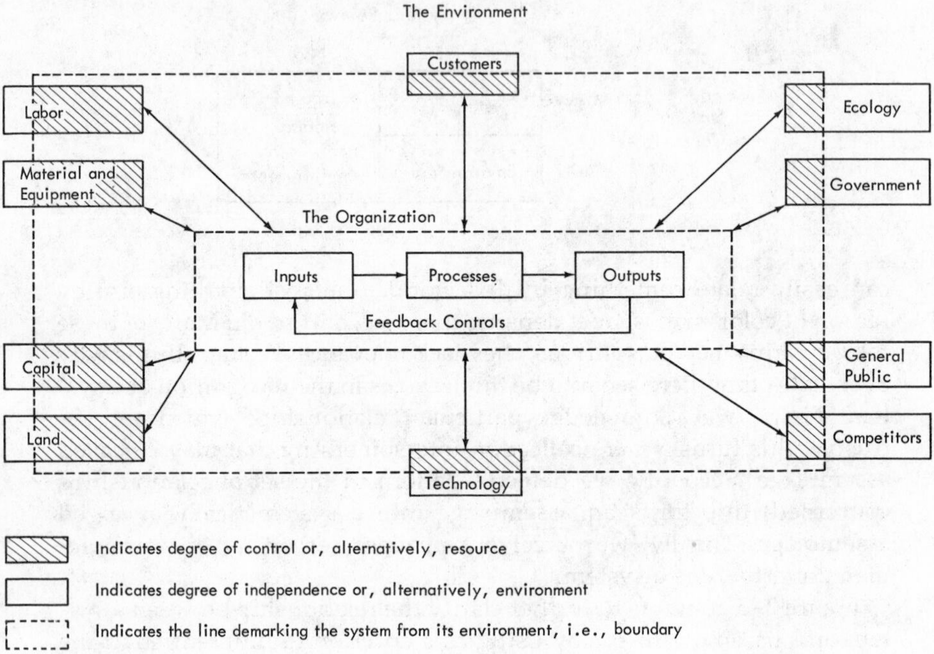

which are relatively less controllable than resources but relatively more controllable than the environment. These factors are consumers and technology.

The degree of controllability reflects the organization's ability to use its resources to influence the external factors or subsystem's behavior. This ability is, in turn, a function of the existence of resources, managerial talent, and the availability of organizational intelligence. Organizational intelligence refers to the organization's ability to recognize the need for control of an external factor, as well as the ability to devise the appropriate influencing strategy.

A few examples should suffice to demonstrate the differences between an organization's resources and its environment. Labor, material, money, and land have always been the exclusive concern of management, primarily because of the necessity of these factors for performing the basic functions of an enterprise and also as a result of the early developments in the discipline of economics. Knowledge of the basic principles of the economics of the firm, or what is usually referred to as micro-economics, enables management to recognize the need of influencing the behavior of these basic subsystems, as well as to develop sophisticated techniques in dealing with them. Thus, labor economics, material- and equipment-handling techniques, money management (finance) and land acquisition and utilization procedures are some of the most highly developed managerial tools.

At the other end of the continuum, knowledge in the fields of ecology, government regulation, social or public responsibility, and competitive strategy development have not advanced enough to enable the construction of a framework for recognizing the need for influence, as well as for enabling management to devise effective techniques for dealing with these external factors. For this reason the degree of controllability available to the organizations is relatively small but not negligible.

The relative degree of control that a particular organization can exert upon these four environmental factors of subsystems will depend on the organization's ability to employ some conventional techniques, as well as to devise some new effective means for dealing with them. Concerning the environmental factor labeled "ecology," the organization can employ sound manufacturing processes to minimize the amount of waste created by its operations. The fact that an environmental crisis has arisen during the last ten years which has forced governments all over the world to adopt governmental policies of the type which is known in the United States as the National Environmental Policy Act indicates that organizations have not been very successful in employing conventional management techniques. Since 1970, companies have been increasingly compelled to devise new techniques for pollution minimization.

Traditionally, organizations attempted to increase their degree of influence over the external factor labeled "government" via conventional lobbying and financial contribution to political parties in an effort to influence favorable legislation or to prevent excessive governmental surveillance of their activities. These techniques are increasingly proving to be ineffective. In the future, organizations will have to devise more sophisticated techniques in dealing with that sector of the external environment in particular, because the degree of government interference with free enterprise is estimated to increase, thereby curtailing the organization's control even further.

Participation in community programs and heavy advertisement are two of the most common and conventional techniques employed by organizations in dealing with the public sector. Consumerism, affirmative action programs, social responsibility demands and other new signs of the public's desire to intervene in the day-to-day activities of an organization have contributed considerably toward a decline in the organization's control of this sector of the external environment. A positive reaction which has been created as a result of XYZ Oil company's contribution toward a clean environment has been nullified by another company's announcement of a 400 percent increase in its corporate profits in a period of gasoline shortages and unemployment which were very visible in the era of the energy crisis.

Conglomeration and vertical integration were the most successful strategies for dealing with the desirable controllability of the external factor referred to as "the competitors." Legal requirements and plain diseconomies of scale caused by swift and excessive external growth (acquisitions, takeovers, mergers, and so on) are rendering these techniques largely ineffective as well. Thus, the organization is gradually experiencing a loss in its ability to control or influence this sector of its external environment.

Customers and technology are two sectors of the external environment which are somewhere between the two extremes of relatively high controllability (resources) and relative uncontrollability (environment). Marketing, which started as a managerial function primarily concerned with salesmanship, has grown into a full discipline employing sophisticated quantitative and behavioral techniques. Research and development, which started as a hit-or-miss type of engineering ingenuity, has developed into a sophisticated organizational entity employing the latest techniques of technological forecasting and the latest developments in quantitative and computer sciences. Thus, these two sectors or environmental variables, although not completely controllable by the organization to the extent that they do not represent its resources, are not outside any influence by the organization.

This brief discussion of the system's environment and its resources was not intended to completely clarify the entire subject. Several subsequent chapters deal with these considerations in greater detail. It should suffice at this point to emphasize that the line separating the system from its environment is indeed not a wall insulating the system from external influences. No open system can survive in such a utopian situation. Since organizations are open systems, their interaction and mutual influence by the environment is indeed a necessary condition for their survival.

BOUNDARIES OF A SYSTEM

Closely allied to the question of environment is that of system boundaries. Chin gives as his operational definition of the boundary of a system "the line forming a closed circle around selected variables, where there is less interchange of energy (or communication, and so on) *across* the line of the circle than within the delimiting circle."[9] One

9 Robert Chin, "The Utility of System Models and Developmental Models for Practitioners," in Warren G. Bennis, *et al., The Planning of Change* (New York: Holt, Rinehart and Winston, Inc., 1961), pp. 201–14.

can readily see that the boundary demarcates the system from its environment.

The boundary of a system is often arbitrarily drawn depending upon the particular variables under focus. One can adjust the boundary to determine whether certain variables are relevant or irrelevant, within the environment or without. A system viewed from two different levels may have different boundaries. This arbitrariness is not necessarily undesirable, since researchers and organizational officials tend to view a particular system from their own intellectual perspectives much as managers tend to evaluate case study problems from the vantage point of their own specialities. It is immaterial that one's system is viewed by others as a subsystem of a larger and different system. What is important in system analysis is that one clearly discriminates between what is in the system and what is in the environment.

NESTING OF SYSTEMS

Whereas there is an obvious hierarchy of systems (the ultimate system being the universe), still almost any system can be divided and subdivided into subsystems and subsubsystems depending on the particular resolution level desired. This nesting of systems within another can be seen in nature as well as in man-made systems. The universe, for example, includes subsystems of galaxies of stars which in turn include the solar system, and so on. The inventory system of a firm is a subsystem of the production system, which in turn is but a subsystem of the firm, which in turn is a subsystem of the industry, which in turn is a subsystem of the economic system.

The amount of nesting of systems within systems employed in any analysis will depend on the nature of the problem being investigated, the depth of analysis sought, and the particular framework employed. Perhaps the reason for failure to adequately solve many organizational and institutional problems may be the tendency to concentrate on too restricted a system. What should be regarded as but a subsystem is taken as the system, with the result that the significant interrelationships of the system with other subsystems are either overlooked or completely ignored. What constitutes the environment of the subsystem should, for practical reasons and for a realistic solution, be part and parcel of the system itself.

OPEN AND CLOSED SYSTEMS

The classification of systems into open and closed rests upon the concepts of boundaries and resources. The resources of a system are all the means available to the system for the execution of the activities necessary for goal realization. They include not only personnel, money, and equipment, but also opportunities (used or neglected) for the aggrandizement of the human and nonhuman resources of the system.

In a closed system all of the systems resources are present at one time. There is no further influx of additional resources across the system's boundary from the environment. In open systems, on the other hand, additional supplies of energy or resources can enter the system across its boundaries.

ISOMORPHIC SYSTEMS

Instances abound in which the structural relationships of one system are similar to or even identical with those of another system. Models in general attempt to represent a correspondence of their structure with the real elements being modeled. Good models correspond point for point with the object modeled. In this case a one-to-one correspondence is said to exist between the elements of the model and the components of the system being modeled. Where a one-to-one correspondence exists of the elements of one system with those of another, the systems are said to be isomorphic ("of like or identical form"). Thus, one can map the 26 letters of the English alphabet on the set of cardinal numbers, 1–26, thus setting up a one-to-one correspondence between the individual numbers and the letters of the alphabet.

The isomorphy most commonly acclaimed is that between mechanical and electrical systems. The two systems exhibit a one-to-one relationship in their structures. The relationship between quantities of the mechanical system and those of the electrical system is expressed by equations of the same form. The corresponding quantities encountered in these equations are force and voltage, speed and current, mass and inductance, mechanical resistance and electrical resistance, elasticity and capacitance.

Of importance in isomorphic mappings is that there is also a correspondence not only of structure but also of operational characteristics. It is this feature of isomorphic systems that has enabled the researcher to investigate and to predict properties of other systems. Cybernetics itself arose out of the realization by Norbert Wiener that the structure and operation of machines which are to be controlled are quite similar to those of animals.

ADAPTABILITY OF SYSTEMS

Adaptability for an organization is its ability to learn and to alter its internal operations in response to changes in its environment. Organization changes are for the most part externally induced. In the present context, adaptability refers to changes in the *kinds* of outputs of the organization rather than to changes within the many subsystems of the organizations. While such internally induced changes in individuals and groups are of interest, these typically do not alter the outputs of the focal system. A prerequisite for adaptability of a system is its familiarity with its environment, whether it be natural or man-made. Just as an individual must first acquire environmental information through his senses before adapting to any changes in it, so too must the organization. Adaptive systems therefore are simply those systems that are cognizant of their environment and are "able and willing" to adapt themselves to it. In the same way, all living systems must possess this characteristic if they are to survive.

In order to provide greater adaptability, firms often employ the strategy of diversification as well as a flexible organization structure. This latter element implies a decentralization policy by which the response time of a firm to react to a rapidly changing environment is typically less than in a highly centralized structure.

As will be shown in a later chapter, the ability and willingness of a firm to acquire information concerning the state of the environment is a critical factor in how well it can adapt. It is only through the acquisition of information that firms can learn of threats or opportunities existing "out there" in the environment. These threats may take the form of new competitors, new regulations, new products, or new processes. Opportunities may take the form of learning of, or creating, new consumers' needs, taking advantage of newly developed technologies of the firm, or a number of other forms. The research of many pharmaceutical firms and the resultant new drugs are illustrative of opportunities for the firm. Such new technology is conducive to increasing the firms' adaptability.

Adaptability can hardly be overemphasized in that some firms that paid little attention to their environment have failed to survive. While the consequences have not been as severe in other cases, market positions have become less secure. One need only to compare the list of Fortune's top 100 companies of 20 years ago with the top 100 of today to be convinced of this fact. Some organizations like du Pont and General Electric have responded to massive economic and social changes of past decades through a transformation process of both structure and outputs. These firms possess adaptability. The earning of many of the so-called progressive firms of today is from products unknown ten years ago. It can be said of such firms that they are in close touch with their environment.

SOME BEHAVIORAL OR FUNCTIONAL CHARACTERISTICS OF SYSTEMS

So far we have attempted to describe a system by enumerating and explaining its major elements. Although this was done in more or less functional terms, we really went about doing this piecemeal, focusing primarily on the functions of the individual parts. We were really concerned with the operations that make up the system. We asked, in other words, "What does an input, process, or feedback do?"

From the individual element's viewpoint this is functional analysis, but from the system's viewpoint this is structural analysis. What one learns about a system through structural analysis is nothing but a list of the items making up the system's structure. This, of course, is unsystemslike thinking, for systems thinking begins with and concentrates on functional analysis. However, one cannot help but ask whether function or behavior is influenced by structure. Since we believe that structure does influence behavior we found this mode of procedure desirable.[10]

Knowing what a system is, we now ask, "What is a system for?" or "What is the system supposed to do and how does it do it?" This, obviously, calls for a description of changes in the attributes of the system's elements as a result of interactions with each other over time. This description of the dynamics of the system can be seen as a two-part endeavor: the study of the reversible processes of a system — i.e., its behavior — and the study of the irreversible processes of a system — i.e., its evolution. We will concentrate primarily on the behavioral characteristics of a system. Evolutionary concepts will be explained only as they appear necessary or useful for an understanding of behavioral concepts.

We begin our discussion of some of the behavioral characteristics of a system by first explaining the concept "system's behavior." By system's behavior is meant a *series of changes in one or more structural properties of the system or of its environment*.[11] More specifically, a system's line of behavior is specified by a succession of states and the time intervals between them.

The state of a system at a moment of time is the set of relevant processes (expressed in numerical values) in the system at that time. What determines the state of a system? *A state of a system is created by the accumulation or integration of the past rates or flows.* Thus, there are two concepts relevant in explaining a system's behavior: (1) states or levels and (2) rates or flows. States or levels are influenced by rates or

---

[10] Our discussion of some behavioral properties of systems is based primarily on Ackoff, "Toward a System of Systems Concepts"; R. Ackoff and F. E. Emery, *On Purposeful Systems* (Chicago: Aldine-Atherton, 1972); Ross Ashby, *Design for a Brain* (Science Paperbacks, Chapman and Hall Ltd., distributed in the U.S.A. by Barnes and Noble, Inc., 1960); J. W. Forrester, *Industrial Dynamics* (New York: J. Wiley & Sons, 1961); and especially on G. Sommerhoff's excellent work, "The Abstract Characteristics of Living Systems," in F. E. Emery (ed.), *Systems Thinking* (Middlesex, England: Penguin Books Ltd., 1970), pp. 147–202.

[11] Ackoff, "Toward a System of Systems Concepts," p. 662.

flows, and rates are influenced by levels, but levels do not interact directly with other levels nor rates with other rates.[12]

These two basic determinants of a system's behavior are key concepts to the study of systems because they trace the movement of the system from one time period to another. Examples of states or levels of systems are the quantity of inventory at a particular time (state), the number of salesmen, and quantity of sales. Rates or flows are the factors which change the state or level from one time interval to another; for example, the production rate, the hiring rate, and the turnover rate. The financial reports implicitly recognize level and rate variables by separating these onto the balance sheet and the profit and loss statement. The balance sheet gives the present financial condition (state) of the system as it has been created by accumulating or integrating the past rates or flows. The profit and loss variables (one overlooks the fact that they do not represent instantaneous values but are averages over some periods) are the rates or flow which cause the state or level variables in the balance sheet to change.

In summary, the individual states of a system are determined by the rates of the different activities. The summation of all the states at a given point of time gives the state of the system at that time. The "movement path" from one state to another represents the behavior of the system. The starting state of a system is the initial state and the last state is the final state. For certain systems, knowledge of the initial state provides a fairly accurate knowledge of the most probable final state. For the majority of systems, however, knowledge of the initial state does not provide any knowledge about the final state. Systems which exhibit behavior are multistate or dynamic systems; one-state systems do not exhibit any behavior: they are static. A table or a house, for example, is a one-state system. On the other hand, a firm or an automobile is a multistate system. Here our concern will be with multistate systems only.

SYSTEM CHANGE PATTERN

What causes rates or flows to change? In other words, what motivates a system's variable or element to change its characteristics from one time to another? According to Ackoff a change in a system's state "can come about either as a result of a reaction, response or an act."[13] A *reaction* is a change in a system's (or environment's) structural properties caused by another change in a more or less deterministic manner.

---

[12] Jay W. Forrester, *Industrial Dynamics* (New York: John Wiley & Sons, 1961), Introduction.

[13] Ackoff, "Toward a System of Systems Concept," p. 664.

Thus, a reaction is a system event for which another event that occurs in the system or the environment *is sufficient*.

For example, changes in the firm's environment along the lines of antitrust, environmental and occupational safety, and health legislation (Antitrust Act, National Environmental Policy Act, and Occupational Safety and Health Act) demand certain rearrangement in the organizational structure. These laws have forced organizations to add to their existing personnel legal, environmental, and safety specialists. Organizational decision-making behavior in all of these instances is reactive because it was initiated by an outside force, namely, the United States government.

A *response* of a system is a system event for which another event that occurs in the same system or in the environment *is necessary but not sufficient*. Here the system does not change its structural properties as a result of the existence of a stimulus. However, should the system choose to change, it needs the stimulus. A response, in other words, is a change in a system's state produced jointly by the system and the stimulus.

The automobile industry's answer to the emission standards imposed by the Environmental Protection Agency (EPA) in the summer of 1973 is an excellent example of responsive behavior. Research and development engineers of the industry have always been trying to make the internal combustion engine less polluting. However, it was the recognition of the EPA's stimulus that resulted in formalized changes in the industry's internal structure, such, for instance, as the addition of new departments or divisions of environmental quality.

The essential difference between reaction and response is that the firm does not alter its structural arrangement ipso facto but rather attempts to co-produce the response by negotiation with that system or environment that causes the response. Finally, an *act* is a system event for the occurrence of which a change in the system or the environment is *neither necessary nor sufficient*. Acts are, in other words, self-determined events, autonomous changes, and are the results of the existence of changes in the internal state of the system. Product diversification, invention, and innovation are mostly the results of self-determined behavior. In other words, the organization's decision-making body decides, after careful assessment of the organization's potentials, to expand and revamp its product or service line. An organization's "social responsibility" programs, such as, for instance, hiring minority or disabled employees, is primarily the result of changes in the management's philosophy or viewpoint.

A system's behavior, then, can be either reactive, responsive, or active. Each mode of systems behavior requires its own method of examination. Thus, reactive and responsive behavior patterns are

TABLE 2–1
System Change Patterns

Change patterns / Characteristics	Reaction	Response	Act
Stimulus or Change	Necessary and sufficient	Necessary but not sufficient	Neither necessary nor sufficient
Examples:	Compliance with the law	Negotiation with employee unions	Product or service diversification

easier to study than are active behavioral patterns. The latter do not have identifiable causes or stimuli but are the results of spontaneous changes in the system's internal elements or characteristics.

SUMMARY

The definition of system adopted here indicated that a system is not just something in which everything is related to everything else. A system was defined in terms of objects (inputs, processes, outputs, feedback), relationships (symbiotic, synergistic, redundant), attributes (defining, accompanying), and environment. The determinants of a system's behavior—states (levels) and rates (flows)—were also discussed. Knowledge of the initial state of a system, it was seen, infrequently enables one to assess its final state. Only multistate systems exhibit behavior; one-state systems do not. Changes in the state of a system were said to be the results of reactions, responses, or acts, the last being the most difficult system behavior to study.

What a system is and how it can behave have been only partially explained. The reason for this is that one essential aspect of systems— self-regulation—has not yet been examined. Because of its importance for communication and control, this element of a system will be treated separately in the following chapter.

REVIEW QUESTIONS FOR CHAPTER TWO

1. Give a definition of the term "system" and illustrate this definition with some examples from your everyday life.

2. Identify the major systems parameter and draw a diagrammatical presentation of a system.

3. Take any kind of organization you are familiar with (e.g., a factory, a bank, a hospital, or a school) and list its inputs, processes, outputs, and feedbacks.

4. In the above system classify its inputs into serial, random, and feedback.

5. What are the main relationships in this system? Classify them into the three categories of symbiotic, synergistic, and redundant relationships.

6. What are "black box and white box processes?" Identify some major such processes in the above system.

7. What is a "system behavior" and what are the two relevant concepts you need to use in order to explain the systems behavior?

8. What are the three main system change patterns?

9. Take a typical manufacturing company or a service organization (e.g., a bank, a hospital, or an insurance company) and identify some of its resources and some of its environmental factors.

10. Choose several environmental variables you identified in the above question and indicate the degree of controllability as well as the most successful strategies which the company employs or should employ.

Part Two

The Science of
Control and
Communication

/

*The central thesis of cybernetics might be expressed
thus: that there are natural laws governing the
behaviour of large interactive systems—in the flesh, in
the metal, in the social and economic fabric. These laws
have to do with self-regulation and self-organization.
They constitute the "management principle" by which
systems grow and are stable, learn and adjust, adapt
and evolve. These seemingly diverse systems are one,
in cybernetic eyes, because they manifest viable
behaviour—which is to say behaviour conducive to
survival.*

Stafford Beer

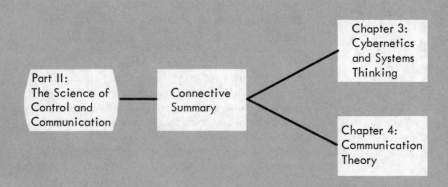

CONNECTIVE SUMMARY

Having considered the origin and nature of systems thinking in Part One, and having developed a vocabulary of the most frequently encountered terms, we now can pass to a closer exposition of the concept of cybernetics.

Chapter Three investigates Cybernetics in general. The overriding objective of this chapter is to expose the student to the science of control and communication, its origin, its logic, its historical developments, and to the tools needed for studying exceedingly complex probabilistic systems which are in the domain of cybernetics.

Cybernetics represents the latest and most sophisticated attempt to tie communication and control—two vital managerial activities—into a meaningful framework. This framework is not, however, fully explained at this early stage of the book. Its understanding will be much easier once the concepts of organizations, environment, information, and decision-making have been explained. A full-scale explanation of this framework is provided in the last part of this book.

There are three main characteristics of cybernetic systems: (1) extreme complexity, (2) probabilism, and (3) self-regulation. The corresponding tools or techniques for dealing with these properties of cybernetics systems are: (1) the black box, (2) information theory, and (3) feedback control. This chapter deals with two of these pairs of characteristics and tools, complexity and the black box, and self-regulation and feedback control. The second pair—i.e., probabilism and information theory—are dealt with in the next chapter under communication theory.

The term "communication" as it is used in cybernetics refers to the transmission of messages between systems. Transmitted messages become information. Thus, communication theory in cybernetics is the study of the behavior of information systems designed to handle information for the purpose of reducing the uncertainty inherent in a system. Information in cybernetic systems can be discussed only in terms of statistical distributions, since not all the factors which must be taken into consideration are known. Thus, the understanding of the basic principles of electronic communication (i.e., amount of information and channel capacity) becomes very relevant to cybernetics.

In applying the basic principles of cybernetics to the management and design of social systems, the electronic handling of information is of secondary importance, since it is provided for by the design of the hardware of the communication device. What is of more importance is the application of certain cybernetic principles to human communication. For this reason, we will briefly present some ideas on nonelectronic communication theory.

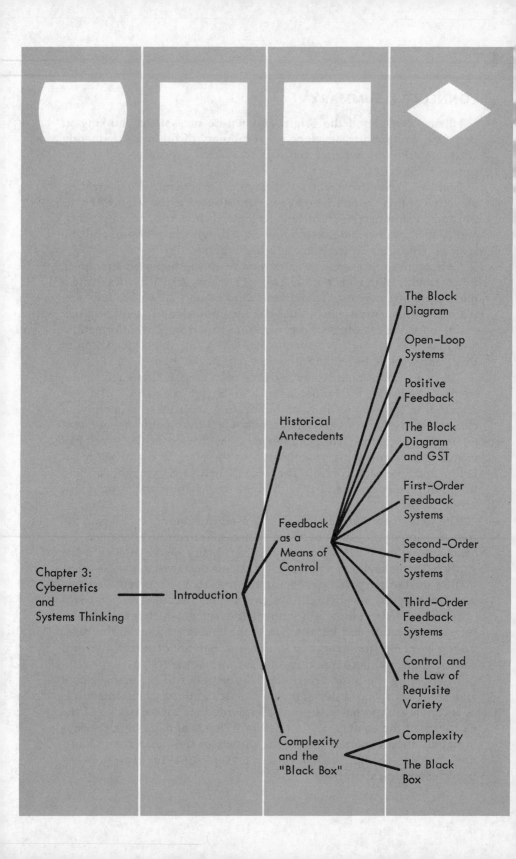

Chapter 3:
Cybernetics
and
Systems Thinking

Introduction

Historical
Antecedents

Feedback
as a
Means of
Control

Complexity
and the
"Black Box"

The Block
Diagram

Open-Loop
Systems

Positive
Feedback

The Block
Diagram
and GST

First-Order
Feedback
Systems

Second-Order
Feedback
Systems

Third-Order
Feedback
Systems

Control and
the Law of
Requisite
Variety

Complexity

The Black
Box

Chapter Three

Cybernetics and Systems Thinking

Benjamin Franklin's reply to a lady who queried the usefulness of his work on electricity: "Madam, what use is a new-born baby?"

Arthur Koestler: *The Ghost in the Machine*

INTRODUCTION

Previously cybernetics was noted as a particularized approach to systems thinking. The cybernetic approach meets the requirements of systems both conceptually and operationally, as the engineering sciences amply testify. However, cybernetics may also be viewed as one of the generalized laws of general systems theory. It merits this appellation since it is concerned with feedback processes of all kinds.

In Chapter One a classification of systems was presented, based upon the criterion of complexity, and nine levels were discussed. As a departure point for the subject of cybernetics, a classification scheme by Beer will be presented which uses two distinct criteria, that of *complexity* and that of *predictability*.[1]

With respect to the first criterion, Beer uses three subclasses: simple, complex, and exceedingly complex. A simple system is one which has few components and few interrelationships; similarly, a system which is richly interconnected and highly elaborate is complex, and an ex-

---

[1] Stafford Beer, *Cybernetics and Management* (New York: John Wiley & Sons, 1964), p. 18.

ceedingly complex system is one which cannot be described in a precise and detailed fashion.

The second criterion concerns the system's deterministic or probabilistic nature; in the former, the parts interact in a perfectly predictable way, while in the latter the system is not predetermined in its behavior although what may likely occur can be described.

The six categories of the two criteria, one threefold and the other twofold, are presented in Figure 3–1. Although Beer is clear in his admonition that these bands are hazy and that they represent merely bands of likelihood, still such a scheme has value since his grouping is done according to the kinds of *control* to which they are susceptible. Not all categories are of equal difficulty and of equal importance.

FIGURE 3–1

A Classification of Systems Based on Susceptibility to Control

Complexity / Predictability	Simple	Complex	Exceedingly Complex
Deterministic (One state of nature)	Pulley Billiards Typewriter	Computer Planetary system	Empty set
Type of control required	Control of inputs	Control of inputs	Control of inputs
Probabilistic (Many states of nature)	Quality control Machine breakdowns Games of chance	Inventory levels All conditional behavior Sales	Firm Human Economy
Type of control required	Statistical	Operations research	Cybernetic

Adapted from Stafford Beer, *Cybernetics and Management*, Science Edition (New York: John Wiley & Sons, 1964), p. 18.

Deterministic systems are of little interest because behavior is predetermined and because they do not include the organization as does an open system. As shown in Figure 3–1, examples of this type of system include the pulley, billiards, a typewriter, most machines in the organizations, the movement of parts on an assembly line, the automatic processing of checks in a bank, and so on. In each of the above examples, the output of the system is controlled by management of the input to the system.

From simple deterministic systems one moves to complex deterministic ones, the singular difference being the degree of complexity involved. The computer is illustrative of this class of system in that it is much more complex than the previously mentioned systems but still operates in a perfectly predictable manner. The point made earlier that the band separating the categories is hazy is demonstrated by the fact that to a computer specialist the computer may not be complex. In a similar manner, the automobile engine is complex for many, but again for a mechanic it is a simple deterministic system. In all of the above examples, there is only a single state of nature for the system which is determined by the structural arrangement of the elements composing it. If these are in the proper configuration, the system will operate in a predetermined pattern.

If one were to introduce a second state of nature in each of the above systems, they would become probabilistic. As seen from Figure 3–1, probabilistic systems can range from the simplest games of chance, such as the flipping of a coin, in which only two possible states can exist, to the organization, in which many multiple states are possible.

In the simple probabilistic system, the additional examples of quality control and machine breakdowns are presented. Because humans are introduced into the production system, and, of course, because humans can exhibit many states of nature, quality becomes a variable factor. It is for this reason that quality control techniques are applied to ensure that a certain state of nature will prevail. Likewise, the wear of parts in a machine necessitates periodic maintenance. The usage rate to a large extent determines the time interval that the machine will be functional (probability of breakdown). In all the above examples, simple statistical techniques can be employed to control the system.

As the complexity of a probabilistic system and the number of states of nature increase, prediction and control of systems behavior become extremely difficult. Thus, while in deterministic systems control of the inputs will provide prediction of the outputs, in probabilistic systems control of the inputs will provide only a range of possible outputs.

The last category of exceedingly complex, probabilistic systems includes the firm, the individual, and the economy, all of which can exhibit variable states of nature. The firm, being composed of multiple subsystems, interacts with other external systems such as the government, competitors, unions, suppliers, and banks. The interaction of the various internal departments and components of an organization and its external subsystem are all so intricate and dynamic that the system is impossible to define in detail.

What, then, is of concern are those systems which exhibit probability and complexity. As noted in Figure 3–1, simple probabilistic systems are controlled through statistical methods while complex probabilistic systems are dealt with through more sophisticated methods of operations research. These tools serve adequately in dealing with systems exhibiting a measure of complexity, but in treating exceedingly complex systems which lack definability they are deficient. Highly complex systems will not yield to the traditional analytical approach because of the morass of indefinable detail; yet these too must be controlled. The technique employed when dealing with extreme complexity is that of the black box. A later section will treat this in detail.

There can be but little doubt that only a few of the systems encountered in the workaday world are of the deterministic type. Most are probabilistic in both structure and behavior. Any system operating within a margin of error is probabilistic and therefore must be treated statistically. Once again the discussion is at an abstract level where it appears that actual organizational situations can be used.

In addition to the two characteristics of probabilism and complexity, Beer includes one additional characteristic of cybernetic systems, and that is self-regulation.

The *self-regulatory feature* of cybernetic systems is essential if systems are to maintain their structure. Control must, therefore, operate from within, utilizing the margin of error as the means of control.

For each of the above characteristics, specialized tools are available for defining, operating, and controlling systems. These, together with the tools of analysis, are presented in Figure 3–2.

FIGURE 3–2
Characteristics and Tools for Analysis of Cybernetic Systems

Characteristics of a System	*Tools for Analysis*
Extreme complexity	Black box
Probabilism	Information theory
Self-regulation	Feedback principle

However, before examining each of these characteristics in detail, it may be useful to trace the development of cybernetics from its inception.

HISTORICAL ANTECEDENTS

For many years now automatic control systems, which have largely been confined to governors, servomechanisms, and the like, have had their greatest impact in the field of engineering. This is not a thing to be wondered at for ever since the 1790s, when James Watt invented his "governor"—the mechanical regulator for stabilizing the speed of rotation of the steam engine—the field of cybernetics has been almost wholly dominated by the mechanical engineer. Even today many of the guidance and control systems employed in missiles are based on fundamentally the same principles enunciated decades ago. While it is true that automatic control systems are used more and more each year, still relatively little application of such systems outside the realm of mechanical devices takes place. Until the recent contributions to cybernetics by such men as Norbert Wiener, W. Ross Ashby, and Stafford Beer, to name a few, the all-important idea of feedback, so vital to a cybernetic system, has only with difficulty been transferred to the political, economic, social, and managerial fields.

Historically, cybernetics dates from the time of Plato, who, in his *Republic*, used the term *kybernetike* (a Greek term meaning "the art of steersmanship") both in the literal sense of piloting a vessel and in the metaphorical sense of piloting the ship of state—i.e., the art of government. From this Greek root was derived the Latin word *gubernator*, which, too, possessed the dual interpretation, although its predominant meaning was that of a political pilot. From the Latin, the English word *governor* is derived. It was not until Watts termed his mechanical regulator a "governor" that the metaphorical sense gave way to the literal mechanical sense. It was this that in 1947 provided the motivation for Norbert Wiener to coin the term *cybernetics* for designating a field of studies that would have universal application. With this the term has now come full circle.

In more recent times the science of cybernetics has been much abused by writers, equating it with electronic computers, automation, operations research, and a host of other tools. Cybernetics is none of these, nor is it a theory of machines, although it derives from a particular type of mechanism (regulators). In his classic text, Wiener defines cybernetics as the science of control and communication in the animal and the machine.[2] It is quite evident that Wiener intended cybernetics to be concerned with universal principles applicable not only to engineering systems but also to living systems.

> In giving the definition of Cybernetics in the original book, I classed communication and control together. Why did I do this? When I communicate with another person, I impart a message to him, and when he communicates back with me he returns a related message which contains information primarily accessible to him and not to me. When I control the actions of another person, I communicate a message to him, and al-

---

[2] Norbert Wiener, *Cybernetics or Control and Communication in the Animal and Machine* (New York: John Wiley & Sons, 1948).

though this message is in the imperative mood, the technique of communication does not differ from that of a message of fact. Furthermore, if my control is to be effective I must take cognizance of any messages from him which may indicate that the order is understood and has been obeyed. . . .

When I give an order to a machine, the situation is not essentially different from that which arises when I give an order to a person. In other words, as far as my consciousness goes, I am aware of the order that has gone out and of the signal of compliance that has come back. . . . Thus the theory of control in engineering, whether human or animal or mechanical, is a chapter in the theory of messages.[3]

This view of control can be profitably applied at the theoretical level of any system and to diverse disciplines in both large and small systems. Wiener further states, "It is the purpose of Cybernetics to develop a language and techniques that will enable us indeed to attack the problem of control and communication in general, but also to find the proper repertory of ideas and techniques to classify their particular manifestations under certain concepts."[4]

Many of the concepts of cybernetics as applied to physical systems are relevant for an understanding of social groups as well. Wiener certainly anticipated this, and, while cautioning against abuse of cybernetics in areas lacking mathematical analysis, he pointed out that the application of cybernetic concepts to society does not require that social relations be mathematicizable *in esse*, but only *in posse*. By clarifying formal aspects of social relations, cybernetics can contribute something useful to the science of society.[5]

Apropos to this, Charles Dechert remarks:

More recent definitions of cybernetics almost invariably include social organizations as one of the categories of system to which this science is relevant. Indeed Bigelow has generalized to the extent of calling cybernetics the effort to understand the behavior of complex systems. He pointed out that cybernetics is essentially interdisciplinary and that a focus at the systems level, dependent upon mixed teams of professionals in a variety of sciences, brings one rapidly to the frontiers of knowledge in several areas. This is certainly true of the social sciences.[6]

Besides the element of control, the other central concept in cybernetics is that of communication. Communication is concerned with in-

[3] Norbert Wiener, *The Human Use of Human Beings: Cybernetics and Society* (Garden City: Doubleday & Company, 1954), pp. 16–17.

[4] Ibid. p. 17.

[5] Norbert Wiener, *God and Golem, Inc.* (Cambridge, Mass.: MIT Press, 1964), p. 88.

[6] Charles R. Dechert, "The Development of Cybernetics," in P. Schoderbek (ed.) *Management Systems*, 2d ed. (New York: John Wiley & Sons, 1971), p. 74. Originally printed in *The American Behavioral Scientist* (June 1965), pp. 15–20.

formation transfer both between the system and its environment and also among the parts of the system. The cybernetic concept of information ranges farther afield than it does in other disciplines. It includes not only electrical impulses as in engineering, signals sent to the brain as in human beings, cardinal values as in mathematics; it embraces all carriers of information. While information theory can be considered as a special tool dealing with quantitative aspects of information, the use here of the term *information* will be much less restrictive. At present, information theory is somewhat limited in application, although attempts to extend it to other disciplines have not been wanting.

While some may find fault with a presentation of cybernetics as a tool for analyzing *all* purposeful behavior, nevertheless, the present writers are convinced that cybernetics can provide a better and a fuller understanding of the system at hand. While social scientists, mathematicians, and other researchers may restrict their definitions to certain domains with the rigidities so necessary for scientific investigation, these writers believe that systems thinking ought to be extended to the many problems encountered in the living firm and not only to those fabricated in the laboratory. The domain of systems thinking and cybernetics should be enlarged as much as possible and not harnessed within narrow and restrictive disciplinary limits.

FEEDBACK AS A MEANS OF CONTROL

Feedback control systems are neither new nor rare. While their historical antecedents date back at least 2,000 years,[7] it was only in the 20th century, and indeed within the past decades, that their underlying principles have been exploited. In some respects it is remarkable that it took so long to unravel the central ideas involved in feedback control systems. Yet the recognition that some systems utilized common principles, and that similar systems could be constructed, was indeed a profound and important discovery. Once this was realized, the first approximation to a theory of feedback systems was soon in coming, coming with quantum leaps in the past two decades.

The feedback control system is characterized by its closed-loop structure. Such a system can be defined as one "which tends to maintain a prescribed relationship of one system variable to another by comparing functions of these variables and using the difference as a means of

---

[7]Otto Mayr, in his extremely thorough book entitled, *The Origins of Feedback Control* (Cambridge, Mass.: MIT Press, 1969) traces the evolution of the concept of feedback through three separate ancestral lines: the water clock, the thermostat, and mechanisms for controlling windmills. The water clock is the earliest description of a feedback device on record and dates from the third century B.C. The thermostat has a more recent history, having been invented in the early 17th century by Cornelius Drebbel. Devices for the automatic control of windmills were invented in the 18th century.

control."[8] The same source also defines feedback as "the transmission of a signal from a later to an earlier stage." For purposes of discussion, a distinction will be made between automatic and manual feedback control systems. In the former, a closed-loop feedback exists which is executed by the system, while in the latter a human operator is needed by the system to close the loop, i.e., to take some course of action on the basis of the feedback information received. Although both types are important, still, because business organizations utilize people in the feedback control system, more attention will be devoted to manual feedback control systems. The automatic closed-loop feedback systems, however, serve as excellent examples for illustrating the feedback concept.

The Block Diagram

Because researchers in different disciplines lack a common language for communicating with one another, attempts have recently been made in the field of feedback control systems for the adoption of standardized symbols and terminology. The block diagram is a basic linguistic tool for illustrating functionally the components of a control system. It is precisely this type of representation that has revealed the underlying similarity of seemingly unrelated systems. The blocks themselves represent things which must be done and not the physical entities, equipment, and the like. In the block diagram, four basic symbols are employed: the arrow, the block, the transfer function, and the circle.

1. The arrow (Figure 3–3) denotes the signal or command which is some physical quantity acting in the direction of the arrow. It is not the flow of energy which is being noted but rather the flow of information which shows the causal relationship that exists. This input is variously termed the command, the signal, the desired value, or the independent variable and is often represented as a mathematical variable.

FIGURE 3–3
The Command Signal

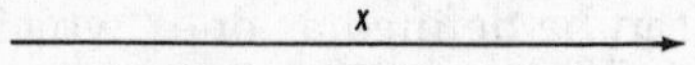

$$x$$
⟶

2. In a block diagram, all system variables are linked to each other through the functional *block* (Figure 3–4). This block is a symbol for the mathematical operation on the input signal to the block which produces

[8] A.I.I.E. Committee Report, "Proposed Symbols and Terms for Feedback Control Systems," *Electrical Engineering,* vol. 70 (1951), p. 909.

FIGURE 3-4
The Block

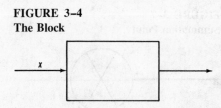

the output. Signals coming into the block are independent, signals leaving the block are dependent, since these are the outputs or effects.

3. Included within the block is the transfer function, which is the mathematical operation to be performed on the block. The output is then the transfer function multiplied by some input. (Figure 3-5)

FIGURE 3-5
Block with Transfer Function

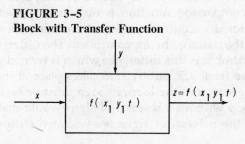

When a signal has two separate effects which go to two different points, this is termed a branch point, as in Figure 3-6.

FIGURE 3-6
Branch Points

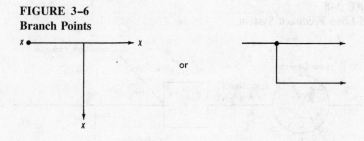

4. The circle with a cross represents a summation point where a comparison is made between two quantities, the command signal or desired state, and the feedback signal or actual state. It is here that the signals are added or subtracted. In Figure 3-7, θ_i is an input signal being fed into the system and θ_o is the output of the system. This summation point is also referred to in the literature as the error detector, the comparator, or the measurement point.

FIGURE 3–7
Summation Point

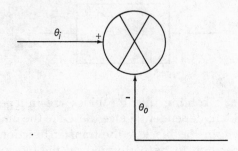

In the simplest closed-loop system, shown in Figure 3–7, the output is fed back to the summation point, where it is compared with the input signal θ_i. This comparison function is obviously one of the requirements of an automatic control system where the command signal is compared with the variable being controlled. The difference is used as the means of control; it is this difference which is termed negative feedback. In negative feedback, subtraction takes place at the comparator. As a signal travels around the loop, its sign must be reversed, since to have a closed loop without a reversal of signs would make the system unstable. Thus, the reversal of signs is associated with *negative* feedback.

Figure 3–8 depicts a closed-loop feedback system. The output in this is obtained by the multiplication of the transfer function (in this case K) by the input to the block (e).

FIGURE 3–8
Closed-Loop Feedback System

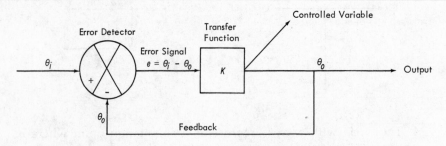

The system in the above figure may be described mathematically by the following set of equations:

$$e = \theta_i - \theta_o$$
$$\theta_o = Ke$$

which can be reduced to the single equation

$$\theta_o = \frac{K\theta_i}{1+K}$$

When K is large, θ_o approximates θ_i

While no one would claim that the following example of an economic system corresponds to reality, still it can provide some insights into the control mechanisms utilized by the government for correcting unstable conditions. Since many econometric simulation models of the economy do in fact deal with real life problems, from this premise the economy is a fitting subject for cybernetic control.

Although several general treatments of the economy from the feedback control approach exist, the following one by Porter[9] serves well for illustrative purposes.

Let

S_d = desired level of spending
S_a = actual level of spending
e = difference = $S_d - S_a$

Assume R = interest level = $(R_o + r)$, in which R_o = standard level of interest and $r = -ke$, is the change in interest, dependent upon e, which provides the control.

The simplified equations of the economic system can now be written

$$S_d - S_a = e$$
$$R = R_o - ke$$

The corresponding block diagram (Figure 3–9) is the following:

FIGURE 3–9
A Simple Economic Control System

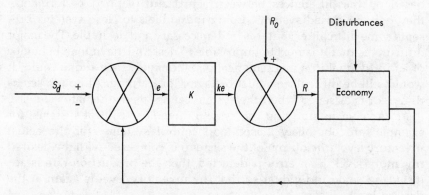

---

[9] Arthur Porter, *Cybernetics Simplified* (New York: Barnes and Noble, 1970), pp. 14–15.

Open-Loop Systems

An open-loop system is one in which the output of the system is not coupled to the input for measurement. Mechanical examples of open-loop systems are the water softener in a home, the washing machine or dryer, the automatic sprinkler system, the traffic light, automatic light switches, and the toaster. In all of these cases the output is not compared with the reference input, but, instead, for each of the reference inputs there exists a corresponding fixed operating condition. Most of the above operate on a time basis. Open-loop systems can be depicted as in Figure 3–10.

FIGURE 3–10
Open-Loop System

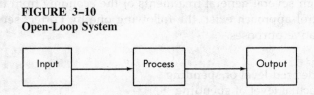

Perhaps all of the above could be made into closed-loop systems if they met the previously mentioned criteria for such systems. The individual examining the dryness or cleanliness of the clothes and comparing it with some standard, the person testing the amount of moisture in the ground and comparing it with a standard, the man checking the amount of daylight and relating it to a standard—all provide for measurement against a goal.

Organizations are basically open-loop systems when viewed without people. However, with the appearance of the human operator as a controller who compares the input and output and makes corrections based on the differences between actual and planned performance, they become closed systems. Using individuals to close systems presents some difficulties, but it is both necessary and desirable. The major difficulty is that it is next to impossible to describe the human behavior of an individual in a mathematical equation and, even if one could, it would still be difficult to adjust behavior for learning. Although scientific rigor is lessened when the system cannot be mathematicized, still there is value in studying the system. Inventory control systems, for example, are obviously closed-loop control systems, for the actual inventory level (the output of the system) is compared with the desired inventory level. If an error is detected, then the production rate is adjusted by some individual so that the inventory level is again at the desired level.

Positive Feedback

Although the preponderance of attention in the literature has been given to negative feedback systems, some mention should be made of positive feedback. Positive feedback systems utilize part of their output as inputs to the same system in such a way that they are, in fact, deviation-amplifying rather than deviation-counteracting systems. All growth processes involve positive feedback systems, since a part of the output is amplified. This is true for both mechanical systems and for human organizational systems. In mechanical systems, power-steering and power brakes are common examples of power-amplifying positive feedback systems.

In social systems such as organizations, the term "positive feedback" is usually interpreted as "good news" as contrasted with negative feedback, which is associated with "bad news." These colloquial expressions are very misleading for the systems man. Positive feedback mechanisms are growth-promoting devices, while negative feedbacks are control-maintaining processes. The activities of the marketing or promotion subsystem of an organization perform growth processes by attempting to enlarge the difference between accomplished usage of the organization's products or services and the aspired goal. On the other hand, the activities of the accounting, quality control, and industrial or public relations subsystems perform control functions in that their purpose is to minimize deviations between set standards (budgets, morale, corporate image, and so on) and actual performance.

The concepts of positive and negative feedbacks are extremely useful in understanding organizational or system behavior. For this reason, a more detailed discussion will be given later in this book after the concepts of organization, the organization-environment interaction, and information technology have been duly dealt with.

The Block Diagram and General Systems Theory

As problems become more complex and the use of computers for simulation and problem solving becomes more widespread, the need for a conceptual framework, both for the definition of the problem and for its solution, becomes more acute. A critical step in this process is the design of the system, which in turn requires a decision regarding the structure of the system. The structural aspects of the system's behavior and operation must therefore be specified. The block diagram serves admirably for the identification of the structural relationships of the system. It is at this point that, as Mesarovic states, the block diagram and general systems theory come together, for general systems theory can assist in the structural considerations by "preserving the simplicity of the block diagram while introducing the precision of mathematics."[10] For complex systems, he holds, a general systems model is a necessary step between the block diagram and the detailed mathematical model as shown in Figure 3–11.

FIGURE 3–11
Relationship of the Block Diagram and General Systems Theory

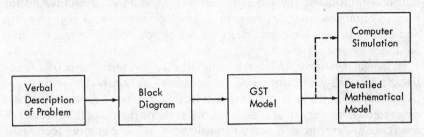

First-Order Feedback Systems (Automatic Goal Attainment)

The closed-loop control feedback systems discussed in the previous section are all feedback systems of the first order because the system is monitored against an external goal (Figure 3–12). It is given one particular command which it is to carry out irrespective of changes in the environment, and so on. In goal-directed systems which operate on the principle of negative feedback, the system is maintained by correcting deviations from the goal. There is no other choice available to the system but to correct the deviation. Thus, the purpose of a first-order feedback system is to maintain the system at a desired state of equilibrium. The system cannot make any conditional response. It has no memory nor does it have available any alternative action. In this type of control

FIGURE 3–12
First-Order Feedback System

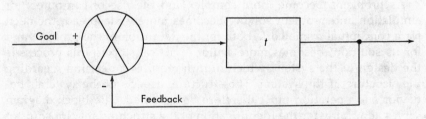

----
[10] Mihajlo D. Mesarovic, "General Systems Theory and its Mathematical Formulation," paper presented at the 1967 Systems Science and Cybernetics Conference, IEEE, Boston, Mass., October 11–13, 1967, p. 13.

system the operation is clearly circular, since, after the comparison against the standard, a recycling must take place. A first-order feedback system always operates in this manner irrespective of changes in the environment. Thus, in the case of the thermostat, a thermal equilibrium is maintained regardless of external weather conditions.

Second-Order Feedback Systems (Automatic Goal Changer)

When a system contains a memory unit and can initiate alternative courses of action in response to changed external conditions and can choose the best alternative for the particular set of conditions, it is said to be a feedback system of the second order. A memory includes all the facilities in the feedback loops available to the system for storing or recalling data from the past. In an organization, such a memory would include the personnel, staffs, policies, filing systems, and so on. This feedback of information from the past is done for decision making in the present.[11] The second-order system has the ability to change its goals by changing the behavior of the system. In other words, goal changing is part of the feedback process itself. In Figure 3–13 Churchman uses a telephone exchange as an example of an automatic goal-changing unit.[12]

FIGURE 3–13
Feedback Circuit with Memory Device

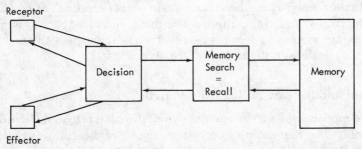

By adding a memory and more complicated feedback loops, an organization can have more control over its own activities. In this case a series of alternatives for action is built into the system if external conditions (detected by the receptor) change. An example is the automatic switching of a telephone exchange.
Source: C. West Churchman et al., *Introduction to Operation Research* (New York: John Wiley & Sons, 1957), Reprinted by permission of John Wiley & Sons.

There are many examples of this type in which, if goal A is attained, priority is shifted to goal B, and so on. When goal B is attained, there is

---

[11] See Karl W. Deutsch, *The Nerves of Government* (New York: The Free Press, 1966), for an excellent exposition of types of feedback. See especially Chapters 11 and 12.

[12] C. West Churchman et al., *Introduction to Operation Research* (New York: John Wiley & Sons, 1957), p. 81.

a shift back to goal A or on to goal C. Any system that can change goals is said to be autonomous. Goal changing is dependent upon memory. When there is no more memory—i.e., when either the system is cut off from all past information or the information has ceased to be effective—the system can no longer change goals. Such a system loses control over its behavior and acts simply as an automaton. The better the memory and the greater the ability to recall past information, the more autonomous the system. The ability to store and recall information, allowing the system to choose alternative courses of action in response to environmental changes, is termed learning. Deutsch[13] defines it as the internal arrangement of resources that are still relevant to goal seeking. This can mean the addition of another channel of communication, a change of information put in the memory, a change in the control process, or any number of similar actions.

Third-Order Feedback Systems (Reflective Goal Changer)

A third-order feedback system is one that can reflect upon its past decision making. It not only collects and stores information in its memory but it also examines its memory and formulates new courses of action. Obviously, such a system refers to both the individual and the organization. The organization as a system can direct its growth by changing its goals, terminating certain activities, initiating new activities, engaging in research, continually searching its memory for vital information, modifying the value system of its personnel, or by changing the firm's operating patterns. The third-order feedback system not only is autonomous, it also possesses a consciousness. Figure 3–14 shows a possible configuration of this type.[14]

Control and the Law of Requisite Variety

The more complex a system, the more difficult it is to understand and control it. The more complex a system, the more difficult it is to define its structure (its interrelationships) and consequently, the more difficult to predict its behavior. As the components of a system increase in number, the interrelationships typically increase and the system is said to possess more variety than it did initially. In moving from a simple organization with few employees having few interdependencies to a more complex one with many employees, the variety of uncertainty increases. When one asks, "How can uncertainty be reduced?" the answer is, "Through information." Information extinguishes variety

[13] Deutsch, *Nerves of Government*, p. 92.
[14] Churchman et al., *Operation Research*, p. 84.

FIGURE 3–14
Additional Memory Refinements

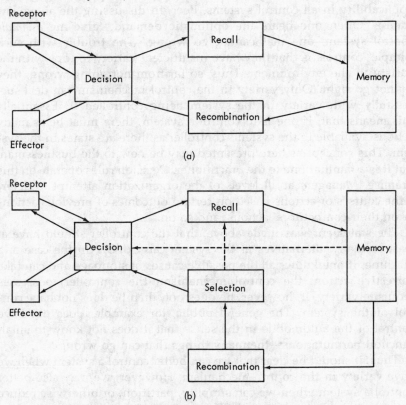

[a] If information in the memory can be recombined and new alternatives produced for action (by the machine or organization itself), the unit becomes more versatile and autonomous. This device makes simple predictions. [b] Development of a consciousness. If many memories can be combined, and if from the many combinations a few can be selected for further consideration, further recombination, etc., the unit will have reached a still higher level of versatility or autonomy. The dashed lines indicate comparisons of what is going on with what has happened in the past and what might occur in the future (second- and third-order predictions). In many organizations, these comparisons are poorly made.

Source: C. West Churchman et al., *Introduction to Operation Research* (New York: John Wiley & Sons, 1957). Reprinted by permission of John Wiley & Sons.

and the reduction of variety is one of the techniques of control, not because it simplifies the system to be controlled, but because it makes the system more predictable.[15] Therefore, what is required is the same amount of variety in the control mechanism as there is in the system being controlled. This important principle Ashby has called the *Law of Requisite Variety*.[16] If there is sufficient permutation variety which will provide for a one-to-one transfer from the control mechanism to the

---

[15] Stafford Beer, *Cybernetics and Management,* Science Edition (New York: John Wiley & Sons, 1964), p. 44.

[16] W. R. Ashby, *Introduction to Cybernetics* (New York: John Wiley & Sons, 1963).

system, then there is "requisite" variety. As Ashby states, "Only variety can destroy variety." This fundamental concept has very general applicability in all control systems. Beer, in discussing the above law states: "Often one hears the optimistic demand: 'give me a *simple* control system; one that cannot go wrong.' The trouble with such 'simple' controls is that they have insufficient variety to cope with the variety in the environment. Thus, so far from not going wrong, they cannot go right. Only variety in the control mechanism can deal successfully with variety in the system being controlled."[17] Essentially this means that, if one is to control a system, there must be as many actions available to the systems controller as there are states in the system. This concept as here presented may be new to the businessman, but it is a familiar one to the practitioner. Decision rules operate on this premise. Managers at all levels of the organization attempt to determine courses of actions based on certain outcomes of previous actions or on their competitors' actions, and so on.

The statement was made above, that the controller should have at least as many alternatives as the system can exhibit. In the case of a machine, if one knows all the possible causes of stoppage and can take corrective action, the control mechanism (the controller) possesses requisite variety. If, however, he does not, then he does not have control of the system. The general public, for example, does not have control of the automobile in the sense that it does not know the malfunction permutations—the many things that can go wrong.

Thus, it should be clear that we can better control a system when we have variety in the control mechanism. However, we may also better control a system when we can simplify, partition, or otherwise reduce the variety in the system itself. This is precisely what policies attempt to do in an organization. Thus, rather than having salesmen deal individually with every instance as it occurs, a credit policy serves as a guideline for all sales inquiries. In this instance the credit policy serves as a regulator whose function is to block the flow of variety into the system. In the following diagram (Figure 3–15), *I* represents some inputs to a system pursuant to a goal. *D* represents disturbances that may occur within or without the system, while *R* is the regulator whose function is to block the transmission of variety to some outcome *O*.

Or—to take another example—in preparing for an important football game, a team will scout its upcoming opponents to determine the pattern of tactics which the team is likely to employ. The proliferation of variety which the opposing team is capable of employing is obviously great, but, nevertheless, the scouting team will deduce some patterns with a high probability of occurrence. This would be especially

---

[17] Beer, *Cybernetics and Management*, p. 50.

FIGURE 3–15
Input-Output Model

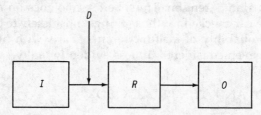

true if the opposing team's "power" is centered around a few individuals. The scouting team, in an effort to control the situation (game), will try to counteract this variety by adjusting its own resources. As is well known, the cybernetician's strategy of requisite variety often works. In this example we are obviously dealing with incomplete information respecting the variety that is possible; however, this is essentially the same for the manager attempting to control a business organization. A firm will typically acquire (scout) environmental information (competition, political factors, the economy, the state of technology, labor activities, and so on) in order to reduce the uncertainty of its operation. It is the task of the manager to scan his environment in order to better control the organizational system.

That the law of requisite variety is of universal application in control systems must not be too obvious; otherwise there would be fewer systems that go out of control each year. One is led to believe that in many cases this principle is not even adverted to. It would be difficult to find systems analysts who claim that federal agencies generally have requisite variety in their control mechanism. Or take the controversy over manned and unmanned space flights. One side states that manned flights provide greater variety for counteracting all possible disturbances to the system. On the other side, the greatly increased costs that the taxpayers would have to pay for unmanned systems with comparable high variety are generally underplayed. The greater the systems variety, the greater the control variety, and the greater the costs.

Ashby's Law of Requisite Variety is at work in many daily situations. One controls a business meeting by limiting the variety of topics to be discussed by the use of an agenda. When one goes on a camping trip, he keeps in mind the motto: Be prepared. Because of the variety of system contingencies (camping situation), he tries to employ requisite variety in the control system (first aid kits, lotions, pills, food, and so forth).

Seldom in the real world does one possess requisite variety; yet one operates as if he had total control. Basically what occurs is this: one attempts to develop variety only for those factors that have a high probability of occurrence. Factors with a low probability of occurrence are given but scant attention. Thus, before one goes on a long trip by car, he has the car checked for those things most likely to go awry (with the highest probability of malfunctioning). Parts with but a low incidence are not even considered, unless evidently faulty.

COMPLEXITY AND THE "BLACK BOX"

Earlier in this chapter the cybernetic system was defined in terms of extreme complexity, probabilism, and self-regulation, and the analytical tools corresponding to each of these systems characteristics were outlined. Thus, self-regulation in a cybernetic system is best understood by employing the analytic tool of the feedback principle. Probabilism is best handled through the vocabulary and conceptual tools of probability theory or its modern equivalent, information theory. Because of its importance in understanding organizational behavior, an entire chapter will be devoted to the logic, principles, and foundations of the subject. This section will be devoted to the first characteristic of a cybernetic system and its corresponding analytic tool; namely, extreme complexity and the black box.[18]

Complexity

The explanation of the term "complexity" can be approached from many different viewpoints. From the mathematical systems viewpoint, complexity can best be understood as a statistical concept. More precisely, complexity can best be explained in terms of the probability of a system's being in a specific state at a given time.[19] From a nonquantitative viewpoint, complexity can be defined as the quality or property of a system which is the combined outcome of the interaction of four main determinants. These four determinants are: (1) the *number of elements* comprising the system; (2) The *attributes* of the specified elements of the systems; (3) The *interactions* among the specified elements of the systems; and (4) the *degree of organization* inherent in the system; i.e., the existence or lack of predetermined rules and regulations which

---

[18] The investigation of the black box approach to complexity is based upon the following works: W. R. Ashby, *An Introduction to Cybernetics*, Science Edition (New York: J. Wiley & Sons, 1963); W. R. Ashby, *Design for a Brain* (London: Chapman and Hall, Ltd., and Science Paperbacks. Butler and Tanner, Ltd., 1960); S. Beer, *Cybernetics and Management*, Science Edition (New York: J. Wiley & Sons, 1969); S. Beer, *Management Science, The Business Use of Operations Research* (New York: Doubleday, 1968); and H. A. Simon, *The Sciences of the Artificial* (Cambridge, Mass.: MIT Press, 1970).

[19] Beer, *Cybernetics and Management*, p. 36.

guide the interactions of the elements and/or specify the attributes of the system's elements.

Most attempts at measuring the complexity of a given system usually concentrate on two criteria: the number of elements and the number of interactions among the elements. This is especially true in classical statistics situations. This kind of measure of complexity is very superficial and, to some extent, misleading. Confining one's self to these two dimensions of complexity will lead one to classify a car engine as a very complex system. There are indeed a large number of elements and an equally large number of interactions among all the parts of a car engine. By the same token, one would be inclined to classify a two-person interaction as a very simple system, for there are only two elements and only two possible interactions involved.

If one were to incorporate the other two determinants of complexity into one's attempt to measure it—namely, the attributes of the elements and the degree of organization—then one would arrive at a different conclusion. Concerning the example of a car engine, one would observe that the interactions must obey certain rules and follow a certain sequence. One would also observe that the attributes of the system's elements are predetermined. By using all these four criteria of complexity one must conclude that the car engine is, in fact, a very simple system.

The seemingly simple system of the two-person interaction is indeed a complex system, since the attributes of each element are not predetermined and since the degree of organization, despite the existence of some rules of human conversation and interaction, is very low. The elements, in other words, have a free will in obeying or disregarding the rules of human conduct. In this case the ultimate outcome of the conversation, that is to say, the degree of predictability of the final state of the interaction is uncertain. One must therefore conclude that this two-person system is indeed complex.

This relationship between the four main determinants of complexity (the number of elements, the attributes of the elements, the number of interactions, and the degree of organization in the system) and the degree of complexity can be easily illustrated by using the so-called span of control principle. This principle states that "no supervisor can supervise directly the work of more than five or, at the most, six subordinates whose work interlocks."[20] The rationale of this principle is the increased complexity which accompanies the increase in the number of the subordinates for each supervisor. This complexity is equated with the number of direct and cross relationships between the different

---

[20] Lynsall F. Urwick, *Scientific Principles and Organization* (New York: American Management Association, 1938), p. 8.

members of the group which increase by a geometrical progression. Thus, a superior interacting with seven subordinates who also interact with each other will generate 490 potential relationships — an enormous complexity indeed.[21]

This is a misleading measure of complexity, however. In order to gain a meaningful measure of the complexity involved in the span of control situation, one must consider in addition to the above two criteria — namely the number of elements (members of the group) and their interactions — the attributes of each member, as well as the organization of the task involved. By considering these two sets of criteria one can arrive at a different set of possible states of the group/system. If the task is highly routinized and at the same time the members of the group are well trained, then, assuming no intentional attempts to overburden the superior, the system will be fairly simple to the extent that most of the possible interactions will not be exercised by the subordinates. In addition, there will be a set of rules and procedures which will tend to reduce the possible number of interactions considerably.

A supervisor attempting to supervise two energy experts, one advocating coal as the most promising future energy source, the other explaining the benefits of fusion as a source of energy, would be confronted with a much more complex system than his colleague who supervises 20 oil engineers. Complexity is indeed a relative concept which is determined by the interaction of all four determinants and not just by the mere number of elements and their interactions.

The Black Box

In explaining the complexity of a system in the above example, no attempt was made to describe or define in detail the elements, processes, interactions, and states of these systems. Only the number of inputs and the number of potential relationships were specified. No speculations were made regarding the nature of the processes responsible for producing one state or another. In other words, as far as the process of the system is concerned, the system was presented as undefinable in detail. This is the same thing as saying that the system has been treated as a black box.

The problem of the black box first arose in electrical engineering.

---

[21] The number of relationships is derived by using the well-known Graicunas formula:

$$C = n \left[\frac{2^n}{2} + n - 1 \right]$$

Where n = number of employees reporting to a superior
C = number of potential relationships.

The engineer is given a sealed black box with terminals for input to which he may apply any voltages, shocks, or other disturbances. The box also has output terminals from which he may observe whatever he can. He is to deduce the contents of the black box.

Although the black box technique originated with electrical engineers, its present range of application is far wider. The physician studying a patient with brain damage may be trying, by means of tests given and responses observed, to deduce something of the mechanisms involved. The psychologist, psychiatrist, or business consultant employs the black box technique whenever he attempts to study anomalies in the behavior of the individual or firm by testing certain input functions of the system and by recording the changes in the composition of the outputs. He manipulates the input and classifies the output.

If the account of the black box technique were to stop here, one might get the impression that what is involved here is but the classical conditioning or stimulus-and-response technique of the early psychologist or the cause-and-effect approach of the analytic thinker. Nothing could be further from the truth than the equating of the black box input-manipulation, output-classification technique with either stimulus and response or cause and effect. Both of the latter techniques assume fairly simple situations consisting of two-term causal relationships that are more often than not fabricated by the observer/experimenter. While here the observer attributes certain responses or effects to certain stimuli or causes, the theory of the black box is simply the study of the relations between the experimenter and the object, as well as the study of what information comes from the object, and how it is obtained.

The black box technique is illustrated in Figure 3–16. From Figure

FIGURE 3–16
The Black Box Technique

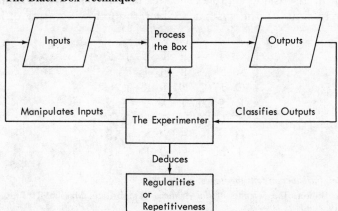

3–16 it can be seen that by thus acting on the box, and by allowing the box to affect him and his recording apparatus (i.e., the protocol), the experimenter is coupling himself to the box, so that the two together form a system with feedback. When a generous length of record has been obtained, the experimenter will examine it for regularities in the behavior of the system represented by the box.

Let us explain the above line of reasoning by a simple example taken from Ashby.[22] Suppose that a system had two possible input states *a* and *b* and four possible output states *f, g, h,* and *j.* Thus, a typical protocol might read as in Table 3–1. As this protocol reveals, the primary data of any black box investigation consist of a sequence of values of the vector with two components: an input state and an output state. One can also note the regularity or repetitiveness in the black box's behavior. For example, the protocol entry *aj* is always followed by either *af* or *bf*—the *j*'s transaction is single-value (*f*) although the *a*'s is not. The more such regularities in the system's behavior the observer or experimenter can detect, the more knowledge he is said to have about the box. In some instances combinational behavior may be exhibited; in others the pattern may be strictly sequential.

TABLE 3–1
The Protocol

Output / Input	*f, g, h, j*									
	1	*2*	*3*	*4*	*5*	*6*	*7*	*8*	*9*	*10*
a *or* b	ag	aj	af	af	aj	bf	ah	bj	bf	af

Source: Adapted from R. Ashby, *Introduction to Cybernetics* (New York: J. Wiley & Sons, 1963), p. 89.

Simon in discussing complex systems uses the aspect of redundancy in order to simplify the system.[23] Given the following array of letters:

```
A B M N R S H I
C D O P T U J K
M N A B H I R S
O P C D J K T U
R S H I A B M N
T U J K C D O P
H I R S M N A B
J K T U O P C D
```

[22] Ashby, *Introduction to Cybernetics*, pp. 86–92.

[23] H. A. Simon, *The Sciences of the Artificial* (Cambridge, Mass.: MIT Press, 1969), pp. 109–110.

Let us call the array $\begin{vmatrix} AB \\ CD \end{vmatrix}$ a, the array $\begin{vmatrix} MN \\ OP \end{vmatrix}$ m, the array $\begin{vmatrix} RS \\ TU \end{vmatrix}$ r, and the array $\begin{vmatrix} HI \\ JK \end{vmatrix}$ h. Let us call the array $\begin{vmatrix} am \\ ma \end{vmatrix}$ w, and the array $\begin{vmatrix} rh \\ hr \end{vmatrix}$ x. Then the entire array is simply $\begin{vmatrix} wx \\ xw \end{vmatrix}$ While the original structure consisted of 64 symbols, it requires only 35 to write down its description:

$$S = \begin{vmatrix} wx \\ xw \end{vmatrix}$$

$$w = \begin{vmatrix} am \\ ma \end{vmatrix} \qquad\qquad x = \begin{vmatrix} rh \\ hr \end{vmatrix}$$

$$a = \begin{vmatrix} AB \\ CD \end{vmatrix} \qquad m = \begin{vmatrix} MN \\ OP \end{vmatrix} \qquad r = \begin{vmatrix} RS \\ TU \end{vmatrix} \qquad h = \begin{vmatrix} HI \\ JK \end{vmatrix}$$

We achieve the abbreviation by making use of the redundancy in the original structure. Since the pattern $\begin{vmatrix} AB \\ CD \end{vmatrix}$, for example, occurs four times in the total pattern, it is economical to represent it by a single symbol a. By recognizing redundancy one reduces the complexity of the system.

This analogy can be usefully applied to the business situation. Say, for example, that a firm in attempting to assess the impact of a possible price reduction in its products can from past experience predict that competitors will do the same. This in effect reduces complexity in the system. Simon also makes the important point that hierarchic systems are composed of only a few different kinds of subsystems but arranged in different ways. If this is known, then again complexity can be reduced. It is precisely this factor that makes the field of cybernetics applicable to all types of systems. The control aspect of a thermostat is similar to those employed in complex space explorations, albeit an obvious difference in degree of complexity is involved.

One of the merits of the black box technique is that it provides the best antidote against the tendency of the investigator to oversimplify a complex phenomenon by breaking it into smaller parts. The black box technique for dealing with complexity represents a selection procedure based on a series of dichotomies. In other words, the investigator of a complex situation manipulates the inputs to the black box and classifies the outputs into certain distinct classes based upon the degree of similarity of the output state. He then converts each class into a "many-to-one" transformation. In this way, the observer obtains a black box with a binary output and a large number of input variables which permute themselves and their interconnections to represent *one* output state.[24]

---

[24] Beer, *Cybernetics and Management;* also, *Management Science.*

To sum up, the black box technique involves the following sequential steps: (1) input manipulation, (2) output classification, and, finally (3) many-to-one transformations.

The input manipulations over an extended number of trials reveal (in the output classification as recorded in the protocol) certain similarities or repetitiveness. These similarities are in turn converted into legitimate many-to-one transformations that act as implicit control devices; these many-to-one transformations account for the reduction of the system's variety without unnecessary simplifications.

Nature is full of examples manifesting the black box technique for dealing with complexity. There are mechanisms common to all life, such as the hereditary apparatus – the genetic structure with its hundreds of genes; occasional variations in the nature of genes through mutation; the distributing and combining of gene variations by sexual recombination. There is the principle of natural selection, whereby favorable (i.e., survival-promoting) mutations gradually become incorporated as normal elements in the gene complex, and so on.

In the industrial world the same process is at work. Were the manager of one of today's complex industrial enterprises to attempt to comprehend all possible combinations of the elementary units of his system, he would simply be overwhelmed by the detail. By assuming the system to be undefinable in detail, and by applying the black box approach, he does succeed in formulating enough many-to-one transformations (policies) that, to use Beer's expression, kill the variety of the system and suppress its concomitant dangers. The modern manager's most indispensable tool, the computer, operates in accordance with the same mechanism of input manipulation, output classification, and many-to-one transformations. Indeed, for the majority of managers the computer is the almost perfect example of the black box both in its figurative and in its literal sense.

SUMMARY

Cybernetics, which deals with feedback processes of all kinds, is characterized by complexity, probabilism, and self-regulation. These elements are analyzable by the black box technique, information theory, and the feedback principle respectively. The concept of cybernetics as the science of communication and control is applicable to highly complex problems, especially those which do not readily lend themselves to traditional analytical approaches.

A feedback control system with its closed-loop structure can employ positive, as well as negative feedback, though negative feedback is the more familiar of the two. The use of block diagramming as a common language for cybernetic applications was illustrated in this chapter, and its suitableness for identifying structural relationships in a system was pointed out. Various kinds of feedback systems were noted: first-order feedback system (the system is monitored against an external goal); second-order (the system can store alternatives in memory and choose the best among them); and third-order (the system can formulate new courses of action).

Ashby's Law of Requisite Variety indicates that, with increases in the complexity of a system, the variety of uncertainties also increases, and, in order to control a complex system, the amount of variety in the control mechanism must equal that in the system itself. A simple control system is possible only where the variety in the system has itself been simplified.

The final section of this chapter concluded with a brief treatment linking complexity with the "black box" technique. The latter, which involves input manipulation, output classification, and many-to-one transformations, finds its almost perfect example in the computer.

Although cybernetics is the science of communication *and* control, only the control aspect was treated here. It is left to the following chapter to take up the element of communication.

REVIEW QUESTIONS FOR CHAPTER THREE

1. What is cybernetics? Briefly present the historical developments or circumstances which led to the formulation of the science of cybernetics.
2. According to Beer there are three characteristics of a cybernetic system. What are these and what are the tools of analysis to deal with each of these characteristics?
3. Explain the classification framework presented in Figure 1 and illustrate the six categories of systems with examples other than the ones provided in the text.
4. How does "feedback" operate and what role does it perform in a control system?
5. Is control possible without negative feedback? Cite some control systems which do not possess negative feedback as a means for control.
6. What is "positive feedback" and what is its function in a system?
7. Show how the concept of "requisite variety" is important in the marketing subsystem of a production-oriented firm.
8. Present several examples of first-, second-, and third-order feedback subsystems of an organization.
9. How would you go about assessing the complexity of a given system?
10. What is the "black box" technique and how is it used in understanding the behavior of complex systems?

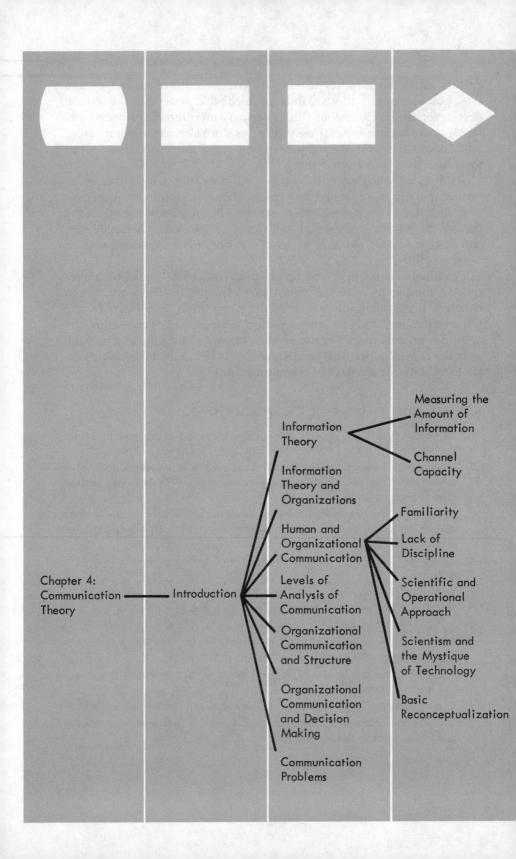

Chapter 4:
Communication
Theory

Introduction

Information
Theory

Information
Theory and
Organizations

Human and
Organizational
Communication

Levels of
Analysis of
Communication

Organizational
Communication
and Structure

Organizational
Communication
and Decision
Making

Communication
Problems

Measuring the
Amount of
Information

Channel
Capacity

Familiarity

Lack of
Discipline

Scientific and
Operational
Approach

Scientism and
the Mystique
of Technology

Basic
Reconceptualization

Chapter Four

Communication Theory

*In giving the definition of Cybernetics in the original book,
I classed communication and control together. Why did I do
this? When I communicate with another person, I impart a
message to him, and when he communicates back with me he
returns a related message which contains information
primarily accessible to him and not to me. When I control
the actions of another person, I communicate a message to
him, and although this message is in the imperative mood,
the technique of communication does not differ from that of a
message of fact. Furthermore, if my control is to be effective
I must take cognizance of any messages from him which
may indicate that the order is understood and has
been obeyed.*

Norbert Wiener

INTRODUCTION

In the previous chapter it was stated that a cybernetic system
possessed three characteristics: (1) extreme complexity, (2) self-regulation, and (3) probabilism. Complexity was equated with system's
variety; i.e., the number of distinct elements in the system together
with the interactions among the elements. It was suggested that complexity can best be dealt with through the black box concept. Self-
regulation is guaranteed by feedback mechanisms built into the system. The third characteristic of a cybernetic state — probabilism — refers
to the determinancy of a system's behavior. Since the first two characteristics of a cybernetic system were described in the previous chapter,
we turn now to the last one, namely probabilism.

Probabilism is defined in philosophy as the doctrine which asserts that "certainty is impossible and that probability suffices to govern faith and practices." As is well known from statistics, probability refers to the likelihood of the occurrence of a certain event and is measured by the ratio of the number of actual occurrences to the number of possible occurrences. It is obvious from this definition that probabilism refers to the degree of knowledge of the system's behavior at a certain point of time. Since the degree of knowledge is closely related to the availability and the obtainability of information, the statistical treatment of information appears as a candidate for dealing with systems whose behavior is not perfectly predictable.

This chapter begins with a brief introduction to information theory, its logic, origin, uses, and shortcomings. Since the subject is somewhat technical and requires knowledge of mathematics, we will confine ourselves to a more elementary treatment, avoiding when possible complex mathematical expressions.

Although the term "communication" is used interchangeably with information theory in the original writings of cybernetics, communication theory has become a discipline, of which information theory is but a small part. For this reason, the last part of this chapter is devoted to other aspects of communication theory. Although a portion of human communication in organization does obey the laws of information theory, one cannot, however, reduce the entire theory of communication to information theory.

INFORMATION THEORY

While the subject of communication has been of interest for many years, it was only within the past several decades that mathematicians were able to treat the subject scientifically. In some quarters, researchers still hold fast to the dictum that unless the object of discussion can be quantified, it lacks suitable description. This was behind the attempt to define the concept of information more accurately and unambiguously through mathematical analysis. Indeed, the endeavor directed to a quantification of information was given a title reflecting the efforts—A Mathematical Theory of Communication.[1]

One should perhaps state here that the mathematical treatment of communication has application only to a specific set of circumstances and that in discussion of information theory the term *information* is used in a very specialized sense. The mathematical theory of information evolved over a number of years as communication engineers

---

[1] Claude E. Shannon and Warren Weaver, *A Mathematical Theory of Communication* (Urbana, Ill.: University of Illinois Press, 1949).

attempted to measure the amount of information that was communicated over telephones, telegraphs, and radios. This is not to say that the mathematical concepts or techniques lack relevance in human communication, for indeed there are similarities, but *direct* application is only to the equipment itself and not to the *users*. As will be noted, while attempts to apply the formal concepts to other disciplines have not been lacking, results have been meager, and after two decades of experimentation the mathematical theory of communications is still dominated by and restricted to the field of telecommunications.

Interest is in the statistical aspect of information, which stems from the view that messages that have a high probability of occurrence contain little information and therefore any mathematical definition of information should be based on statistical analysis; i.e., the probability of that particular message being chosen from a given set of messages.

According to the classical theory, information is viewed as an entity that is neither true nor false, significant nor insignificant, reliable nor unreliable, accepted nor rejected.[2] As such, it is concerned neither with meaning nor with effectiveness. This is so because in the transmission of signs or words it is the signs or physical signals that are transmitted and not their meaning. Thus, information theory is associated only with quantitative aspects, the *howmuchness* of the uncertainty or ignorance reduction.

A communication system will consist of the following five elements: (1) an information source, (2) an encoder/transmitter, (3) a channel, (4) a detector, and (5) a decoder. (See Figure 4–1.)

FIGURE 4–1
Communication Model

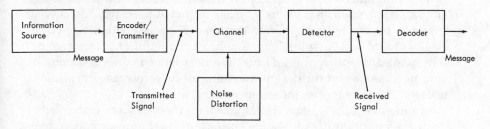

The information source selects the desired message out of a finite set of possible messages (verbal, written, and so on). The message is then transformed into a signal (encoded) and sent over the channel.

---

[2] See T. F. Schouten, "Ignorance, Knowledge, and Information," in *Information Theory*, Colin Cherry (ed.) (New York: Academic Press, Inc., 1956), pp. 37–47.

The channel is the medium used for sending messages from the source to the receiver. The detector picks up the transmitted signal. The signal is finally decoded into a message.

When one wishes to communicate there must obviously be common agreement as to the language (symbols, phonemes, and so on) to be used. Specifically, both the sender and the receiver of information must agree on the set of symbols available in the language to be used. When the symbols or words that the sender selects are unknown to the receiver no information is transmitted. This merely says that if I transmit a signal which has meaning for me but not for the intended receiver, obviously no information is transmitted. What is in my information bank is not in the receiver's. Thus, it is necessary that the sender and the receiver both know the set of possible messages from which a particular one will be selected.

It is precisely the restriction that both sender and receiver have the same information bank (source) that makes possible quantification of information. The larger the information bank of the sender and the receiver is, the greater the number of choices available, and the more information needed to resolve the uncertainty. Put very simplistically, if a child's vocabulary is limited to ten words, there is less uncertainty as to what word the child will say as compared to a grown-up whose vocabulary is less restricted. Thus, it can be seen that information, selection, and uncertainty are all interrelated.

Measuring the Amount of Information

One way to measure the amount of information in a statement is to enumerate the number of possible outcomes the statement eliminates. If only one outcome is possible, no information is required. For example, it is known that in the English language the letter q is always followed by the letter u. If we were attempting to spell any specific English word in which the letter q appears, no additional information is conveyed when we are told that the next letter is u. When there is zero uncertainty, no further information can be conveyed. Put simply, uncertainty decreases as information increases.

In measuring the amount of information, the unit used is the "bit," short for *bi*nary dig*it*. A bit is the smallest amount of information possible, and it represents a single selection between two alternatives, as between ON and OFF, YES and NO, OPEN and CLOSED, 0 and 1.

A central issue in information theory is the number of bits that are required (given a set number of alternatives) for the sender to communicate certain information. One example to illustrate this is to take your local telephone book and have a person choose a particular name from somewhere in the white pages. The question is how many bits of information are required for you to choose that same name or, to put it another way, how many bits are required for the sender to be certain that the receiver is getting the message. It was said that if only two alternatives were possible one bit of information would be conveyed. Obviously there are many alternatives possible in this example, but still it is not difficult.

In the following figure, the relationship between the number of alternative choices and the number of bits is depicted. Figure 4–2 shows that if the number of alternative choices is 4,000 then it will take 12 bits of information to completely specify the particular choice. If the telephone book has 500 pages it would take no more than 9 bits of information to find out what page the particular name is on. This is simply done by continually halving the number of alternatives (decreasing uncertainty) by taking the following line of questioning.

FIGURE 4–2
The Relationship between the Number of Alternative Signs and the Number of Bits Needed to Move from Uncertainty to Certainty

Bits	0	1	2	3	4	5	6	7	8	9	10	11	12
Powers of 2	2^0	2^1	2^2	2^3	2^4	2^5	2^6	2^7	2^8	2^9	2^{10}	2^{11}	2^{12}
Number of alternative signs	1	2	4	8	16	32	64	128	256	512	1,024	2,048	4,096

"Is the name in the front half of the book (pp. 1–250)?" If the answer is "yes," then the next question is, "Is the name in the front fourth of the book (pp. 1–125)?" It will take nine yes or no answers to specify the page number that the name is located on. The additional number of questions required will vary with the number of columns in the telephone book and the number of names vertically listed. If there are four columns, then two more additional questions are required. "Is the name in the left half of the page (first two columns)?" Likewise, most telephone books have approximately 100 names listed vertically, which means that it will take another seven bits of information to completely specify the name. Thus, given the above conditions, 18 bits of information are required to specify the particular name.

It was mentioned previously that at least two alternatives must be present before any information is conveyed. This is the simplest type of communication system and is termed the binary system, since for each bit two choices are required. It is for this reason that logarithms to the base 2 are used to measure the amount of information. The number of bits per alternative is then $\log_2 N$, where N is the number of alternative signs available in the entire repertoire of the sender and receiver. In the situation where only one sign (outcome) is possible ($N = 1$) it was stated earlier that no information can be conveyed. It takes at least two alternative signs to convey one bit of information. This is shown in the following manner: the amount of information $(H) = \log_2 1$ which is equal to 0. With two alternatives such as YES and NO, $N = 2$, and the number of bits (see Figure 4–2 for verification) equals 1 ($H = \log_2 2 = 1$) This is so because 2 raised to the first power is equal to 2. When there are four alternatives ($N = 4$) the number of bits is 2.

Let us assume that each letter in the English alphabet has an equal chance of being selected. Let us also consider the blank space as another character, thus making the list of possible alternative signs employed in a message 27 (26 letters of the alphabet plus one space). The amount of information would be $H = \log_2 27$. A log table would show this to be 4.75 bits, or a quick examination of Figure 4–2 will show that for 27 alternative signs five bits are required. Now if we were to let n represent the number of actual signs in the message, then the number of bits in the entire message (H_m) thus becomes $H_m = n \cdot \log_2 N$, which is the information rate measured in bits.

In the following message, PLAY BALL, the number of bits would be $H_m = 9 \cdot \log_2 27$. PLAY BALL is eight digits plus a space, which is nine. Thus, in this message 42.75 bits are required (9×4.75) or 43.

So far we have assumed that each outcome has an equal probability of being chosen. This, of course, is an oversimplification of the actual situation. Of the 26 letters in the English alphabet, not all have the same probability of appearing in English words. One would expect that the vowels would appear more frequently than some consonants. In fact, the letter *e* appears about 60 times more often than the letter *z*.[3] Consequently, the probability of the letter *z*'s occurrence is considerably less than that for the letter *e*. But note. When *z* does occur, its informational content will be greater than when *e* occurs because it does more to identify a word.

To take account of differences in the *probability* of messages, a message is assigned a probability p when it is selected from a predetermined set of $1/p$ messages. The *amount of information* that must be transmitted for that message is then $\log_2 1/p$ or $-\log_2 p$. If all N messages have an equal probability of being chosen, then the probability of any

[3] Schouten, "Ignorance, Knowledge, and Information," p. 36. The frequency probabilities associated with letters in the English messages are the feature of Edgar Allen Poe's classic story, "The Gold Bug."

of these is $p = 1/N$ and then $-\log_2 p = -\log_2 N$. If the probabilities assigned to N messages are $p_1, p_2, \ldots, p_n$, then the amounts of information associated with each message are $-\log_2 p_1, -\log_2 p_2, \ldots, -\log_2 p_n$.

Very seldom, however, is one concerned with a single message. Generally what is of interest is the capacity of the channel for generating messages and the average amount of information per message per channel. The average amount of such information from a particular source is generally given by the equation:

$H = -(p_1\log_2 p_1 + p_2\log_2 p_2 + \ldots + p_n\log_2 p_n)$, or more simply

$$H = -\sum_{i=1}^{n} p_i \, log_2 p_i$$

Channel Capacity

When information is transferred from one location to another, it is necessary to have a channel of some sort over which the information can travel. The measurement unit used when describing channel capacity is again the bit. The challenge is to devise efficient coding procedures that match the statistical characteristics of the information source and the channel. The upper limit of the amount of information that can be transmitted over a channel is termed the channel capacity.

For example, in a 100-words-per-minute teletype channel, it is possible to transmit 600 letters or space characters per minute, or ten characters per second. Since the maximum information associated with one such character is 4.75 bits, the capacity of this channel is 47.5 bits per second.[4] Up to this point we have been concerned only with signals made up of discrete characters. Although dealing with continuous communication such as musical tones, video signals, speech waves, or color is somewhat more difficult and complicated, still it is not essentially different. Information theory is so general that it can accommodate any type of symbol.

INFORMATION THEORY AND ORGANIZATIONS

In recent years the many claims regarding the importance of information theory and its applicability to the theory of business organizations have bordered on the extravagant. Since 1949, when the classic information theory was first formulated by Shannon and Weaver, the literature has grown by leaps and bounds. More recently, the number of articles purporting to relate information theory to the business organization has noticeably increased. This association of information

---

[4] Gordon Raisbeck, *Information Theory, An Introduction for Scientists and Engineers* (Cambridge, Mass.: MIT Press, 1963), p. 45.

theory with business organizations undoubtedly arises from the frequent use of the word information in the business context. Here and there one speaks of management information systems, accounting information, the information explosion, information for decision making, information control systems, and so on. In most of these instances, the term *information* is equated with mere data acquisition, with the quality of data, its flow through the system, its functional characteristics, and so on. However, as stated earlier, information theory as originally developed has little to do with these connotations.

It should not be forgotten that information theory was initially developed for application in telecommunications, where it is both possible and feasible to compute the amount of information that can be transmitted over a wire or a radio band. For determining channel capacity it has indeed been of significant benefit. However, its utility when applied to other disciplines has been of doubtful value and the not infrequent attempts to apply it to the business sector have been equally disappointing. In recent years some have endeavored to apply the formal theory to the fields of experimental psychology,[5] sociology,[6] decision making,[7] accounting,[8] and to many other diverse situations.

Since information theory is concerned with reducing the uncertainty associated with many possible outcomes, its focus on the amount of information is understandable and its prescinding from the semantics problem is justifiable. Perhaps, it is too early to state that information theory will never find application outside its present arena. Although the results in settings other than engineering are indeed modest, this is perhaps no different from other scientific concepts of today that also lacked a noble and auspicious origin. But for the present Rapoport states:

> However, one must admit that the gap between this sort of experimentation and questions concerning the "flow of information" through human channels is enormous. So far no theory exists, to our knowledge, which attributes any sort of unambiguous measure to this "flow." . . .
> If there is such a thing as semantic information, it is based on an entirely different kind of "repertoire," which itself may be different for each

---

[5] Colin Cherry, *On Human Communication* (New York: Science Editions, 1961). Also F. C. Eric, "Information Theory in Psychology," *A Study of a Science,* Sigmund Koch (ed.) (New York: McGraw-Hill, 1959).

[6] See Walter Buckley, *Sociology and Modern Systems Theory* (Englewood Cliffs, N.J.: Prentice-Hall, 1967).

[7] Russell L. Ackoff, "Toward a Behavioral Theory of Communication," in *Management Science,* vol 4 (1957–58), pp. 218–34.

[8] Norton M. Bedford, and Mohamed Onsi, "Measuring the Value of Information— An Information Theory Approach," *Management Services* (January–February, 1966), pp. 15–22.

recipient. . . . It is misleading in a crucial sense to view "information" as something that can be poured into an empty vessel, like a fluid or even like energy.[9]

Summary

Since there is little direct application of information theory to business situations, does this mean that cybernetics cannot profitably be applied to these areas? The answer is obviously no. Just as it is necessary to have humans in many cybernetic systems to provide the feedback function, communication must also often rely on the human element to operate and control the system. Granted that the statistical aspects of information are inappropriate, this simply means that information must be treated from some other dimension to make it applicable to organizations. The remainder of this chapter will be concerned with human and organizational communications.

HUMAN AND ORGANIZATIONAL COMMUNICATION

Although Wiener's principal concern with communications was a quantitative one, he was keenly aware of the importance of other modes of communications that exist in the organization. He states:

> Communication is the cement that makes *organizations*. Communication alone enables a group to think together, to see together, and to act together. All sociology requires the understanding of communication.

> What is true for the unity of a group of people, is equally true for the individual integrity of each person. The various elements which make up each personality are in continual communication with each other and affect each other through control mechanisms which themselves have the nature of communication.

> Certain aspects of the theory of communication have been considered by the engineer. While human and social communication are extremely complicated in comparison to the existing pattern of machine communication, they are subject to the same grammar; and this grammar has received the highest technical development when applied to the simpler content of the machine.[10]

Rightly suggestive of the above quotes is that society's very existence is dependent upon communication. And yet, in spite of the in-

---

[9] Anatol Rapoport, "The Promise and Pitfalls of Information Theory," *Behavioral Science*, vol. 1 (1965), p. 303.

[10] N. Wiener, *Communication* (Cambridge, Mass.: MIT Press, 1955).

tensive study of communication processes and the voluminous literature in existence, few substantive theories have evolved. There are several fundamental problems which serve to explain this situation. Thayer notes five conceptual difficulties which have impeded progress in the development of communication theory.[11]

Familiarity

The more familiar a concept the more difficult to develop a sound empirical base. This fact can also be noted in the numerous definitions and connotations of the word "system" discussed throughout this text. The more popular a concept typically the more ambiguity is possible. Another concept falling into this category is the very substance of this chapter—information. This word is used daily in a multiplicity of ways. For the accountant, the receipt of cost figures represents factual information; knowledge of a competitor's strategies may be construed as valuable information. Still there are many other connotations given to the word. It is precisely this ambiguity surrounding the word information that led to efforts to define it more rigorously.

Lack of Discipline

Thayer notes that a second difficulty in the development of a theory of communication is that it is discipline-less. No single discipline exists that purports to study communications in the main. He notes "There are 'loose' professional associations of persons having some part interest in communication, of course, as well as academic programs built upon some special orientation; and there is undoubtedly an 'invisible college' of scholars whose scientific interests and pursuits with respect to communication do overlap to some degree. But there is nothing like the discipline foundation one sees in physics, for example."

This point is also made by Cherry: "At the time of writing, the various aspects of communication, as they are studied under the different disciplines, by no means form a unified study; there is a certain common ground which shows promise of fertility, nothing more."[12]

Because communication is so basic to each discipline and is studied within its own disciplinary boundaries, the fragmented results do not extend beyond its walls and remain for the most part segmented, never adding up to more than the sum of the parts.

---

[11] Lee Thayer, "Communication—*Sine Qua Non* of the Behavioral Sciences," *Vistas in Science*, 1968, pp. 48–51.

[12] Colin Cherry, *On Human Communication* (Cambridge, Mass.: MIT Press, 1967), p. 2.

Scientific and Operational Approach

The haziness of the line separating theoretical aspects of communication from the pragmatic ones is still another difficulty. While research in other disciplines rarely alters social behavior, this is not always the case with communications. The point being made here is that much of the work designed for the scientific inquiry into communications inevitably becomes "rules or ways to communicate better," thus enhancing the practical knowledge of communicating. A clear delineation is required since the operational aspects of communication differ from the scientific aspects.

Scientism and the Mystique of Technology

Another barrier, states Thayer, is the incompatibility of our blind faith in scientism, on the one hand, and on the other the nature of the communications phenomenon itself. The caution expressed here is that employment of "scientific techniques" in the study of this subject area does not imply that the subject will necessarily yield. It is quite likely that such approaches will reveal only that which is scientizable in the first instance. The application of new technology to old problems will not ipso facto solve these problems; indeed the problems plaguing organizations today are the same ones that plagued them many years ago. The communication problems of today are identical to those centuries ago.

Basic Reconceptualization

The remaining difficulty noted is that once such an ubiquitous subject as communications becomes conceptualized it is nigh impossible to reconceptualize it. The initial conceptualization is one of the largest obstacles to conceiving the subject in new and different ways. Thayer cites the oft-used formula $A \longrightarrow B = X$ in which A communicates something to B with X result. Thayer maintains that the "thing" communicated is just as much a product of the receiver as it is of the sender and in effect the message is coproduced. What is required is a basic reconceptualization of the underlying phenomena.

These, then, are some of the obstacles which stand in the way of a universal body of knowledge of communication. This is not to say, however, that there has not been any progress, but rather that the progress has been fragmented. Developments have come from many resolution levels — from the communications engineer who is basically concerned with the transmission of signals, to the behavioral scientist who is concerned with the behavioral aspects of communication.

LEVELS OF ANALYSIS OF COMMUNICATION

Although various authors present alternative approaches to the study of human communications, nearly all of them categorize them at differing levels. The particular scheme we wish to present is that of Thayer, who notes five levels of an analysis from which one can approach the study of communication:

(a) the intrapersonal (the point of focus being one individual, and the dynamics of communications as such);

(b) the interpersonal (the point of focus being a two or more person interactive system and its properties—the process of intercommunication and its concomitants);

(c) the multi-person human enterprise level (the point of focus being the internal structure and functioning of multi-personal human enterprise);

(d) the enterprise $\longleftrightarrow$ environment level (the point of focus being the interface between human organizations and their environments); and

(e) the technological level of analysis (the focus being upon the efficacy of those technologies—both hardware and software—which have evolved in the service of man's communication and intercommunication endeavors).[13]

The first level of analysis, that of the intrapersonal level, concerns itself with the individual's own physiological and mental processes. The individual acquires, processes, and consumes information about himself and other events in his environment. The individual may acquire this information through reading, observation, speaking, writing, and so on. Emphasis at this level is in the inputting and processing of information by the individual, since it is here that communication occurs in the individual.

The next level of analysis is the interpersonal one, often termed intercommunication since a minimum of two people is required in the system. In this type of system, attention is focused on how individuals affect each other through intercommunications (influential level). Most of the standard texts on communications treat each of the above types in depth and therefore attention will not be devoted to these.

The third analysis level is that of organizational communication and one of particular interest to us. It is to this area that we will shortly direct our attention.

The enterprise-environment level is concerned with the ways that the organization communicates with its environment. There have been

---

[13] L. Thayer, "Communication," pp. 56–75.

a number of recent attempts to incorporate environmental variables into theories of organization, since the environment can be treated as "information" which either becomes routinely available to the firm or which the firm actively seeks. It is only through information that an organization can learn of and adapt to changes "out there" in the environment. Because this subject area is of critical importance in the study of organization, a separate chapter is devoted to this environment-organization interaction.

The final level of analysis (technological) obviously refers to computerized management information systems. While the mathematical theory of information of Shannon and Weaver can be included under this level, major attention must be devoted to the advances in technology which facilitate communications in organizations. To this end, information technology and its attendant vehicles (the computer) will be given special consideration in two of the following chapters.

ORGANIZATIONAL COMMUNICATION AND STRUCTURE

That there is a relationship between the efficacy of the communication system of the organization and the structure of the organization is not to be denied. Horti[14] views the process of communication as "the dynamics of the organization structure" and proceeds to show that the communication system and organization structure are interdependent. He also states that the uniqueness of any organization is reflected in its structure upon which the communication system is based. Thus, in order for a firm to remain viable the communication system must be in balance with the organization structure.

Deutsch, in commenting on the relationship of communication and control, gives an opposing viewpoint:

> Communication and control are the decisive processes in organizations. Communication is what makes organizations cohere; control is what regulates their behavior. If we can map the pathways by which communication is communicated between different parts of an organization and by which it is applied to the behavior of the organization in relation to the outside world, we will have gone far towards understanding that organization. . . .[15]

The above suggests that organization structures follow the development of the communication system. Perhaps Deutsch was taking account of the informal channels of communication that exist and precipitate the informal organization structure. This is not to say that com-

---

[14] Thomas Horti, "Organization Structure and Communication: Are They Separable?" *Systems and Procedures* (August 1968), pp. 6–10.

[15] Karl Deutsch, *The Nerves of Government* (New York: The Free Press, 1966). Chapter 5.

munications must follow the formal established channels, for, if this were the case, decision making would entail a long drawn out process. The existence of informal communication systems as well as informal power structures must be acknowledged and indeed it is often through this vitalizing force that the organization succeeds. Although current organizational practice is hardly reason to state that the communication system should follow the organizational structure, this is what occurs. In the design of information systems, managers are asked what information is required for decision making in their particular functional and hierarchial position. In other words, an implicit relationship exists between structure and information. Likewise, when a reorganization occurs in the organization, seldom if ever is this based upon an analysis of the existing communications network. Suffice to say that the organization communication system and the organization structure are intertwined.

ORGANIZATIONAL COMMUNICATION AND DECISION MAKING

The relationship of communication and decision making is inseparable, since decisioning must rely on information. Likewise, if decisions are to be carried out, they must be communicated to the people in the organization. Irrespective of how one approaches the subject of decision making—i.e., from a mathematical, statistical, psychological, or any other viewpoint—decision making always involves a choice among alternatives. Choosing implies that the alternatives are known and this obviously involves communications. The range of alternatives is limited by the variety and the amount of information available to the decision maker.

Decision making always takes place in an environment, which can have a weighty impact on the process. As mentioned previously and as will be discussed more thoroughly in a following chapter, an organization constantly adapts to changes in its environment. In its adaptation process the organization makes a series of decisions based on events in the environment. In order to make such decisions the firm must first engage in soliciting and/or receiving information depicting the events. The entire decision-making process can be conceived of as a series of communication events. A decision is made based upon the receipt of communications from a variety of sources including a memory bank of the organization, and is then communicated to others in the organization.

Simon defines three steps in the decision-making process: (1) the listing of all the alternative strategies, (2) the determination of all the consequences that follow upon each of these strategies, and (3) the comparative evaluation of these sets of consequences.[16] Suffice it to say that each of these involves information and communication.

COMMUNICATION PROBLEMS

1. The problem most cited in the communication literature is that of meaning, the semantic aspect of communication. A major point regarding meaning is that it is not inherent in messages and, therefore, it is not conveyed by the sender but rather meaning is imposed by the receiver. The problem encountered, then, is to give the precise meaning to the message that the sender intended. The relationship between a word and the object that the word stands for exists only in the minds of the people using the word. That meaning rests with the individual is an important concept in all communication theory. It is through the commonality of experiences that meanings take on the same connotations. However, since each person's experiences vary widely, so does the meaning of objects and reference points. Boulding, in his oft-quoted work *The Image,* states that each experience encountered becomes a part of the individual's image of the world:

> The image is built up as a result of all past experience of the possessor of the image. Part of the image is the history of the image itself. At one stage the image, I suppose, consists of little else than an undifferentiated blur and movement. From the moment of birth, if not before, there is a constant stream of messages entering the organism from the senses. At first, these may merely be undifferentiated lights and noises. As the child grows, however, they gradually become distinguished into people and objects. The conscious image has begun. . . . Every time a message reaches him his image is likely to be changed in some degree by it, and as his image is changed his behavior patterns will be changed likewise. We must distinguish carefully between the image and the messages that reach it. The messages consist of information in the sense that they are structured experiences. *The meaning of a message is the change which it produces in the image.*[17]

Thus meaning is subject to change because of changes in our experiences. Even if two people were to experience the same situation, the meaning cannot be the same since the experience is essentially a private affair. Of course, what we attempt to do is to utilize certain words that will stimulate recall of common experiences. But even though two people may share the identical stimuli, the response will be somewhat different. It is through words that mental associations with past expe-

---

[16] Herbert Simon, *Administrative Behavior* (New York: The Macmillan Company, 1957), pp. 80–84.

[17] Kenneth Boulding, *The Image* (Ann Arbor, Michigan: University of Michigan Press, Ann Arbor Paperbacks, 1956), pp. 6–7.

riences are recalled. Presumably if you have indulged in the contents of this text up to this point, your viewpoint of "systems" will have been altered. This alteration will be different for each individual, depending on one's particular experiences. In any case, the reader's "image" of systems will have changed.

2. A second problem noted in the literature is that of a participant's concern for the other participant's intentions. It is often heard that "It is our intentions in the use of language, not the language itself, which hinder or facilitate communication."[18] While this can be useful in the communication process it can also call forth behavior which tends to distort the intended meaning. Strategies are often encountered between superiors and subordinates which make it difficult, if not impossible, to assess true statements. We can all name an individual who we know tends to exaggerate, and in order to put related facts into a proper perspective (from our standpoint) we tend to discount some or much of such a person's statements. The individual aware of communication bias may in turn introduce a counterbias.

3. The interpersonal level of communication is not to be confused with the organizational communication system. Although individuals may be effective in their communications with other individuals in the organization one should not mistakenly identify these interpersonal communications with a viable organization communication system. The danger is in thinking that other communication problems which may arise are soluble at the interpersonal level when in reality they are communication systems problems.

4. Since organization structure and communication systems are interdependent, an elimination of the structural problems in the organization will also eliminate the communication problems. An inevitable fact is that the hierarchial structure in an organization must terminate in one focal point of authority. It does not follow that such a situation is problematical. Simon states: "Organization is innately hierarchical in structure; to the extent that one looks upon certain communicative consequences of that structure as 'problematical,' he is destined to deal with symptoms rather than causes."[19]

These are but a few of the organizational issues and problems encountered in the work-a-day world. Many other texts treat these and other communication issues in greater depth. Our main purpose here has been to elucidate several of the major problems encountered in organizations and to set the stage for the interaction of the organization and the manner that the organization adapts to changes via an "information" system.

---

[18] P. Meredith, *Instruments of Communication* (London: Pergamon Press, 1965), p. 36.

[19] Herbert Simon, *The New Science of Management Decision* (New York: Harper & Row, 1960).

SUMMARY

The purpose of this chapter was twofold, first to present the concept of information theory and to discuss its main elements, and its area of applications; secondly, to discuss the more encompassing field of communication theory. The relationship between organizational structure and the communication system was treated, as well as the inseparability of decision making and the communicative system.

Finally, some organization issues and problems were briefly mentioned.

REVIEW QUESTIONS FOR CHAPTER FOUR

1. Discuss how information is basically a statistical concept.
2. How many bits of information are required to specify the following messages? (Assume equal probabilities for each letter.)
   you passed the course
   do not quit
   be prepared
3. What factors limit the use of information theory in human communications?
4. In many textbooks on communication problems, a frequently cited one is the semantic (meaning) problem. Show how this is related to the communication model presented in this chapter.
5. Using an 8 × 8 matrix, ask a friend to specify a square (a number) within the matrix. After he has done this in the conventional way, reverse the procedure and show him how you would do it.
6. International symbols are used in all Olympic games and international traffic signs are also increasingly used. Explain this in terms of communication theory.
7. Some people might argue that rather than the field of communication being contentless, it is multidisciplinary. How would you harmonize these two statements?
8. What are some typical gestures employed by the following people which have a special meaning as acceptable as the popular?
   a baseball coach
   a football referee
   a truck driver
   a student
9. How might the structure of an organization inhibit communications within it? Are there any organizational structures which are more suitable for communicating or is communication an individualized problem of the person communicating?
10. Thayer notes that the more familiar a word or concept, the more ambiguous it is. Name five words or concepts that are very ambiguous. Give words or concepts in which there is little ambiguity.

Part Three

Organizations and Systems

Whenever I draw a circle, I immediately want to step out of it.

R. B. Fuller

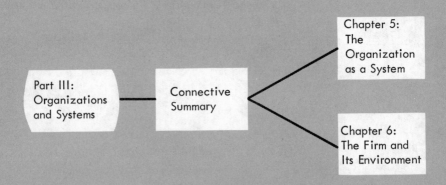

Part III:
Organizations
and Systems

Connective
Summary

Chapter 5:
The
Organization
as a System

Chapter 6:
The Firm and
Its Environment

CONNECTIVE SUMMARY

The two preceding parts have provided enough insight into the subject of general systems theory and cybernetics to enable us to propose a conceptual framework for viewing an organization or enterprise from a different perspective. From GST we know that an organization must be conceptualized as an open system which is in constant interaction with its environment. The central issue of maintaining controlling systems is largely the field of cybernetics.

Chapter 5 addresses itself to the task of developing a conceptual framework for organizations based on the tenets of GST and cybernetics by tracing the main concepts of various theories of organizations. After a brief review of the so-called classical and neoclassical organization theories, focus is placed on modern organization theory, upon which a modern systems theory of organizations will be built.

In this modern systems theory of organizations, enterprises are conceived as hierarchical structures possessing feedback systems of three orders: first-, second-, and third-order feedbacks.

An organization so designed exhibits two main characteristics: (1) it is open to the environment; and (2) it feeds on information. To further enhance the understanding of these complex enterprises and how they cope with their environments, a model will be presented in Chapter 6. The ways that the firm identifies and adapts to its environment will be examined.

The process of acquisition of external information, known as "scanning," constitutes the cybernetic machine for dealing with the extreme uncertainty surrounding the organization's environment. It is only through this information gathering that firms learn of changing factors in their environment and gain adaptability.

105

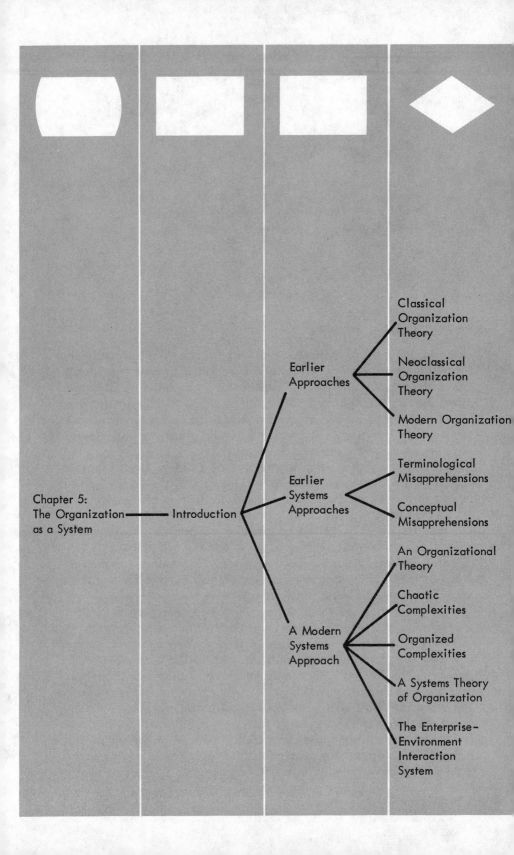

Chapter 5:
The Organization
as a System

Introduction

Earlier
Approaches

Classical
Organization
Theory

Neoclassical
Organization
Theory

Modern Organization
Theory

Earlier
Systems
Approaches

Terminological
Misapprehensions

Conceptual
Misapprehensions

A Modern
Systems
Approach

An Organizational
Theory

Chaotic
Complexities

Organized
Complexities

A Systems Theory
of Organization

The Enterprise–
Environment
Interaction
System

Chapter Five

The Organization as a System

No man is an island—he is a holon. A Janus-faced entity
who, looking inward, sees himself as a self-contained unique
whole, looking outward as a dependent part. His self-
assertive tendency is the dynamic manifestation of his
unique wholeness, his autonomy and independence as a
holon. Its equally universal antagonist, the integrative
tendency, expresses his dependence on the larger whole to
which he belongs; his "part-ness." The polarity of these two
tendencies, or potentials, is one of the leitmotives *of the*
present theory. Empirically, it can be traced in all phenomena
of life; theoretically, it is derived from the part-whole
dichotomy inherent in the concept of the multi-layered
hierarchy; the self-assertive tendency is the dynamic
expression of the holon's wholeness, the integrative
tendency, the dynamic expression of its partness.

Arthur Koestler, *The Ghost in the Machine*

INTRODUCTION

Man is by nature inquisitive; with his sensory organs he searches the world about him and with his mind he attempts to organize his observations into coherent schemes. His curiosity has led to theories which modern sophisticated technology has validated so that they have become part of his "common sense."

Modern man is man in organization. He not only spends one-half of his waking day contributing to the cooperative effort of an organization, but he also occupies the other half watching television, reading books, or going to a theater to be entertained—all output of the cooperative effort of men in organizations! It is no small wonder then that

107

man should find the genesis, growth, and evolution of organization a fascinating study.

Contributions to the study of organizations have come from individuals with varied backgrounds. The classical economist, the historian, the lawyer, the sociologist—all have added their insights to this intriguing subject. Even the novelist and satirist have deepened interest in this field.

The contemporary student of organizations encounters difficulties in comprehending all of this. For him, organizations are more like life: they display no one outstanding regularity nor any singularly striking anomaly. Instead, a variety of events and processes present themselves, waiting to be understood and incorporated into a coherent conceptual framework.

Complex phenomena, such as organizations, require equally complex methods of inquiry if one is to understand and appreciate their genesis, growth, and evolution. Modern systems thinking does provide such a framework. What one hopes to accomplish here is to lay the foundations for a systems-oriented theory of organizations. Under the heading "Earlier Approaches" we will examine the traditional viewpoint according to which organizations were regarded either as mechanisms for efficiently converting certain resources into finished products and services, or as social enterprises in which the human factor occupied the center of attention.

The end of World War II marked the beginning of the Second Industrial Revolution, namely information technology. Information technology (roughly the application of computerized techniques to industrial and business problem solving) added a new dimension to the organization's complexity. It is not accidental that the "systems viewpoint" first appeared at about the same time as the computer. Under the heading "Earlier Systems Approaches," the first attempts to transplant systems thinking to the study of organizations will be recounted. The final section of this chapter, "A Modern Systems Approach," will sketch the most promising contemporary attempts at conceptualizing organizational systems.

While systems-oriented theory of organizations is far from becoming part of our "common sense," still we are beginning to realize that the inputs, outputs, processes, feedback loops, and computers are really the stuff of which organizations are made.

EARLIER APPROACHES

A complete account of the various approaches to organizations will not be undertaken in the limited space devoted to this topic. The interested reader can find a fuller treatment of the subject in many standard textbooks on management or organization theory.[1] The brief discussion presented here is intended only as an aid for understanding the theoretical synthesis to be offered in the latter sections of this chapter.

Classical Organization Theory

Two main themes occupied the attention of the earlier thinkers on organization: (*a*) the use of men as adjuncts to machines in the performance of routine productive tasks, and (*b*) the formal structure of the organization. Contributions to the first theme came from engineering-minded thinkers, primarily F. W. Taylor and his disciples. The second area, while initially the exclusive domain of military experts (Urwick and Gulick) and industrialists (H. Fayol), later became the concern of sociologists (M. Weber), practicing executives (C. Barnard), and others. F. W. Taylor's *Scientific Management* was an attempt to investigate the effective use of man in industrial organizations. Although Taylor's initial intention was to develop a general organization theory by which to analyze interaction between the human factor and the social and task environments of the organization, scientific management focused its attention on man-machine interaction in the performance of routine productive tasks.

Scientific management's investigation of man centered on the physiological variables affecting human productivity. Thus, scientific management identified three physiological variables that related to task accomplishment: capacity, speed, and durability. Capacity of a human being refers to the limit of his productive potential, speed to the time required to accomplish a task, and durability to muscle fatigue.[2] Thus, scientific management, originally proclaimed to be the application of scientific methods to the management of organizations, never went beyond the time-and-motion study techniques it employed.

While scientific management focused on the development of standardized methods at the production (operation/shop) level, classical administration theory focused on the firm as a whole. Realizing, on the one hand, that the accomplishment of organizational objectives requires the division of larger tasks into smaller units, and, on the other, that the grouping of these tasks into larger classes is needed to form

---

[1] See, for example, A. C. Filley and R. J. House, *Managerial Process and Organizational Behavior* (Chicago: Scott, Foresman and Company, 1969); J. G. March and H. A. Simon, *Organizations* (New York: John Wiley & Sons, Inc., 1958); P. M. Blau and W. R. Scott, *Formal Organizations* (San Francisco: Chandler Publishing Company, 1962); F. E. Kast and J. E. Rosenzweig, *Organization and Management* (New York: McGraw-Hill, 1970).

[2] J. G. March and H. A. Simon, *Organizations* pp. 15–16.

coherent and matching series of operations (coordination), administrative theorists attempted to discover the ideal design for organizational structure that would facilitate this division of labor and coordination.

Some of the major questions that the classical administrative theory dealt with were associated with the number of administrative units (departmentalization); the number of subordinates that an executive can effectively supervise (span of control); the formation of hierarchical processes through which the coordinating authority operates (scalar principle); functional differentiation between various types of duties (functional principle); and the differentiation of authority into command authority and staff authority.[3]

To sum up, classical organization theory did have relevant insights into the nature of organizations, but its value was diminished by its overconcentration on the formal structure of the organization and its continuing assumption that the behavior and development of organizations were but corollaries of their structural properties.

Neoclassical Organization Theory

Neoclassical organization theory's point of departure was the observation that, in the classical organization theory, the human element was either oversimplified or ignored. Thus, neoclassical organization theory offered the human dimension as the focal point as contrasted with the impersonality of classical theory. For this reason the movement initiated by this school was termed the human relations movement.

Neoclassical theory accepts the basic postulates of the classical theory but modifies them by superimposing changes in operating methods and structure evoked by individual behavior and the influence of the informal group. Here, individual and group behavior became the center of focus.

Contributions to this theory came primarily from psychologists and sociologists. The human relations movement started when Elton Mayo and his colleagues from Harvard University were invited to participate in the now classic Hawthorne studies. The research concentrated on the effects of certain variables like illumination, heat, fatigue, and machine layout upon productivity. A major finding of the experiments was that the output of human effort is actually a form of social behavior.[4]

Research inspired by the Hawthorne experiments led to some important findings. Thus, the relationship between division of labor, specialization, and productivity, assumed by the classical theorists to

---

[3] J. D. Mooney and A. C. Reiley, *Onward Industry* (New York: Harper & Row, 1931), pp. 5 ff.

[4] Filley and House, *Managerial Process and Organizational Behavior*, p. 18 ff.

be desirable because a high degree of division of labor is associated with a high degree of productivity, was now called into question. A high degree of specialization could in certain instances lead to a low degree of productivity because of boredom, monotony, and so on. Informal organizations (the natural groupings of people in the work situation not prescribed by the formal structure) act as agents of social control.[5]

Modern Organization Theory

Modern Organization Theory is characterized by three parallel developments: (*a*) the extension of earlier classical and neoclassical theory; (*b*) the emergence of behavioral science research; and (*c*) the emergence of operations research. Of these three developments, only the first two will be discussed here. Operations research will be dealt with in the next section.

The extension of classical and neoclassical theories was inspired by the work of C. Barnard. Although his book, *The Functions of the Executive*, appeared several decades ago, his ideas were not current until H. Simon popularized them. Barnard-Simon's conceptualization of organizations and their emphasis on decision making and communication can be regarded as the onset of systems thinking. Other developments along classical and neoclassical lines, especially the much talked-about management process, were refinements of Fayolian management theory.

The behavioral science approach attempted to study observable and verifiable human behavior in organizations by means of social science research methods (surveys, lab experiments, and field studies). This approach draws heavily upon psychology, sociology, economics, and to some extent, upon the exact sciences. Since the 1950s, behavioral science has focused on three levels of analysis. The first level of research deals with the *individual* in an organization (personality, learning, motivation, attitudes, and leadership patterns); the second level focuses on the *group* (social norms, communication patterns, group conflict, and problem solving); the third level is concerned with *complex organizations as institutional units*. Most of the research at this level deals with the empirical testing of Max Weber's theory of bureaucracy.[6]

The behavioral science approach contributed to improvements in the methodology used to test and validate classical and neoclassical theory as well as in the conceptualization of organizations resulting from the interdisciplinary nature of the research.

---

[5] H. A. Simon, *Administrative Behavior* (New York: The Macmillan Company, 1945).

[6] Hilley and House, *Managerial Process and Organizational Behavior*, p. 9.

EARLIER SYSTEMS APPROACHES

The earlier approaches to organizations tended to focus on the functional departmentalization as well as on the person-to-person relationships within the organization. If one were to add to the knowledge gained through the anatomical and behavioral studies the knowledge of the microeconomist about the product and market, then one ought to have a fairly good understanding of what the organization is and what it does. One would then have a universal theory of organizations. However, such a theory does not exist. What we have is a multiplicity of theories, such as the economic theory of the firm, the behavioral theory of the firm, the decision theory of the firm, and the social systems theory of the firm.

Despite the diversity of theories, all of them seem to have been formulated with just one purpose in mind: to supply the ingredient or viewpoint missing in the other theories. For the human relations theorist, the classical and economic theory of the firm neglected human behavior. For the modern organization theorist, the emphasis on human behavior by the neoclassicist was done at the expense of productivity. The modern organization theorist holds that the set of variables affecting human satisfaction is closely related to the set of variables affecting productivity.

This plausibility of the interrelatedness of sets of variables determining organizational behavior supplied the initial impetus to systems theory. The systems approach maintains that the best way to approach the study of organizations is to view them as systems, with emphasis on the interrelationship and interdependency of parts.

Earlier systems approaches to the study of organizations were characterized by various misapprehensions. These can be categorized as philosophical, terminological, or conceptual. The philosophical misapprehensions of the earlier systems approaches have already been detailed in Chapter 1 and can be summarized in the phrase: an adherence to mechanistic and reductionistic views of the world.

Terminological Misapprehensions

The vagueness of terms used by the systems theorists was obvious from the very beginning. Enamored of the success and prestige of the exact sciences, systems enthusiasts were quick in casting off their own terminology for that of the physical scientist. The precision and clarity of the physicist's terms made the price seem just right. Earlier opponents of this casting-off process were subdued to silence for fear that their own ignorance would be exposed.

The discipline that supplied the technical vocabulary was, for the most part, that of engineering. The process of borrowing engineering terms, begun with F. W. Taylor's scientific management, gained momentum in the mid-50s as a result of increased employment of engineers in management positions and of the so-called Second Industrial Revolution; namely, information technology.

In less than a decade, words such as system, feedback, information, entropy, and steady state became part of the systems enthusiasts' standard vocabulary. These terms were and still are used in either their original or altered meaning to fit the new technology. In any case, the use of terms still remains the prerogative of the user with all the concomitant possibilities of confusion. Thus, even today, after more than two decades of struggling with these terms, most systems theorists of organizations will either refer to an engineering text for definitions of the terms used or will assume that the terms have already become part of our common inheritance.

Evidence of this confusion can be seen in the term system. While the term does have a clear and definite meaning in engineering, in the social sciences it is often clothed in considerable vagueness. In engineering, a system is defined as "an arrangement, set, or collection of things connected or related in such a manner as to form an entirety or whole."[7] A mathematical definition of the above verbal definition might read as follows: $S = (E, R)$, meaning that a system (S) is a set of elements (E) together with a set of relationships (R). There can be many elements in the system so that $E = e_1, e_2, \ldots e_n$, as well as many relationships among these elements so that $R = r_1, r_2, \ldots r_m$.

The social scientist, viewing the organization as a system, may either not define the term at all or may define it ambiguously. For him it could simply be an "assemblance or combination of things or parts forming a complex whole."[8] It is easy to see how controversy would be generated, were one to inquire into the meaning of such terms as assemblances, combinations, parts, and whole. Similar problems, no doubt, would be associated with terms such as feedback, entropy, and the host of others subsumed from the vocabulary of the hard sciences.

Conceptual Misapprehensions

The major conceptual misapprehension is associated with the inadequacy of the models of the organization employed by most sys-

---

[7] J. J. DiStefano and A. R. Stubberud, *Feedback and Control Systems*, with an application to the Engineering, Physical and Life Sciences, Schaum's Outline Series (New York: McGraw-Hill Book Co., 1967).

[8] E. W. Martin, Jr., "The Systems Concept," *Business Horizons*, Spring, 1966; reprinted in D. I. Cleland and W. R. King, *Systems, Organizations, Analysis, Management: A Book of Readings* (New York: McGraw-Hill Co., 1969).

tems theorists. Were one to examine a sampling of books and articles entitled "Organizations and/or Management: A Systems Approach," one would find that the organization is depicted as an input-transformation-output device,[9] a communication network, a decision-making mechanism,[10] a control system,[11] an operation/activity system, or finally, a combination of the above.

In all of these conceptualizations, an identical routine is followed: the author sketches out an engineering input-process-output system or a communication, decision, or control system, and then proceeds to match his conception of the organization with that model.

The first false step that the conceptualizer is most likely to make in this modeling process is to use an unwarranted analogy between the engineering model and his conception of the organization. Analogizing is not an attribute or technique peculiar to the systems enthusiast. The technique has its roots in the use of metaphors by writers of literature the world over, as well as by writers in the exact sciences. But analogies are analogies and some are better than others. Some may be practically useless, only confounding the issue.

Obviously, there can be no quarrel with the description of the systems discussed above. Even a superficial look into the first few chapters of a basic engineering text will support this. Difficulties arise, however, when one attempts to fit the organization into this simple input-process-output-feedback-environment scheme. What are the inputs, processes, and outputs of an organization? Is the labor force of an organization an input or a process? What is the environment of an organization? Who draws the boundary? What is within and what is without? Are the environment and the organization parts of the same whole, or are they two different and separate wholes? Is the whole different from the algebraic sum of the parts? Is the organizational system a causal system, one whose outputs depend only on the present and past values of the inputs?

Earlier theorists who studied the organization as a system concentrated their efforts on a few evident analogies. In general, one can identify three main viewpoints: (a) an organization is a communication network resembling, to a remarkable degree, the telecommunication system of Shannon and Weaver; (b) an organization is a collection of

---

[9] See for example: A. K. Rice, *The Enterprise and Its Environment: A System Theory of Management Organization* (London: Tavistock Publications, Ltd., 1963); D. Katz and R. L. Kahn, *The Social Psychology of Organizations* (New York: John Wiley & Sons, Inc., 1966).

[10] See for example: R. W. Cyert and G. March, *The Behavioral Theory of the Firm* (Englewood Cliffs, N.J.: Prentice-Hall, Inc., 1964).

[11] See for example: Jay Forrester, *Industrial Dynamics* (New York: John Wiley & Sons, Inc., 1961).

activities aimed at transducing certain material and energy inputs into certain material and energy outputs; and (c) an organization is a social system consisting of networks of roles and relationships.

The main conceptual difficulty of the first approach lies in the analogy between an electrical or electronic communication network and a human communication system. While the model may be adequate for the design of an electronic system, it is insufficient when applied to human communication,[12] and totally lacking when employed with organizational communication. True, some communication of the electronic type does take place within every organization, but to reduce all organizational communication to the exchange of signals and messages between the telephone sets of two executives would indeed be disastrous. The same holds true for attempts to study behavioral characteristics (leadership patterns and attractiveness) by measuring the amount of information sent and received by the member of a small group exchanging cue cards.[13]

The second systems approach to organizations is operations research (OR). Although launched auspiciously, in its subsequent development it was reduced to an in-depth analysis of certain observable and measurable activities of the organization. Recent developments are but further quantifications, mathematizations, and computerizations of the same operations, not a broadening of OR's conceptual foundations. Thus, after over two decades of considerable intellectual effort, today's typical OR problems are still centered about inventory, allocation, waiting-line, replacement, and competitor models, the only difference being that new developments in mathematics, computer languages, and simulation have enabled the operations researcher to make his problems even more complicated than before.

Admittedly, there is nothing wrong with quantification, mathematization, or simulation, provided that the problem has been placed in proper context and that mathematization and computerization do not constitute *the* problem. One would expect little opposition to operations research if it had managed to get out of both traps by now. Churchman, a pioneer in the field, has himself expressed serious doubts similar to the ones presented above.[14] Others insist that OR is not to be identified with a heterogeneous assortment of mathematical

---

[12] For a critical view on the subject see: Lee Thayer, *Communication and Communication Systems* (Homewood, Ill.: Richard D. Irwin, Inc., 1968), and his "Communication: *Sine qua non* of the Behavioral Sciences" in D. L. Arm (ed.), *Vistas in Science* (Albuquerque, N.M.: New Mexico University Press, 1968, pp. 48–77).

[13] See for instance: B. E. Collins and H. Guetzkow, *A Social Psychology of Group Processes for Decision Making* (New York: John Wiley & Sons, Inc., 1964).

[14] C. West Churchman, "Operations Research as a Profession," *Management Science*, vol. 17, no. 2, October 1970, pp. B 37–53.

techniques, but is rather a continuum of methods embedded in concepts. This continuum of methods begins with useful but crude qualitative models and reaches to the most highly refined forms of mathematical expression.[15]

The third systems approach to organizations represents the sociologist's attempt to provide a theory of human organizations that would be free of the mistakes of past social scientists. The point of departure here is the observation that human beings do not live in a vacuum nor are they the exclusive product of the most global of all influences, that of culture; rather they are members of organizations and institutional settings.[16]

A MODERN SYSTEMS APPROACH

Here the point of departure is that organizations come into existence, change, and disappear and that man's role is basically that of a controller, a steersman of the structure, the function, and the evolution of these organizations. To fulfill that role, he needs a logically consistent and generalizable set of concepts which will make intelligible the changing structure and behavior of organizations, as well as their effective control.

In tracing the origins of systems thinking, we have already dealt with some of the basic philosophical, terminological, and conceptual considerations. It will be recalled that the general philosophical and conceptual predisposition underlying modern systems thinking is "organicism." Organicism is the philosophy or viewpoint that puts the organism at the center of one's conceptual scheme. The term "organism" has often been replaced by the term "organized complexities" or "organized systems," defined as entities composed of many subentities which are interrelated and interconnected with respect to each other and, more importantly, with respect to their environments and to the whole. In his attempt to understand these organized complexities, the systems-oriented researcher employs the holistic method. This approach forces him to acquire an adequate knowledge of the whole before he proceeds to an accurate knowledge of the workings of its parts.

These are the major premises that govern the modern systems thinker's approach to the structure of organizations and the activities of the human beings associated with them. These organized complexities, although sharing certain common characteristics, are substantially

---

[15] D. W. Miller and Martin Starr, *Executive Decisions and Operations Research,* Prentice-Hall International Series in Management (Englewood Cliffs, N.J.: Prentice-Hall, Inc., 1969).

[16] D. Katz and R. L. Kahn, *The Social Psychology of Organizations* (New York: John Wiley & Sons, Inc., 1966); F. E. Kast and J. E. Rosenzweig, *Organization and Management: A Systems Approach* (New York: McGraw-Hill Book Co., 1970).

different, the differences in systems being associated with the way they are *organized*.

An Organization Theory

While theories of organizations have during the last two decades grown at an accelerating rate within the social sciences, there have been but few serious attempts to construct a theory dealing with organization as an abstract principle. Whatever contributions have been made toward the formation of such a theory have come from the biologists, who at the beginning of the 19th century introduced the concept of *organization* primarily as a substitute for the then waning concept of *vitalism*. Starting with some provocative, though crude, ideas about organization formulated by de la Mettrie in the 18th century, Claude Bernard equated organization with self-regulation of the body.[17] The two developments — the biologist's study of organization as an abstract principle and the social scientist's study of organization as an institution — have progressed independently of one another.

The last decade witnessed a surge in interdisciplinary research that made possible some progress toward a unified theory of organization. Research in biology and physiology has redefined the foundations of organization theory, revealing its applicability to realms outside these disciplines. Although the fundamental principles and concepts derived from this research are still not universal enough to become part of our common heritage, still they do serve as guides for theory formulation and empirical research. Despite the logic of the arguments presented by its opponents,[18] there is reason to hope that the indictment of general systems theory as a discredited theory is at present much too premature. Enough evidence appears to be piling up to support the belief that Rapoport and Horvath's dream for a unified organization theory may become a reality in the not too distant future.[19]

---

[17] S. E. Toulmin and J. Goodfield, *The Architecture of Matter*, Harper Torchbooks, The Science Library (New York: Harper & Row, Publishers, 1962), Ch. 14; also, Julien de la Mettrie, *Man a Machine* (La Salle, Ill.: The Open Court Publishing Co., 1961).

[18] See, for example: D. C. Phillips, "Systems Theory — A Discredited Philosophy," in P. P. Schoderbek (ed.), *Management Systems*, 2d ed. (New York: John Wiley & Sons, Inc., 1971); also his "Organicism in the Late Nineteenth and Early Twentieth Century," *The Journal of History of Ideas*, March 1970, pp. 413–32.

[19] The authors concluded their classic article with the rather optimistic observation: "In totality, we have today [1959] a variety of approaches to the study of organization (as an abstract principle) and a variety of approaches to the study of organizations (i.e., human aggregates with certain specified relations of interdependence among the members). The two developments are destined to travel along separate roads for a while. Occasionally, a connecting path will be discerned, along which ideas can trickle from one stream to the other. Eventually, it is hoped, the two streams of ideas will actually merge." In W. Buckley, *Modern Systems Research for the Behavioral Scientist* (Chicago: Aldine Publishing Co., 1968), p. 75.

Chaotic Complexities

When confronted with populations of elementary units, one generally has two ways of dealing with them: he can begin by specifying the attributes of each individual unit of the population, or he can derive the overall statistical averages of the individual attributes of the population. When the number of the individual elementary units contained in the population is small and the attributes under consideration are few, the first method is appropriate. In a class of five pupils ($N = 5$), for instance, the determination of the age attribute of each of the individual elementary units (students) would be made using the first method. Were class size (N) or the number of attributes (X) to increase to, say, $N = 100$ and $X = 10$, the method becomes impractical.

The phenomena that today's researchers study are nearly all complex: the number of the elementary units is large and their attributes many. It would seem therefore that the second method is most applicable here. This would, of course, be true if all phenomena were *chaotic complexities* which are characterized by a large number of elementary units and/or a large number of attributes and very little organization. However, the majority of phenomena that modern researchers are confronted with belong to another category, namely *organized complexities*.[20] A theory of organization appropriate for the present would have to be a theory of organized complexities.

Organized Complexities

Organized complexities are phenomena which are composed of a very large number of parts which interact in a nonsimple way. However, this interaction of the parts is arranged or organized into an orderly scheme and is guided by a purpose. In other words, organized complexities have a specific structure and exhibit a purpose or goal-directiveness. The structure of an organized complexity most commonly found in organic systems is that of hierarchy. The purpose most commonly pursued by organic systems is that of goal attainment or teleology. Thus, the adjective "organized" preceding the term "complexities" refers to both the existence of a hierarchical structure and goal setting.

Teleology (from Greek *telos*, "end"), or goal-directiveness and goal-attainment, was held by earlier philosophers as the specific characteristic of life organisms and the one which differentiates them from inorganic matter.[21] In modern systems nomenclature, teleology is a

---

[20] Ibid., p. 73.

[21] Ernest Nagel, "Teleological Explanations and Teleological Systems," in H. Feigl

cybernetic concept and refers to "behavior controlled by negative feedback."

It has been shown in cybernetics that teleological behavior is not the result of a vital force peculiar to biological phenomena, but rather the result of the operation of an error-activated and error-correcting mechanism found in machines and animals alike. Thus, a cybernetic system regulates itself by constantly comparing its actual performance to a goal, measuring the deviation from the goal and taking corrective action to minimize the difference between the two states. This is what is meant by teleological behavior in modern systems thinking.[22]

Teleological behavior is the result of a hierarchical structure. In its conventional usage relative to formal organizations, the term implies a superior-subordinate (authority) arrangement. Here we will be using the term hierarchy or hierarchic systems as defined by Simon. He states, "By a hierarchic system, or hierarchy, I mean a system that is composed of interrelated subsystems, each of the latter being, in turn, hierarchic in structure until we reach some lowest level of elementary subsystem."[23]

In hierarchic systems, absolute subordination among parts does not exist. In fact, the division between absolute "parts" and "wholes" is arbitrary, if not meaningless. What we do find in such systems are intermediate structures on a series of levels in an ascending order of complexity: subwholes which display some of the characteristics commonly attributable to wholes and some of the characteristics commonly attributable to parts.[24] Hierarchic systems of this kind exhibit certain characteristics which, if understood correctly, can become powerful tools in the hands of a competent holist.

In every organized complexity there are *interactions* within subsystems and interactions *among* subsystems. The particular structure of the hierarchy is determined by the degree of interaction among the different subsystems; however, it is possible that in some instances the degree of interaction is determined by spatial arrangements of the subsystem. For example, this would be true of mechanical systems with direct coupling of parts. Normally, however, the geographic dispersion of the divisions of a firm has little effect on the structure. Some

and M. Brodbeck, *Readings in the Philosophy of Science* (New York: Appleton-Century-Crofts, Inc., 1953), p. 539.

[22] A. Rosenblueth and N. Wiener, "Purposeful and Non-Purposeful Behavior," in W. Buckley, *Modern Systems Research*, p. 232.

[23] H. A. Simon, *The Sciences of the Artificial* (Cambridge, Mass.: The MIT Press, 1969), p. 87.

[24] A. Koestler, *The Ghost in the Machine* (New York: The Macmillan Company, 1967), Ch. 3.

firms, however, do take on a particular organization structure because of the spatial factor. This is less important today than formerly when telecommunication systems and computers were used less frequently.[25]

In organized complexities, coded relationships seem to predominate. In such cases, the degree of interaction and the shape of the hierarchy are determined by the existence and quality of the scanning devices, channels of communication, and decision-making devices. Thus, an organized complexity with a potent scanning device, efficient channels of communication, and accurate decision making would have a hierarchy with a larger span of subsystems than one that does not have these properties. However, most hierarchies would be expected to have only moderate spans, given their limitations.

Interactions within and among subsystems take the form of feedback loops.[26] Thus, an organized complexity, viewed as a hierarchy, can be described as a series of feedback loops arranged in an ascending order of complexity.[27] While these feedback loops were discussed previously in the chapter on cybernetics, here they will be treated from the viewpoint of complexity.

Figure 5–1 depicts such a hierarchy. The foundations of this hierarchy consist of nonfeedback simple transformation and sorting units.[28] In a simple transformation unit continuous outputs are produced by a continuous series of inputs. No goal as such is involved in a simple transformation unit. In a simple sorting unit, a given input is converted into several outputs. In the limiting case in Figure 5–1 one input is converted into two outputs.

Although a simple sorting system does make a decision regarding the proper ratios of the outputs, the decision rule or criterion is built into the system by a higher hierarchy. Simple transformation and sorting units can be complex insofar as they consist of a large number of elements. However, their simplicity lies in their lack of choice regarding the inputs and outputs as well as in the lack of goals. For this reason one can say that this level is occupied by organized simplicities.

The first level of the hierarchy of an organized complexity is occupied by simple goal-maintaining units. It is at this level that the simplest self-regulation begins. These feedback systems attain their goals via negative feedback. The degree of goal maintenance reflects

---

[25] Simon, *Sciences of the Artificial*, p. 98.

[26] C. W. Churchman, R. C. Ackoff, and E. L. Ansoff, *Introduction to Operations Research* (New York: John Wiley & Sons, Inc., 1957), Ch. 5.

[27] Cf. Chapter 3 of this book.

[28] See, for example: Jay Forrester, *Industrial Dynamics* (New York: John Wiley & Sons, Inc.), 1961.

FIGURE 5-1
A Hierarchic Arrangement of Feedback Loops Within an Organized Complexity

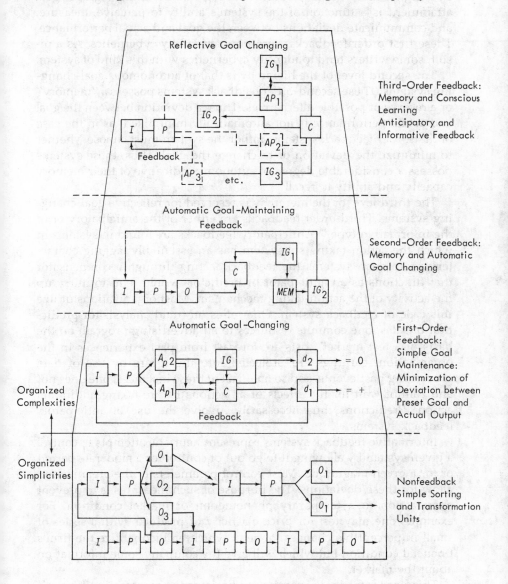

Legend: I = Input, P = Process, O = Output, FB = Feedback, I_G = Intended Goal, A_p = Actual Performance, C = Comparison, MEM = Memory, D = Deviation = $I_G - A_p$

Note: The importance of the external environment and the position/role of the observer vary directly with the degree of complexity (i.e., order of feedback). Thus, environmental disturbances and observer-observed uncertainty (i.e., product space) are less important for first-order feedback loops than for third-order feedback loops. Environment-hierarchy arrangements at third-order feedback level become "buffers" for second- or first-order feedback loops. That is to say, adaptation of a system on one level is coordinate functioning of the system on another level.

the system's degree of control. Although the goal of this system is set by a higher hierarchy (someone in the organization), the degree of goal attainment is a function of the system's ability to perceive, measure, and communicate deviations between the goal and actual performance. These first-order feedbacks were popularized by cybernetics. As a result, some writers tend to identify cybernetics with this kind of system.

The second level of the hierarchy is that of autonomous goal-changing systems. These second-order feedback systems possess a "memory" or a reserve of possible alternatives. Thus, a deviation between the goal and actual performance is not automatically minimized, as in the case of first-order feedback systems; rather the system must choose whether to minimize the deviation or to change the goal. Thus, these systems possess a considerable degree of autonomy indicative of their memory capacity and ability to recall.

The third level of the hierarchy is reserved for reflective goal-changing systems. Third-order feedbacks are either of the anticipatory or of the informative type.[29] Anticipatory feedbacks are found in systems in which the action-taking subsystem has an essentially lagging characteristic. In such systems the feedback, acting through a compensator that functions as an anticipator or predictor, would tend to hurry up the activity of the action-taking mechanism. Most organizations utilize this type of feedback system. One relies on trend analysis for prediction of sales, one commits resources to an advertising program on the basis of a test market, costs are forecast from past experiences in the organization, earnings are projected into the future by use of data concerning past events, and so on. In all of these instances, one does not and cannot wait for the effects of all actions before taking additional action. Predictions thus necessarily involve the use of anticipatory feedback systems.

Informative feedback systems represent heuristic attempts to control a given system by allowing it to go out of control for a moderate period or for a given magnitude. Most control parameters of the organization allow for these deviations. The purpose of such measures is to prevent overreaction to a temporary phenomenon or market condition. For example, one may test for price elasticity of products with a series of small imperceptible changes, none of which will jeopardize the firm's financial condition but which will give the company some information about the market.

Third-order feedback systems can be regarded as conscious learning processes in which past experiences in similar situations are recalled and used to revise the methods of control.

---

[29] N. Wiener, *Cybernetics* (Cambridge, Mass.: The MIT Press, 1961), Ch. 4.

Let us illustrate the above conceptual considerations with some familiar examples of organizational activities. Any organization from a typical manufacturing business enterprise to an educational institution can be visualized as a hierarchical arrangement along the lines of Figure 5–1. In a manufacturing firm, for instance, the bottom of the hierarchy is occupied by organized simplicities. For example, each machine operator is engaged in a simple sorting and/or transformation process whose standards or goals and actual performances are fixed either by the mechanical or technical configurations of the machine or by management's plans. The individual machine operator has little freedom regarding the inputs and outputs of this transformation and/or sorting process. The bank teller or accountant, the keypunch operator, the lab technician of a hospital, or the sanitary personnel of a university perform simple transformation and/or sorting activities.

At the first level of an organized complexity a minimal amount of freedom is introduced in the form of maintaining a given level of performance by regulating the actual performance to a fixed goal or standard. A foreman of a production shop, for example, has the exclusive responsibility of maintaining a given level of output. His goals or standards are set by the immediate higher level of feedback arrangements. His only strategy is to adjust the inputs so that his subsystem will achieve the desired output. The head nurse of a hospital must see to it that the nursing personnel attain a given level of performance. A supervisor of a bank's accounting department or the university's sanitary personnel are confronted with the same task.

Second-level hierarchies characteristically have more freedom to the extent that the performers of these activities may choose to adjust either the actual performance or the goals or standards. Although a given set of standards or goals is set by the upper hierarchy of a manufacturing company's production departments or divisions, or by a university or hospital, middle management personnel have some freedom in choosing certain goals or sets of goals and actual performances. The amount of data collection and information generation that takes place will somehow affect the amount of freedom.

By using data that support his position, the middle manager may be able to alter goals or expected performance, as when top management sets goals for middle management that are unattainable.

The line of demarcation between second- and third-order feedback systems is thin indeed. In the third-order system, top management of the organization provides a set of goals for all divisions and departments as well as the necessary feedback systems. It is this latter element that is more developed in this instance. If an organization is to survive, it must possess third-order feedback mechanisms, which means that it needs information about its environment, it needs current information about the state of internal operations, and it needs information concerning past events retrievable through its memory system (organization policies, sales records, production records, students records, patient records), all of which foster learning.

Such consciousness, if it exists in an organization, obviously becomes a determining element in the behavior of the system. If the sources of information mentioned above are cut off from the organization, it soon loses control of its own behavior. An organization simply cannot operate if data on past events are unavailable to it. Similarly an organization cannot function if the information streams of its various parts are severed. The company simply cannot suitably adapt and survive if it does not know what is occurring out there in the environment. Such information nets are crucial for the behavior and survival of a system.

It should be clear that all three feedback systems operate in a hierarchical fashion. Outputs of lower-level systems are inputs to higher-level systems. This in turn allows the focal system to set goals for the subsystems and the cycle repeats itself.

A Systems Theory of Organization

With this in mind we can now attempt to define organization. Such a definition should include both structural and functional considerations. From the structural viewpoint, the degree of organization reflects the degree of hierarchic arrangements, as well as the number of subsystems under one hierarchy (span). Hierarchies with moderate spans facilitate understanding because, by and large, they are nearly decomposable or dissectable.[30] From the functional viewpoint, the degree of organization reflects the degree of self-regulation involving control (negative feedback) as well as evolution or growth (positive feedback). Thus, from both structural and functional viewpoints, the degree of organization of an organized complexity is defined by fixed rules (hierarchic arrangements) and flexible strategies (orders of feedback loops).

There remain two additional concepts to be dealt with: the impact of the external environment upon the operation of the feedback loops, and the position and role of the observer of organized complexities.

Some of the inputs of the feedback system at any of the three levels of the hierarchy of an organized complexity might be in the form of "outside" disturbances. In certain instances accommodation of the system to these external disturbances would be accompanied by a certain

---

[30] Simon, *Sciences of the Artificial*, p. 99, and Koestler, *Ghost in the Machine*, p. 52.

loss in internal organization; some of these disturbances could be accommodated with no extra effort.

Hierarchy and teleological behavior are not intrinsic properties of organized complexities but, rather, are extrinsic, based on the relationship of the observer and the observed organized complexity. For a given observer, a certain organized complexity may appear to have few hierarchies with a large span or may exhibit first- or second-order self-regulation. For another observer the same organized complexity may have a large number of hierarchies with moderate spans or may exhibit third-order teleology.

Taking into consideration both of these additional concepts, one can say that *organization* is a relative concept, depending upon the relation between the real thing (the organized complexity), the environment, and the observer. Ashby is concerned with another kind of relativity: good vs. bad organization and high vs. low level of organization. The goodness or badness of organization refers to its self-regulative property. In any case, Ashby admits that there is no such thing as a "good organization" in any absolute sense. It is always relative. An organization that is good in one context may be bad in another. High and low degrees of organization are associated with degrees of fitness of the organization to certain environments.[31]

To appreciate the relevance of a system-oriented theory of organizations, a review of the traditional ways of studying a business firm may be in order here. It is axiomatic that any conception of a phenomenon (here, the firm) constitutes, by necessity, an abstraction of its relevant aspects and a subsequent rearrangement of these aspects into a coherent and logical unit less complex than the original phenomenon. This, of course, lies at the very heart of man's way of coping with complexity: by building conceptual models. These models are later used to construct a theory.

The most traditional model of an organization is the financial or accounting model, one that has stood the test of time exceedingly well. The financial or accounting model, although originally devised for the management of profit-oriented business enterprises, has been adapted by other enterprises such as education, health, and government and by some nonprofit organizations. The rationale behind this model is that each activity within an organization has its equivalent monetary representation. Thus, the monetary surrogate (model of the firm) is treated as an adequate representation of the firm's human and material processing activities. Such a model includes both static and dynamic

---

[31] W. Ashby Ross, "Principles of the Self-Organizing System," in Buckley, *Modern Systems Research*, p. 108.

aspects of a business enterprise. A balance sheet is, for instance, a static model of the firm, indicating the state of its different activities at a given moment of time—usually at the end of the firm's economic year. Dynamic aspects of the firm's behavior are expressed in the profit and loss statement. The accounting model is a relatively closed-system model to the extent that environmental considerations are limited to the stock market and financial institutions.

Another traditional and relatively closed-system model is the organizational chart approach to understanding a business firm. Here the organization is pictured as a pyramidal arrangement of different sectors of the enterprise called departments, districts, sectors, and so on. This very popular way of viewing an organization is as old and as universal as organizations themselves. Its openness to the environment is relatively limited and, to some extent, inconsequential as far as understanding and managing the firm is concerned. As a matter of fact, one would not be severely criticized were he to characterize this model as completely closed to the environment.

It should perhaps be pointed out that no model of itself asserts that the organization *is* a closed system; rather the model-builder considers the interaction between the organization and its environment as an irrelevant and unnecessary complication in explaining the behavior of the firm. Thus, the human behavior model assumes that an understanding of small group dynamics (i.e., the behavior of employees of a particular department or a group), will suffice for an understanding of the organization as a whole. The same applies to the market, economic, operational, and legal models and theories of organizational behavior.

All conventional models and theories of organizations constitute relatively closed-systems approaches to the study of organizations: they assume that the *defining* characteristics of an organization are its internal aspects while the external environment and the organization's interaction with it are, for the most part, inconsequential and constitute its *accompanying* characteristics. Thus, a study of an organization focuses primarily on its internal aspects and makes certain allowances insofar as a particular sector of the external environment has a bearing on a particular internal activity of the organization. For example, the financial/accounting theory takes into consideration the behavior of the stock market exchange because the relationships between the stock market and the firm affect the latter's ability to provide the monetary means for its survival.

A systems theory of organizations will, however, focus on the interface between the organization and the totality of its environment, simply because neither the strictly internal nor the strictly external aspects of the organization constitute its defining characteristics. Thus, the basis of an open-systems theory of organization will be a model depicting the exchange of energy and/or information between the internal and the external environment of the whole system.

The Enterprise-Environment Interaction System

Every manager is conditioned to think of a business enterprise as a system that *creates wealth*.[32] In fact, this seems to be the main theme in most standard textbooks on microeconomics: the firm, via the entrepreneur, converts disorganized resources into useful goods and services that consumers can acquire to satisfy their needs. The manager's chief task in this conversion process is to come up with the most efficient (least costly) combination of factors involved in the production and distribution of the goods or services. The manager associates consumption with the outputs of the enterprise's endeavor and not with its inputs. The inputs are considered to be abundant, though not free. Scarcity of inputs is not generally associated with the quantity of physical resources, but with the price to be paid for the use of these resources and for the alternative uses to which these can be put.

Contrast this view with the systems view. In systems nomenclature, every input to a system is the output of another system to which the latter is serially or randomly connected. This is essentially the problem of interdependencies. Thus, production that depends upon the importation of certain resources that are the output of another system can be considered as the consumption of these outputs. A manufacturing enterprise, for example, that produces automobiles is a consumer of steel, tires, and so on; steel, tires, and so on, are the outputs of a steel mill or tire factory; the inputs of a steel mill or tire factory are the outputs of iron-ore and rubber-producing systems, and so on. From the systems viewpoint then, it is really erroneous to refer to production as the creation of wealth and not as the consumption of wealth. As Friedrich Georg Juenger succinctly put it, "What is euphemistically called production is really consumption."[33]

In Figure 5–2 the firm is depicted as an open system which functions by importing the necessary resources from the environment and by exporting the product of the combination of these resources into the environment. The input side of the environment, which economists call the "factor market," can be regarded as a reservoir of both nonrenewable and renewable resources. The output side of the environ-

---

[32] A more detailed explanation of this line of reasoning can be found in A. G. Kefalas, "The Environmental Invariant and the Limits to Growth," paper presented at the conference "Environmental Protection: A Dialogue," October 10–11, 1973, The University of Georgia, Athens, Georgia.

[33] F. G. Juenger, *Die Perfektion der Technik,* English translation, *The Failure of Technology* (Hinsdale, Ill.: Henry Regnery Co., 1949).

FIGURE 5-2
An Enterprise as an Open System

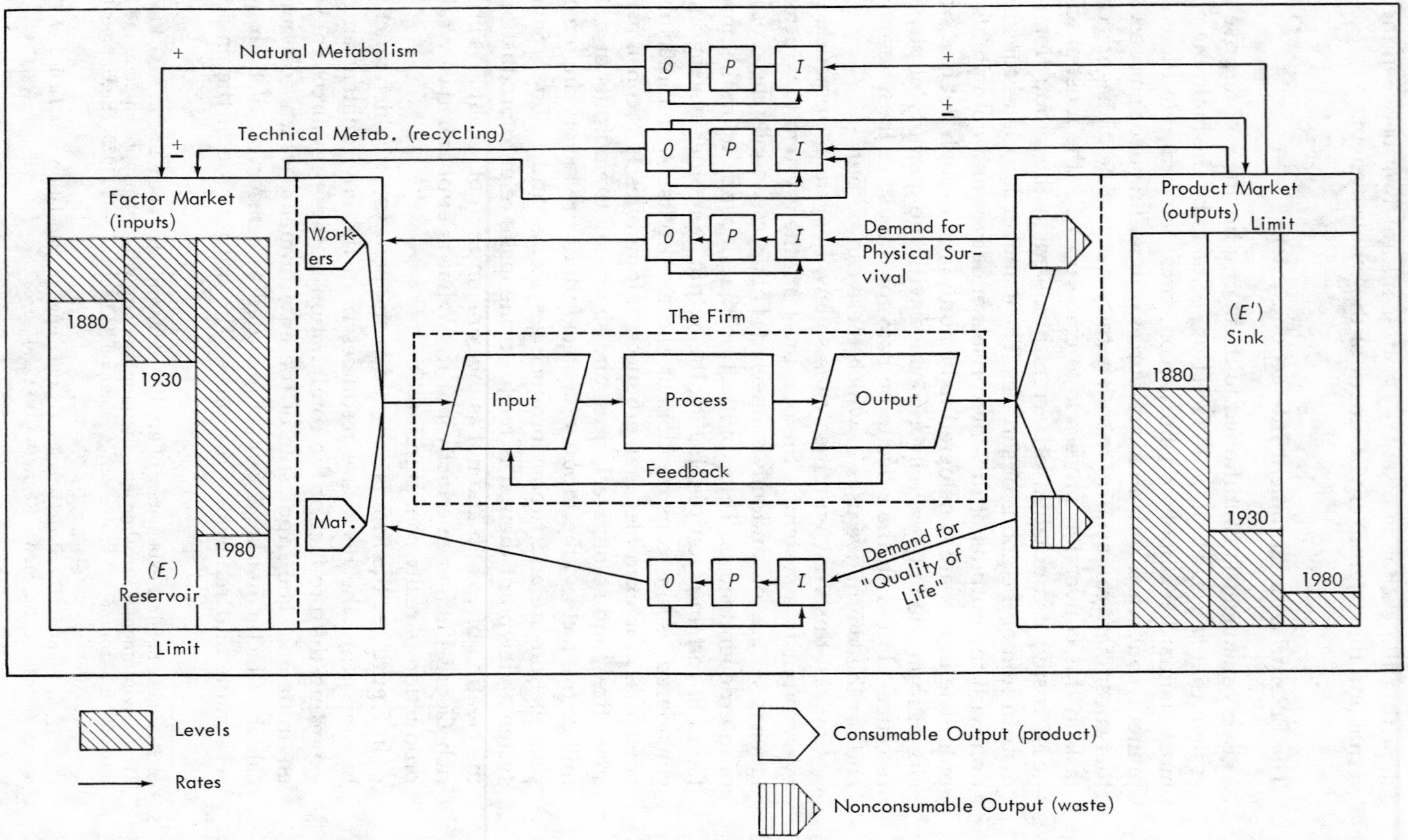

ment is identified with what the economists call the "product or consumer market." Local equilibria in these two markets guarantee a general equilibrium.

A closer look at Figure 5–2 reveals that the output of the open system actually consists of two outputs: consumables and nonconsumables. These terms refer to the output's demand upon the carrying capabilities or tolerances of the physical environment. Thus, every process of transformation of inputs into outputs results in primary products and by-products. Although both primary products and by-products may eventually become the inputs to other systems, they ultimately leave a residue that can become nobody's input. This will be relegated to the physical environment, which is here called a "sink."

A simple example should suffice to clarify this point. A typical meat-packing plant is an assembling point for live animals, which are disassembled into their component edible and inedible parts for further disposal. In carrying out this disassembling process, the plant produces a certain amount of solid and nonsolid waste which, for all practical purposes, is nonconsumable. The consumable portion of the output is "packaged" and shipped for consumption. However, a portion of this consumable (marketable) output is also biologically nonconsumable. That portion consists of nonmetabolizable tin cans, synthetic wrappings, and so forth, here called garbage (e). This is then forced to become an input to that portion of the environment labeled sink (E').

In general, the model depicted in Figure 5–2 can be interpreted as follows: the firm's environment can be thought of as two interconnected vessels — a reservoir and a sink. The levels of these vessels are the aggregation or integration of certain incoming and outgoing rates or flows. The magnitude and direction of these flows determine the level of the reservoir and the sink at a given moment of time. The capacities of both vessels are finite. Whether or not these limits will be reached will depend on the decisions made by the two governors of the flows: (1) nature — i.e., the subsystem that governs the levels of the reservoir and the sink via natural metabolism — and (2) the firm — i.e., the subsystem that governs the levels via technical metabolism (recycling). While the first governor is a cybernetic system that keeps the difference between system imports and exports to a minimum, the second decision-maker is a noncybernetic system, since it extracts more resources and deposits more residues than the optimal rate of renewal and absorption allows.

Only recently have business enterprises begun to realize the importance of the symbiotic relationship between themselves and their environments. As long as the physical environment appeared to be the "horn of plenty" for supplies and "the bottomless pit" for waste, business enterprises succeeded in carrying out their goals despite their ignorance. Recently, however, through the efforts of the ecologists, man has been made increasingly aware that the physical and man-made environment is but an aggregate of finites, and so is itself finite. Both the reservoir of resources and the sink for waste are bounded. One can no longer "burn the candle at both ends."

SUMMARY

In this chapter a review of organizational theories was undertaken, beginning with the classical school of scientific management and ending with the modern school with its stress on behavioral science principles and operations research. Following this we considered the philosophical, terminological, and conceptual misapprehensions of the earlier approaches to the study of organizations. Multiple theories of the firm arose in the hope of supplying a viewpoint missing in the others. The modern systems approach, however, first attempts to acquire an adequate knowledge of the whole and only then an accurate knowledge of the several parts.

A systems approach takes cognizance of the various organized simplicities, chaotic complexities, and organized complexities of the phenomena under study. In looking at organized complexities to which the modern organization corresponds, the properties of teleological behavior and hierarchic structure were stressed. These are extrinsic properties and not intrinsic, and are based on the relationship of the observer and the observed organized complexity. Consequently, in hierarchic structure there is no absolute subordination, while the teleological behavior is that of a third-order feedback loop. Conventional organizational theories assume that the environment is merely an accompanying characteristic of the organization, while the system approach takes the environment as a defining characteristic and focuses on the organization-environment interface and the exchange of energy/information. The environment is viewed as both a sink and a reservoir, the capacities of which are finite.

With this in mind, one needs to know the mechanism whereby one can successfully cope with the environment. The heuristic information acquisition and processing mechanism for dynamic exchange of information between the environment and the organization and between the subwholes is the subject of the following chapter.

REVIEW QUESTIONS FOR CHAPTER FIVE

1. Contrast the three organizational theories touched on in the chapter, noting their similarities and dissimilarities.

2. Discuss some of the "misconceptions" of the earlier systems approaches. Note whether or not these misconceptions also apply to GST.

3. The modern systems approach conceives of organizations as "organized complexities." What does this term really mean and how does it manifest itself in pragmatic business examples?

4. What are "organized complexities" and "chaotic complexities" and what tools or methods are available to deal with these?

5. In Figure 5–1, an organization is depicted as a "hierarchic arrangement of feedback loops." Give some examples of organizational activities—i.e., management tasks which in your opinion fall into each of these feedback loops.

6. Analyze a bank, a savings and loan association, a retailer, a manufacturing organization, a public utility, an educational institution, a stock brokerage office, and so forth, from the standpoint of feedback loops. Are feedback loops more discernible at the various levels of organizations? Why or why not?

7. What are some of the shortcomings of conventional financial models commonly used to study a business firm? What are the shortcomings of the attempt to study an organization from the systems viewpoint?

8. What type of feedback system is utilized in the classes you are presently taking?

9. Take a firm in your local community with which you are familiar and draw the input-output model as depicted in Figure 5–2.

10. Briefly discuss the energy situation in terms of input-output analysis.

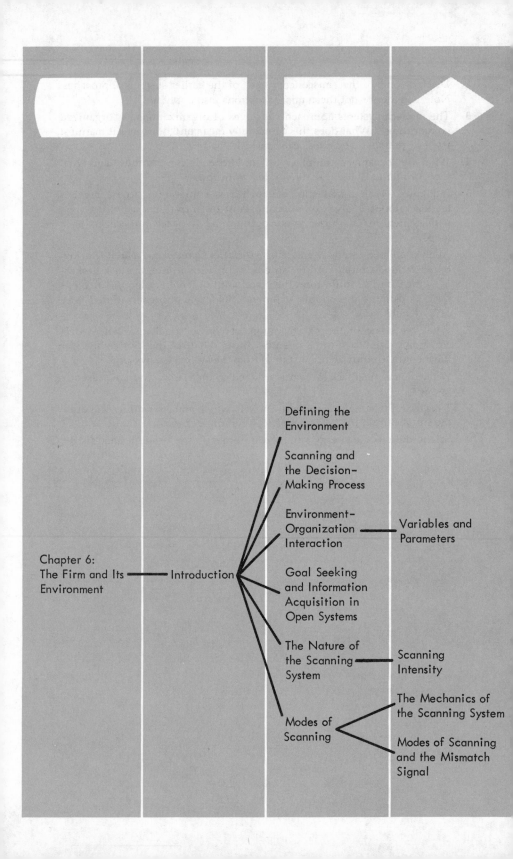

Chapter 6:
The Firm and Its ——— Introduction
Environment

Defining the
Environment

Scanning and
the Decision-
Making Process

Environment-
Organization ——— Variables and
Interaction   Parameters

Goal Seeking
and Information
Acquisition in
Open Systems

The Nature of
the Scanning ——— Scanning
System   Intensity

Modes of   The Mechanics of
Scanning   the Scanning System

                   Modes of Scanning
                   and the Mismatch
                   Signal

Chapter Six

The Firm and
Its Environment

*The world is not made up of empirical facts with the
addition of the laws of nature: what we call the laws of nature
are conceptual devices by which we organize our empirical
knowledge and predict the future.*

R. B. Braithwaite

INTRODUCTION

During the last decade a good deal of discussion has concerned the
need to incorporate environmental variables more explicitly into the
study of organizations. Organizational researchers admit that different
environments impose unlike demands and provide varying opportunities. An organization gains knowledge of these demands and
opportunities by gathering data about environmental events and by
the subsequent analysis and evaluation of the data. This information is
then utilized in organizational decision making for determining appropriate adjustments in strategies.

As used here, the term environmental factors refers to a set of
measurable properties of the environment, perceived directly or indirectly by the organization operating in that environment. These are
assumed to influence its operations.

For some purposes, the term environment is not difficult to con-
ceptualize. When the researcher's purpose is simply to describe the
environment of an organization, a dictionary definition of the term, as
"the aggregate of surrounding things, conditions, or influences," may
suffice. However, when the researcher's goal is not simply to define the
organizational environment but to analyze its properties or its role in

133

the functioning of the organization, the concept becomes a bit more complex.

In management literature, the term environment is generally loosely defined as "those things surrounding the organization." Typical factors noted are government, organized labor, competition, technology, the economy, and, of importance now, ecology. This definition also lacks sufficient discrimination in that it refers to an aggregate of all these factors and applies indifferently to one's perceptions.

DEFINING THE ENVIRONMENT

One way to overcome this shortcoming is to define the environment not as an objective "fact" but rather as an "image" in the entrepreneur's mind.[1] However, this, too, can lead to misunderstanding.

Between the two extremes of the highly objective—those things that surround the organization—and the highly subjective—the executive's image—lies a middle ground incorporating both viewpoints.

Churchman defines environment as those factors which not only are outside the system's control but which determine in part how the system performs.[2] Things that are within the control of the organization are, according to Churchman, resources or means that the organization may use in whatever way it finds appropriate. These, therefore, do not constitute environment. Things that have no direct impact upon organizational performance are also not in the actual environment. These may become organizational environments only if there is a change in the organization's objectives or goals.

For present purposes, environmental information will be treated as information which becomes available to the organization or as that to which the organization acquires access through its scanning activity. Environmental information flows either are routinely communicated to the organization or are deliberately sought out.

This definition of environment coupled with the "image" concept provides a clearer idea of what "that over there" actually is. The mere enumeration of the things surrounding the organization does not provide the organization with any specific information about the environment; it only hints at potential sources of data that the organization should monitor. What constitutes external environmental information is analysis and evaluation of the properties of these sources of data, together with the executive's "image" of the environment, that is, his *Weltanschauung.*

---

[1] E. T. Penrose, *The Theory of Growth of the Firm* (New York: John Wiley & Sons, Inc., 1959), p. 215.

[2] C. W. Churchman, *The Systems Approach* (New York: Delacorte Press, 1968), p. 36.

Most investigations of organizations and their environment rely heavily on the assumption that environmental demands or opportunities are presented to the organization in the form of "the problem" (constraints, threats, opportunities). However, once this assumption is challenged, one is then forced to ask how the organization becomes aware of these problems. How does it learn of impending threats? Organizational decisions are made as a result of the organization's ability and willingness to *scan* its external environment for the purpose of identifying environmental problems. Scanning, then, becomes the first element to be investigated here.

SCANNING AND THE DECISION-MAKING PROCESS – A DEPENDENCY

At the outset, it will suffice to define scanning as the activity or the process of acquiring information for decision making. Simon describes the decision-making process (DMP) as comprising three principal phases,[3] "(1) finding an occasion for making a decision, (2) finding possible courses of action, and (3) choosing among courses of action."

The first phase of the decision-making process (DMP) – searching the environment for conditions calling for a decision – is termed by Simon the "intelligence activity." The second phase, designated the "design activity," refers to inventing, developing, and analyzing possible courses of action. The third phase, referred to as the "choice activity," consists of selecting a particular course of action from those available.[4]

Initially, one may be tempted to suggest that a relationship between scanning and decision making exists only for the first phase. Indeed, the relationship here is very explicit, for the terms "searching the environment" and "intelligence" imply gaining knowledge through active information-acquisition behavior. Intelligence has, in fact, been defined by some scholars as "data selected and structured such as to be relevant in a given context for a decision."[5]

A more careful examination, however, will reveal that the second phase of DMP, the design activity, is also dependent on scanning. One can hardly deny that inventing, developing, and analyzing possible courses of action are influenced by the kinds and amounts of informa-

---

[3] H. A. Simon, *The Shape of Automation for Men and Management* (New York: Harper Torchbooks, The Academy Library, 1965), pp. 52–54.

[4] Simon, *Shape of Automation,* p. 53.

[5] O. H. Poensgen and Z. S. Zannetos, "The Information System: Data and Intelligence," Alfred P. Sloan School of Management, MIT, Working Paper no. 404–69, July 1969.

tion acquired through the scanning process. Similarly, whether or not a certain event or situation will be considered as a possible course of action will depend upon the degree of knowledge that the observer has about the event or situation. It will depend upon how well the decision maker is informed.

Finally, the dependence of the choice activity on scanning follows logically from the above. The less the environment is scanned, the fewer the possible courses of action, and therefore, the more limited the final choice.

If one were to add, as a fourth phase of DMP, the implementation and evaluation of the chosen courses of action (decision), then one would be forced to admit that the link between the final phase of DMP and scanning is quite significant. Implementation of a decision will require information about the system that is prior to or concurrent with it (feed-forward). The evaluation of a decision will require information for determining how effective the course of action was (feedback).

Each step in this process has as its inputs the outcome of activities in preceding steps. While DMP is depicted here as a chain of sequential activities, very often it takes the form of recurring chains with feedbacks.

If one accepts the proposition that "man's judgment is no better than his information," one would expect to find the same amount of attention given to scanning as to decision making per se. Furthermore, if the four phases of DMP are of a sequential nature, and if the dependence of the intelligence activity on scanning is substantially high, then one would expect that every scientific treatment of DMP would be preceded by an extensive investigation of the scanning process.

A search of the literature, however, shows that this is not the case. Not only does most of the literature on decision making concentrate on the last two phases (the comparison of alternatives and making the choice), but in nearly all cases the scanning process has been assumed to have already taken place.

When empirical research on the subject includes some kind of controlled experimentation, the information necessary for making the choice is provided to the subjects in the form of an "information survival kit," called "data bank," "information base," or "information structure." As Lanzetta and Kanareff state:

> Empirical studies of decision-making have typically provided the decision-maker with an information base in terms of which a choice among alternatives must be made. . . . The information base includes specification of the alternatives, the possible consequences of a choice, the probabilistic data on the relationship between alternatives and out-

comes. The information-acquisition processes preceding decision are assumed to have occurred, in essence, are simulated by the experimenter.[6]

As a result, most of our knowledge of scanning is far too incomplete and based upon implicit or explicit recommendations in the literature that deal only incidentally with information-acquisition behavior. It is our purpose to present a conceptual framework of the scanning process. In this framework it is assumed that factors outside the formal organization's boundaries do have a significant effect upon the functioning of the enterprise. Although opening the organization to its environment unduly complicates organizational behavior, still it is the only way one can "get a feel" of what the firm, the enterprise, the organization, is all about. Just because such a procedure necessitates an analysis of extremely complex interactions does not justify its omission. It is precisely when the organization interacts with its environment that its amazing complexity is revealed.

ENVIRONMENT-ORGANIZATION INTERACTION

In the study of the environment-organization interface, we are obviously interested in the types of relationship existing among the members of the set. The point scarcely needs belaboring that, in studying organizational change, the environment in which the organization exists is *also* in a state of change, but before examining the changing organization in a changing environment, one should look at the processes the organization employs for adapting to environmental change.

The two-way exchange between the environment and the organization can be depicted in the following matrix (Figure 6–1), presented by Emery and Trist.[7] The Ls in the matrix indicate some potentially lawful connection; the suffix 1 refers to the organization, the suffix 2 to the environment. Thus, L_{11} refers to the processes solely within the organization; L_{12} and L_{21} refer to exchanges between the organization and environment; L_{22} refers to processes solely within the environment. Of the four relationships depicted in the matrix, the two that fall outside our scope of investigation are relationships L_{11} and L_{22}.

In systems terminology, the environment-organization interactive system is the superordinate system — it is the Whole. On the other hand,

[6] J. T. Lanzetta and V. T. Kanareff, "Information Cost, Amount of Payoff, and Level of Aspiration as Determinants of Information Seeking in Decision Making," *Behavioral Science,* vol. 7 (November 1962), pp. 459–73. See also C. W. Churchman, "Operations Research as a Profession," *Management Science,* vol. 17, no. 2 (October 1970), pp. B37–53.

[7] F. E. Emery and E. I. Trist, "The Causal Texture of Organizational Environments," *Human Relations,* vol. 18 (1965) pp. 21–32.

FIGURE 6–1
The Environment-Organization Interaction Matrix

Inputs \ Outputs	Organization	Environment
Organization	L_{11}	L_{12}
Environment	L_{21}	L_{22}

the environment and the organization are the parts whose positional values relative to the whole need to be identified and studied. This investigation is best carried out in terms of the information flow of the transactional interdependencies between the environment and the organization (L_{21}) and between the organization and the environment (L_{12}).

In order to identify and trace the L_{12} and L_{21} transactions, one must be able to draw a line of demarcation between the two parts of the superordinate system, here called the environment-organization (EO) interactive system. In other words, one must be able to define the boundaries around both systems.

If one approaches the study of the EO system from the morphological (anatomical) viewpoint, then the boundary-definition problem becomes critical, for anatomically there is a unique and obvious distinction between the two parts of the EO system. However, if one views the system functionally, ignoring purely anatomical facts, then the division of the superordinate system into "organization" and "environment" becomes extremely vague. Supportive of this reasoning is the remark of Ashby that when "the organism and the environment are to be treated as a single system, the division between organism and environment becomes partly conceptual, and to that extent arbitrary."[8]

Variables and Parameters

When viewing the EO interactive system from the functional viewpoint, it is necessary that the observer be in a position to initially designate the system. The very first step involves the identification and allocation of all possible elements into two classes: (1) those within the

---

[8] W. R. Ashby, *Design for a Brain* (London: Science Paperback, Chapman & Hall, Ltd., 1954), p. 40.

system, and (2) those outside the system. Elements within the system are termed *variables* while those outside the system are called *parameters*.

In Figure 6–2 the two subsystems are depicted as being "located" next to each other. *A, B,* and *C* are the elements of the subsystem *environment* and 1, 2, and 3 are the elements of the subsystem *organization*. Elements *A* and *B* of the environment and 1 and 2 of the organization are variables of the two systems, respectively (*A* and *B = L_{22}* and 1, 2 = L_{11}). Variable *C* of the environment affects the organization in a marked way. This variable of the environment is an effective parameter of the organization. The same applies to variable 3 of the organization. It is a parameter of the environment.

FIGURE 6–2
Variables and Parameters of Two Systems

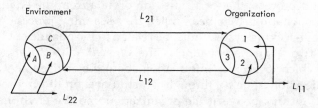

Traditionally there has been a tendency to consider the parameters of a system as constants. In systems thinking, however, there are but few situations in which a parameter can be considered to be constant, and that only for a limited time.

Figure 6–1 indicates that the behavior of the EO system is determined by two sets of *variables* (L_{11} and L_{22}) and two sets of *parameters* (L_{12} and L_{21}). For system survival it is imperative that both systems behave in such a way as not only to keep their essential variables (L_{11} and L_{22}) within certain desirable limits, but also to maintain their effective parameters (L_{12} and L_{21}) within certain predetermined limits.

Open systems depend, for their survival, on their ability to arrange their variables to accommodate outside disturbances. This characteristic of open systems to cope with the effects of changes in their parameters is referred to as adaptability. In the EO interactive system considered here as an open system, adaptability, stability, and survival belong to the combination of both subsystems; they cannot be related to either separately. Dynamic systems, like the EO system here under consideration, ensure their survival through stability in both their effective variables and their effective parameters.

GOAL SEEKING AND INFORMATION ACQUISITION
IN OPEN SYSTEMS

In general, the transactional interdependencies of EO systems (L_{12} and L_{21}) can take the form of either *energy* and/or *information*, as is true for all open systems. As Thayer states,

> The basic processes of an organization, which are the basic processes of *all* open living systems are:
>
> 1. *Importing* from the environment certain raw materials and resources for conversion into products or services which are *exported* for consumption by the same or other parts of the environment; and
>
> 2. *Acquiring* data from the environment, and from its internal parts, to be "consumed" in problem-definition and decisioning in the service of its attempts to alter its intended-states-of-affairs, its internal structure or function, or some aspect or domain of its environment.[9]

The second process, that of acquisition of data from the environment, constitutes what has here been called the "scanning process."

Any goal-seeking system must be related to the outside environment through two kinds of channels: the *afferent* (a scanning system), through which it receives information about the environment; and the *efferent* (decision system), through which it acts on the environment.[10] The relationships between the organization and the environment are shown in Figure 6–3. For logical completeness, a third subsystem has been added to Simon's two subsystems here labeled Scanning System (#1) (afferent system) and Decision System (#3) (efferent). This third subsystem is the Intelligence or Internal Organizing System (#2).[11] The addition of this system reflects the authors' understanding of the role of information in the decision-making process. Organizational actions which are the outputs of the efferent (decision) subsystem are not based upon outputs of the afferent (scanning) subsystem, but rather upon the outputs of the intelligence (internal organizing) subsystem which acts as an evaluator of the receptor or scanning subsystem. Data received by the scanning (receptor) system are eventually fed into the decision (efferent) system where they are utilized for problem-solving purposes. Contrary to popular belief (at least in the business literature) evaluated data do not constitute information unless and until they

---

[9] L. Thayer, *Communication and Communication Systems* (Homewood, Ill.: Richard D. Irwin, Inc., 1968), p. 101.

[10] H. A. Simon, *The Sciences of the Artificial* (Cambridge, Mass.: The MIT Press, 1969), p. 66.

[11] D. MacKay, "The Mechanization of Normative Behavior," in *Communication: Theory and Research*, Proceedings of the First International Symposium, Lee Thayer (ed.) (Springfield, Illinois: Charles C Thomas, 1967), p. 228.

FIGURE 6-3

The Environment-Organization (EO) Interaction System

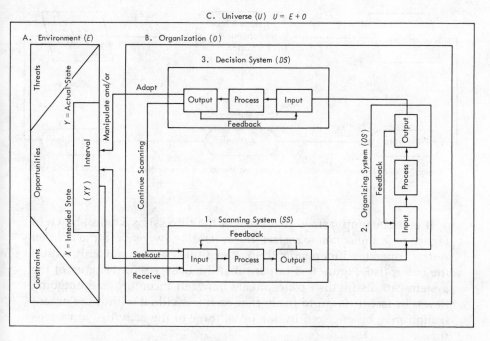

C. Universe (*U*)   *U* = *E* + *O*

enter the decision system. What the scanning system receives from the environment are raw sensory data about some aspects of the external environment.

The next step in this sequence is a selective conversion of the scanning data into a form suitable for consumption. It is only through this conversion process that the data become information.

It is this information which serves as a basis for decision making. The proportion of the potentially available data to be converted into immediately consumable information will be determined by the problem at hand. This will be indicated by the discrepancy between the intended state (X) and the actual state (Y) of the system; or in terms of Figure 6–3, the *existence, direction,* and *magnitude* of the (XY) interval.

Thus, what actually goes into the decision system are evaluated data, but not yet information. Information is *formed* in the mind of the problem solver or decision maker as an outcome of a comparison between the problem and the data. Adrian McDonough has presented this in Figure 6–4.[12]

---

[12] A. M. McDonough, *Information Economics and Management Systems* (New York: McGraw-Hill Book Co., 1963), p. 71.

FIGURE 6–4
Information

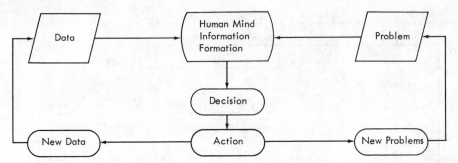

Source: A. M. McDonough, *Information Economics and Management Systems* (New York: McGraw-Hill Co., 1963).

It must be emphasized that none of the three subsystems shown in Figure 6–3, the *scanning system,* the *organizing system,* and the *decision system,* operates independently or constitutes a separate unit within the enterprise. Figure 6–3 implies that the observer or designer of the system can distinguish conceptually between incoming and outgoing flows. In fact, it is quite likely that each individual within the organization may be engaged in one or all three of the activities at various times.

Thus far the environment-organization interaction system has been treated conceptually as a hierarchical system. A hierarchical system is one that is composed of interrelated subsystems, each subordinate to the one above it.[13] At the bottom of the rank order is some lowest level of elementary subsystem to which no other is subordinate. This "boxes-within-boxes" way of looking at complex phenomena is not a mere partitioning of elements but one joined to the relationships of the several parts with one another and with the whole.

The EO interaction is regulated by the decision-making process. More precisely, the decision maker coproduces the future of the system along with the environment, which he does not control. The decision-making process in its entirety (i.e., sensory system + organizing system + decision system) becomes a subsystem within the total system of the EO interaction.

The organization through these three subsystems attempts to build associations between the states and changes of the environment and the organizational actions that will bring the EO system into a harmonious relationship. Messages are sought out or received by the

---

[13] Simon, *The Science of the Artificial,* p. 87.

organization through the scanning system. These are then transmitted to the organizing system, where they are evaluated. Finally, these are fed into the decision system where they are converted into information to be consumed in action.

Event data are gathered by the scanning system in terms of the deviations between an intended state and an actual state, referred to in the diagram as the XY interval. Assessment of the existence, direction, and magnitude of the XY interval is the function of the *scanning* system. Minimization of the existing deviation between the intended state and the actual state is the function of the decision-making process.

THE NATURE OF THE SCANNING SYSTEM

The organizational task of building associations between particular environmental changes and accommodative organizational actions was referred to above as the "strategic problem."

The term "strategic problem" alludes here to the basic feature of every open system: *maintenance and regulation of flow of information between the system and its environment.* Adaptability of the system to environmental demands and opportunities constitutes a *conditio sine qua non* for survival of the open system. It is in this sense that the term "strategic" is used.

Scanning Intensity

One might hypothesize a priori that at the organizational level the degree and intensity of scanning will depend, among other things, upon:

1. the availability of organizational economic resources,
2. the perceived nature of the relationship between the organization and its environment, and
3. the frequency and magnitude of changes in the states of the environment as it is related to the organization—i.e., the XY interval.

For the sake of simplicity, prescinding from the availability of economic resources that the organization can muster for information-acquisition purposes, the degree and intensity of scanning will be dependent on the two remaining factors.

Suppose there were only two kinds of relationship possible between the enterprise and its environment: (*a*) symbiotic and (*b*) synergistic. A symbiotic relationship is of the functionally necessary type: the relationship between the two systems is necessary for survival of *both* systems. A synergistic relationship, on the other hand, is not functionally necessary, but its existence enables the two systems to achieve

Figure 6–5
Determinants of Scanning Intensity

Degree of Change of the Environment \ Nature of the Relationship	Symbiotic	Synergistic
Stable	1	2
Dynamic	4	3

a performance that is greater than the sum of the two individual performances taken separately.

Disregard of a symbiotic relationship increases the probability that the relationship will eventually get out of control, leading temporarily to undue exploitation of each system, and ultimately to disintegration of both systems. In the EO interaction system, certainly the ultimate loser will be the organization, for that is the system that has to adapt.

Disregard of a synergistic relationship also increases the probability that the relationship will eventually get out of control, but the consequences here are much less severe than in the previous case. For what is at stake is not the whole relationship but, rather, the additional increment in the system's ability to survive, attributed to the synergistic effect.

In an open system the degree of scanning of the environment will depend upon the importance of the relationship between the two systems. This in turn will be influenced by the degree of interaction. The test for the intensity of interaction is the same as the test for the dependence of a system on certain parameters. To test whether a parameter is effective, one observes the system's behavior on two occasions when the parameter has different values.

The relationship between scanning and environmental states would be considerably more complicated were one to take into consideration the third factor mentioned above, that is, the frequency and magnitude of changes in the environment, as evidenced in the XY interval.

Were one to classify a slowly changing environment as relatively stable and a frequently changing environment as relatively dynamic, then one would expect to find the degree of scanning to be higher in the dynamic environment than in the stable. This is so because the variety inherent in a frequently changing environment can only be handled or controlled through equal variety in the system designed to monitor it. Since "information kills variety," dynamic environments call for "busier" scanning systems.

Figure 6–5 combines both determinants of scanning intensity. The numbers in the cells represent scanning intensity in ascending order.

The environmental sector that is perceived as being symbiotically related to the organization and that has a high frequency and magnitude of changes requires the most monitoring. The opposite is true of the symbiotic/stable combination.

MODES OF SCANNING

Scanning was defined as the process whereby the organization acquires information for decision making. This certainly must include human activity. As with all human activity, scanning is subject to all the biological, psychological, social, cultural, and economic laws governing human behavior.

Biologically, the process of information acquisition is today fairly well understood. From the psychological and, to some extent, the cultural and social viewpoints, scanning is considered as part of the process of thinking and problem solving.

From the economic viewpoint the acquisition of information is said to be subject to the law of efficiency, which states that the cost of acquiring information should not exceed the benefits to be derived from the acquired information.[14] Despite the obvious soundness of this principle, empirical research has thus far failed to provide any substantial evidence confirming its operationality.

The principle of efficiency, however, provides a useful conceptual tool. Every human activity is an economic activity to the extent that it requires the allocation and expenditure of scarce resources that have alternative uses. Since many different activities compete for the limited resources that the individual can devote to each activity, every individual will develop some scheme for rationing his resources among the different activities. In addition, the individual will develop some sort of modus operandi for each activity that will either minimize the expenditure of effort or maximize the returns.

As with most aspects of human behavior, scanning covers a broad continuum of possibilities that merge imperceptibly into one another.

---

[14] J. March and H. A. Simon, *Organizations* (New York: Wiley & Sons, Inc., 1958); H. Simon, *Models of Man* (New York: J. Wiley & Sons, Inc., 1957); J. Marschak, "Towards an Economic Theory of Organization and Information," in R. M. Thrall et al. (eds.), *Decision Processes* (New York: J. Wiley & Sons, Inc., 1954); J. March, *Handbook of Organizations* (Chicago: Rand McNally, 1965).

For purposes of analysis, however, it may be necessary to establish some recognizable, even if arbitrary, reference points within the continuum.

Since the organization and the environment are viewed here as parts of the same system, the different modi operandi of scanning utilized by the organization will depend upon the states of the variables defining the system organization (L_{11}), and upon the state of the parameters determining the exchange activity between the organization and its environment (L_{12} and L_{21}).

Although the two sets of determinants of the modes of scanning are not independent of each other, only the second set (L_{12} and L_{21}) will be examined here. As Ruesch puts it, "In our modern technological society, environmental change is so rapid that modern man's way of adaptation consists of holding the internal surroundings stable."[15] One may perhaps safely hypothesize that since "the outer environment determines the conditions for goal-attainment," and since data about the degree of goal attainment are gathered by the scanning system (Figure 6–3), *the mode of scanning is for the most part determined by the external environmental stimuli.*

In general, one can distinguish between two basic methods of scanning: *surveillance* and *search*.[16] The term *surveillance* refers to "a watch over an interest." The term is similar to what is termed "current awareness," the function of which is to give the information seeker some *general* knowledge.

Search, on the other hand, aims at finding a *particular* piece of information for solving a problem. The meaning of this term is familiar enough and has been dealt with in a number of works.

For present purposes, the difference between the two basic modes of scanning may be considered to be one of degree and not of kind. The difference lies in the degree of involvement of the scanner and in the formalization of the scanning procedures as measured by the degree of commitment of scarce resources in time.

It was mentioned above that the degree of involvement of a receptor will be influenced by the degree and frequency of the environmental changes that it is designed to monitor. Depicted in Figure 6–6 is the continuum that begins with surveillance at the extreme left end and terminates with search at the extreme right. The degree of involvement in time spent also runs in the same direction.

---

[15] Jurgen Ruesch, "Technology and Social Communication," in Lee Thayer (ed.), *Communication: Theory and Research* (Springfield, Ill.: Charles C Thomas, Publisher, 1967), p. 466.

[16] F. J. Aguilar, *Scanning the Business Environment* (New York: MacMillan Co., 1967); W. Keegan, "Scanning the International Business Environment," unpublished dissertation, Harvard Business School, June 1967.

FIGURE 6-6
Modes of Scanning

Surveillance	Search
Involvement time, etc.	Involvement time, etc.
Low	High

The two basic scanning modes can be further subdivided into view-ing and monitoring for surveillance, and investigation and research for search. Figure 6–7 shows the complete "scanning tree," consisting of the two basic branches and the four smaller but more detailed sub-branches.

FIGURE 6–7
The Scanning Tree

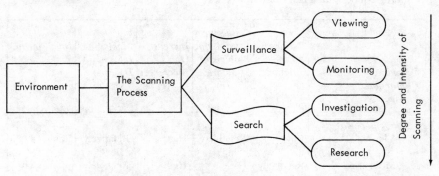

All scanning modes can be viewed as processes for:

(a) seeking a problem solution,
(b) gathering data about problem structure that will ultimately be used in discovering a problem solution,
(c) increasing one's awareness or familiarity with an environment, and
(d) making an "information decision."[17]

---

[17] The term "information decision" refers to the decision that a decision maker has to make with respect to (a) making the decision with the existing information or (b) decid-ing to acquire more information. If the decision maker chooses the second alternative, he is said to have made an information decision. The term could approximately be equated with "continuation and termination of search." See, for example, J. C. Grayson, Jr., *Decisions Under Uncertainty: Drilling Decisions by Oil and Gas Operators* (Cambridge, Mass.: Harvard Business School, Division of Research, 1960), Ch. 11.

The scanner assigns different values to each branch of the scanning tree. The assigned values obviously are not "true" values but rather estimates of the gain to be expected from further scanning along the same branch of the tree. Figure 6–8 presents the four modes of scanning and their relative values or utility as (*a*), (*b*), (*c*), and (*d*).

From Figure 6–8, it can be seen that surveillance (viewing + monitoring) has high value as an information-gathering process rather than as a means for finding solutions to problems. This relationship is reversed in the search (investigation + research) situation. Most research is, indeed, aimed at finding a satisfactory solution to a specific problem. Research activity, like all human problem-solving activities, is a varying mixture of trial, error, and selectivity. The selectivity derives from various rules of thumb, or heuristics, that suggest which pattern should be tried first and which leads are promising.

FIGURE 6–8
Scanning Modes and Their Values or Uses

Scanning Continuum (*modes*) \ Values or Uses	Seeking a Problem Solution (*a*)	Gathering Data About Problem Structure (*b*)	Increasing One's Awareness (*c*)	Making an Information Decision (*d*)	Involvement. Time
Surveillance					
Viewing	Low	Relatively high	High	Relatively low	Low
Monitoring	Relatively low	High	High	Relatively high	Relatively low
Search					
Investigation	Relatively high	High	Relatively high	High	Relatively high
Research	High	Relatively low	Low	Relatively high	High

The Mechanics of the Scanning System

Organizational information-acquisition activity cannot be investigated apart from its function in the survival of the system. Organizations, as goal-seeking or goal-guided open systems, depend for their survival upon their ability to adapt themselves to environmental states. Assessments of the states of the environment are made through the receptor or scanning system.

Figure 6–9 below is a simplification of Figure 6–3 illustrating the system's dependence upon scanning for its adaptability and ultimate goal attainment.

FIGURE 6–9
A Goal-Guided Open System

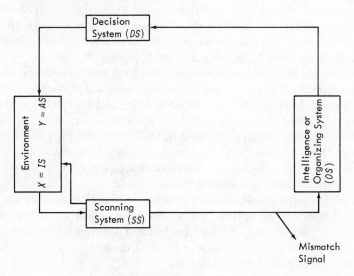

From the environment, the scanning system acquires data about the XY interval ideally in terms of:[18]

1. Existence of the XY interval (i.e., $X - Y > 0$)
2. Direction (i.e., $X - Y = \pm$) and
3. Magnitude (i.e., $X - Y = c$).

Signals about the XY interval can be viewed as feedback emanating from the environment. It will be convenient to view these as negative feedbacks which may be either "direct" or "real" or "anticipatory."[19] Thus, assuming that the XY interval can vary from Y_1 to Y_2 for the actual state and from X_1 to X_2 for the intended state, then it follows that the scanning system is confronted with a 3×3 matrix of signals which it has to watch. For each of these nine signals, the scanning system will have to acquire and transmit to the organizing system (1) existence, (2) magnitude, and (3) direction of X and Y movements. In addition, the system will have to monitor each state movement separately. Thus,

---

[18] D. MacKay, "Towards an Information-Flow Model of Human Behavior," in W. Buckley (ed.) *Modern Systems Research for Behavioral Scientists* (Chicago: Aldine Publishing Co., 1968), p. 359 ff.

[19] N. Wiener, *Cybernetics* (Cambridge, Mass.: MIT Press, 1967), p. 95 ff.

there are four additional feedbacks that have to be taken into account: $X_0 - X_1$, $X_0 - X_2$, $Y_0 - Y_1$, $Y_0 - Y_2$.

Feedback information about these changes in the external environment of the simple system with *one* known intended state and *one* known actual state is conveyed to the organizing system by way of the mismatch signal. The capability of the scanning system to transmit in the mismatch signal not only data about the existence of the XY interval (mismatch) but also its direction and magnitude will affect the number of trials that the decision system will have to make in order to adapt itself suitably to the external environment.[20]

Obviously, the organization will try to minimize the number of trials needed for perfect adaptation and survival. However, keeping the XY interval as small as possible will depend on the organization's willingness and ability to maintain a receptor or scanning system whose mismatch signal is capable of indicating not only the existence but also the direction and magnitude of the XY interval. Gains to be derived from minimizing the number of trials would be proportional to the costs for such a scanning system.

As with most maximization or minimization problems, the organization will seek not an optimal but a satisfactory solution, that is, a solution that is good enough to the extent that it satisfies all the constraints. Such a solution will consist of a flexible multistage scanning system like the one depicted in Figure 6–7.

Modes of Scanning and the Mismatch Signal

The four modes of scanning can be associated with different *degrees of completeness of the mismatch signal* in terms of the *existence* of a mismatch ($X - Y > 0$), *direction* ($X - Y$ at $t = 0$ is greater or less than $X - Y$ at $t = 1$), and *magnitude* ($X - Y = c$ where c is a number greater than zero).

Figure 6–10 presents the expected relationships between the four modes of scanning and completeness of the mismatch signal. Viewing, for example, the weakest form of scanning, does virtually nothing in terms of producing a complete mismatch signal and therefore contributes almost nothing to the reduction of the number of trials needed to achieve suitable adaptation. This mode, however, does provide information about the existence of the XY interval and can therefore relieve the higher modes of searching for this kind of information. It is only because of this contribution that the value assigned to viewing is positive. A system engaged in viewing should be considered a search-directing rather than a search-performing system.

---

[20] MacKay, "Towards an Information-Flow Model," p. 362.

On the other end there is research. The output of the research phase of the scanning process ought to be complete in the sense that statements about the XY interval should also indicate its direction and magnitude. Here the decrease in the number of trials will be the greatest. It should be emphasized, however, that research is the end-phase of a multistage process and not a totally independent activity.

Since all four phases of the scanning process belong to the same overall system, values assigned to lower steps can be regarded as savings of effort for the higher steps. Ideally, search should begin with assessing the direction of the XY interval, since its existence should have already been detected through one of the surveillance modes.

FIGURE 6–10
Modes of Scanning and Completeness of Mismatch Signal

Scanning Continuum (*modes*) \ Completeness of the Mismatch Signal	Existence of XY Interval $(X - Y \neq 0)$	Direction of XY Interval $(X - Y)_0 < (X - Y)_1$	Magnitude of Interval $(X - Y = c)$	Number of Trials
Surveillance				
Viewing	Maybe	No	No	Very large
Monitoring	Yes	Maybe	No	Large
Search				
Investigation	Yes	Yes	Maybe	Relatively small
Research	Yes	Yes	Yes	Very small

The direction and magnitude of the XY interval will depend on two factors: (1) upon the organization's internal capability to correctly formulate the intended state X and to actually produce the state Y that approximates the desired state, and (2) upon the nature of the environment.

Although the first factor definitely pertains to information acquisition, it is exclusively a matter of goal setting, goal seeking, and goal attaining. The second factor more directly touches upon the nature of the environment.

If the magnitude of the feedback is proportional to the intensity of the stimulus picked up from the environment, then one should expect a large deviation (a large $X - Y$) to be accompanied by a correspondingly strong feedback. (Considering that an undetected signal from the environment will have an impact which is proportional to the signal's original intensity, since it is not filtered anywhere, then one should expect this kind of environment to be monitored fairly intensively.)

The particular mode of scanning, however, will depend on the direction of the XY interval (i.e., its sign). When $X_1 - Y_1 > X_0 - Y_0$, that is, when $FB_{t=1} > FB_{t=0}$, the negative feedback indicates an unfavorable development in the L_{12} or L_{21} relationship. The discrepancy between the goal and the achievement is now larger than before. This certainly constitutes a problem, and, following Cyert and March's thinking, one would expect search to be the predominant mode of scanning. According to the above authors, search is always stimulated by a problem; and therefore all organizational search is "problemistic search."

When $X_2 - Y_2 < X_0 - Y_0$, that is, when $FB_{t=2} < FB_{t=0}$, then the negative feedback indicates an improvement in the L_{12} or L_{21} relationship. The discrepancy between the goal and the actual state of the system is now less than before. In this case one would expect surveillance to be the predominant form of information-acquisition behavior. In J. D. Thompson's thinking, surveillance is *the* scanning mode for the discovery of opportunities. According to him then, all surveillance by an organization is "opportunistic surveillance."

Although both movements—movement away from the intended state and movement toward it—can be perceived as problems, the "not-reaching-the-goal" situation (i.e., $X_1 - Y_1 > X_0 - Y_0$) is much more of a problem than the "doing-better-than-expected" situation (i.e., $X_2 - Y_2 < X_0 - Y_0$).

The particular scanning mode will also be affected by the importance of the environmental source that generates the feedback signal. If the environmental feedback affects a fairly large number of the system's variables, the system's ability to control any discrepancy is diminished and the amount of information and the number of trials needed to reach the desired (terminal) state are increased. In this kind of dynamic environment one needs fairly close monitoring. Again, the sign of the feedback signal will determine the mode of scanning.

SUMMARY

In this chapter an attempt was made to present a conceptual framework of the firm and its environment. The scanning process was viewed as the process of linking the organization to its environment.

The enterprise, viewed here as an open system, was related to its environment through two mechanisms commonly found in all open systems: the afferent or sensory system, and the efferent or motor system. These two systems are here called the scanning and decision systems, respectively.

Enterprises are man-made systems, often referred to as "artificials" whose actions are less integrative than biological systems. For this reason, to these basic systems a third was added. This system has been referred to here as the intelligence or internal organizing system.

These three subsystems are viewed as mechanisms enabling the organization ultimately to choose the correct action to diminish the difference between an intended state and an actual state. The achievement of this goal will depend, in part at least, on the state of the variables defining the system (organization) as well as upon certain parameters of the environment that significantly affect the organization.

The fate of the organization and of the more comprehensive EO system will depend on keeping both the *effective variables* (L_{11} and L_{22}) and the *effective parameters* (L_{12} and L_{21}) within certain desired limits. Survival is a quality inherent in both subsystems taken together and not in either of the two systems taken separately.

The function of the scanning system is the assessment of the relation between the intended state and the actual state. Data on these states are communicated via a mismatch signal to the internal organizing system for evaluation, and from there to the decision system for the necessary information formation. By means of these three systems the organization tries to build associations between the states and changes of the environment and the organization actions that will bring the EO system into a harmonious relationship.

Different environments will call for different scanning modes. In a stable environment in which the differences between the actual and the intended states are not very large and do not vary often, the amount of scanning required will be relatively small. However, scanning here will be well organized and considerably formalized. A dynamic environment, on the other hand, requires considerably more scanning, although the degree of formalization will be much less than in the case of the stable environment.

Specific modes of scanning the environment are determined by the magnitude and by the direction of the discrepancy between the goal and its realization.

Environmental sectors that in the past revealed an opportunistic type of feedback are expected to evoke more surveillance than search. Problem sectors, on the other hand, are expected to trigger problemistic search.

In the following chapter we will examine the development and impact of information technology in the 20th Century. We have looked at information acquisition; now we must turn to the significance of the acquired information in terms of the individual, the organization, and society.

REVIEW QUESTIONS FOR CHAPTER SIX

1. In the previous chapter an organization was said to interact with its environment. Just how does this occur? Is it possible for a firm to have more than one environment?

2. If you were hired as a "scanning manager" in an organization what do you suppose your job would be? Write a job description for your new job.

3. Most firms would acknowledge that they do informal scanning but do not have any such department solely concerned with this activity. In reality, many firms do scan on a formal basis. How is this so and what department in a large organization might be noted as doing most of the scanning? Could this vary with the type of organization? Give some examples.

4. Utilizing the organization-environment interaction matrix presented in the chapter, sketch out the matrix of an organization which you are familiar with.

5. Discuss the environment for your university. Be careful to note things which are in the system but perhaps controllable at a higher level of the organization.

6. The transactional dependency of L_{22} implies that some elements of the environment are affecting other elements of the environment. Give some examples of this.

7. Speculate on what some of the determinants of scanning would be for an insurance company. Contrast this with a firm in the computer industry.

8. How do data differ from information and why is this distinction necessary? What does "information formation" mean?

9. Name several industries which you think have a dynamic environment and several which have a stable environment. Does it make any sense to talk about a stable company in a dynamic environment or a dynamic company in a stable environment? Discuss fully.

10. Draw up a list of questions which attempt to discriminate between the various modes of scanning, administer this questionnaire to your students, and then determine whether they can distinguish between the various modes of scanning.

Part Four

Information and Decisioning

To live effectively is to live with adequate information.
N. Wiener

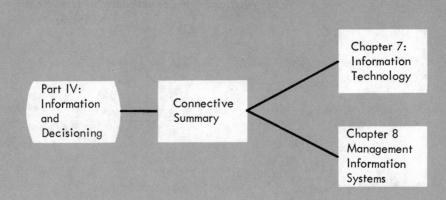

Part IV:
Information
and
Decisioning

Connective
Summary

Chapter 7:
Information
Technology

Chapter 8
Management
Information
Systems

CONNECTIVE SUMMARY

The main thesis that emerged from the previous two chapters of Part Three is that interactions between an organization and its environment can best be understood as exchanges of information. Considering the uncertainty of the environment as well as the immense variety of organizational activities, the two-way information flows constitute an information network of great complexity.

Information technology represents man's newest discovery for dealing with informational complexity. Information technology devices can easily be visualized as extensions or prostheses of man's cognitive capacities. The nature, uses, and impacts of information technology have been greatly misunderstood and misinterpreted. For this reason, Chapter 7 focuses on the logic, uses, and impacts of information technology. Thus, after a brief exposure to the rationale behind man's discovery of information technology and its basic concepts, we will concentrate on its potential and actual uses in managing human organizations.

The use of information technology for gathering, evaluating, and disseminating information in organizations is usually referred to as management information systems or MIS. Management information systems have been heralded as the latest panacea for effective decision-making. Recently, some scholars have argued and some practitioners have painfully found out that an MIS does not necessarily lead to better complex decision making per se. In fact, the entire sphere of MIS is saturated with many erroneous assumptions which will be examined.

Some of the more publicized criticisms about the inadequacy of long established techniques for designing, implementing, and managing management information systems are presented at the end of Chapter 8. It is hoped that cognizance of these pitfalls in MIS design and implementation, along with a better understanding of information technology provided in Chapter 7, will enable future designers to do a more efficient job.

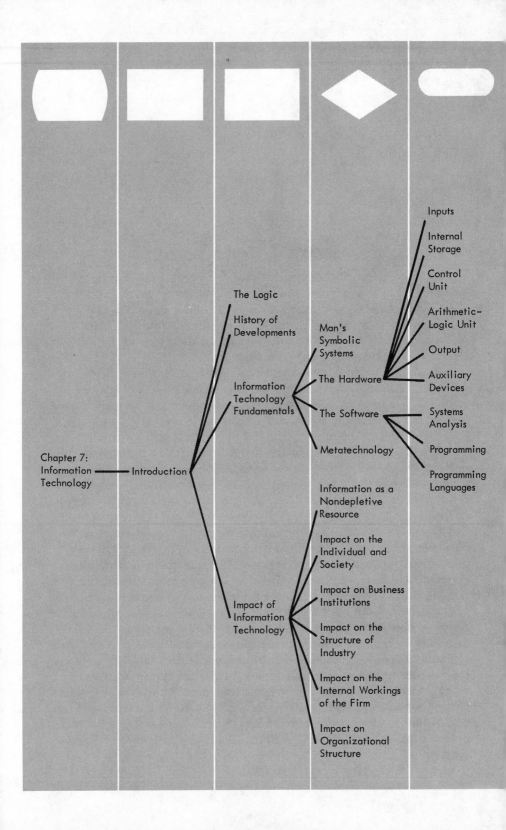

Chapter 7:
Information ———— Introduction
Technology

The Logic

History of
Developments

Information
Technology
Fundamentals

Man's
Symbolic
Systems

The Hardware

Inputs

Internal
Storage

Control
Unit

Arithmetic–
Logic Unit

Output

Auxiliary
Devices

The Software

Systems
Analysis

Programming

Programming
Languages

Metatechnology

Information as a
Nondepletive
Resource

Impact of
Information
Technology

Impact on the
Individual and
Society

Impact on Business
Institutions

Impact on the
Structure of
Industry

Impact on the
Internal Workings
of the Firm

Impact on
Organizational
Structure

Chapter Seven

Information Technology

Information technology is the nervous system of society.
Max Ways

INTRODUCTION

The second half of the 20th century, which ushered in the Second Industrial Revolution, is characterized by both the systems approach and information technology.

The increase in the size and complexity of 20th-century organizations has led many students and managers to adopt the analytical method employed traditionally by researchers of the so-called hard sciences. The analytic thinker, attempting to understand a complex phenomenon, breaks it down into smaller and less complex parts, then studies the parts separately, and then finally puts his findings together to gain an understanding of the whole.

Systems thinking, on the other hand, represents the "Age of Synthesis." Here the researcher's approach to the understanding of complex phenomena is to put together the findings of various disciplines with the aim of developing a technique applicable to many different phenomena.

Thus far, we have dealt with the origins, developments, criticisms, applications, and logic of the systems approach. The following two chapters are devoted to the second pillar of the "Age of Systems," namely, information technology. In this chapter we examine its logic, origin, and developments, and its impact on individuals, institutions,

159

and society. The next chapter focuses on the applications of information technology to the management of organizations. This application of information technology is commonly known as management information systems (MIS) and electronic data processing (EDP).

THE LOGIC

An idea running through the writings of systems thinkers is that all phenomena feed on information. As the biologist's maxim is "Life feeds on energy," so, too, that of the systems thinker is "Systems feed on information." Those aspects which tie all systems (biological, physical, social, and so on) together are the inflows, processes, and outflows of information. The biologist studies a cell or an organism by observing what it does (function) and what it becomes (evolution) as a result of supplying or withholding certain information from the organism's environment. The stock market analyst studies the behavior of the economy by tracing the impacts of certain kinds of information. The firm, too, feeds on information. The organization depends upon an efficient information system that can scan the environment to acquire data; after classification, evaluation, and transformation of the data into information, the information system transmits it to decision makers for more effective managing of the enterprise.

Data are best handled when expressed in a language other than that of ordinary communication. The most efficient way of compiling data is by numerical codification. While not everything can be so treated, a great deal of that which the decision maker needs to perform his task can be expressed in numbers.

Human beings are sometimes surprisingly inefficient in dealing with numerical expressions. They can barely manipulate numbers consisting of more than five or six digits. In the period immediately following World War II, a solution was sought in the development of a technology that would aid man in dealing with numerical expressions. Right from the start it was recognized that such a technology would have to go far beyond the existing hardware. If it were to aid man in overcoming his computational deficiencies, it should not only enable him to manipulate numbers by the four basic number transformations (adding, subtracting, multiplying, dividing), but should also enable him to arrange, organize, classify, transmit numerical information, and thus to solve problems. Furthermore, this technology should require *no* human intervention, once the data were received by the machine.

The requirements for this new technology were succinctly stated by Norbert Wiener in his classic book on cybernetics.

1. That the central adding and multiplying apparatus of the com-

puting machine should be numerical, as in an ordinary adding machine, rather than on a basis of measurement, as in the Bush differential analyzer.

2. That, in order to secure quicker action, these mechanisms, which are essentially switching devices, should depend on electronic tubes rather than on gears or mechanical relays.

3. That, in accordance with the policy adopted in some existing apparatus of the Bell Telephone Laboratories, it would probably be more economical to adopt the scale of two for addition and multiplication, rather than the scale of ten.

4. That the entire sequence of operations be laid out on the machine itself so that there should be no human intervention from the time the data are entered until the final results are taken off, and that all logical decisions necessary for this should be built into the machine itself.

5. That the machine contain an apparatus for the storage of data which should record them quickly, hold them firmly until erasure, read them quickly, erase them quickly, and then be immediately available for the storage of new material.[1]

Nowadays nearly every organization employs the new technology to some degree. Hence, the need to understand the logic, potentials, shortcomings, and impacts of this new technology.

HISTORY OF DEVELOPMENTS

Man is perhaps the most physically handicapped of all the primates. During the course of evolution he learned that the satisfaction of his needs and desires was limited chiefly by his physiological capacities. His desire to move bigger and bigger boulders in order to build ever larger pyramids, castles, churches, and roads was constrained by his physiological capacities. Material things, Newton told us, are subject to the universal laws of inertia: they resist displacement and seek states of equilibria. Thus, man, if he is to satisfy his desires, is compelled to find ways of overcoming these physical limitations.

Man is not only a *homo sapiens* — a creature capable of forming abstract ideas and of reasoning — but also a *homo faber*, a toolmaker. Man, the *homo sapiens*, conceived of novel artifacts while man, the *homo faber*, constructed tools and used them to produce even more elaborate implements, enabling him to substitute mechanical power for "muscle power."

Man's substitution of mechanical for human power reached a peak

---

[1] N. Wiener, *Cybernetics: or Control and Communication in the Animal and the Machine,* 2d ed. (Cambridge, Mass.: MIT Press, 1961.)

with the First Industrial Revolution. Mechanization now enabled man to use low-energy sources to trigger off high-powered operations. His role in the man-machine system became one of controlling these low-to-high-power transformation processes.

The consequences of the First Industrial Revolution are well known. On the positive side, mechanization enabled man to concentrate on more efficient power substitution. As a result, inexpensive goods were plentifully produced. On the negative side, some social and economic dysfunctions arose that man could not deal with. Mechanization also imposed another limit on man's ability to survive. This limit was man's ability to control technology.

Once again man was forced to do some rethinking; only this time it was not the relationship between his desires and the limits of his physical capabilities, but, rather, the relationship between the artifacts he himself had created and his cognitive capacities to comprehend and control them. Man's rethinking of this relationship resulted in a complexity/sophistication continuum: advanced mechanization-automation-cybernation.[2]

Advanced mechanization, characterized by mechanical handling between transfer machines, represents the linking together of mechanical devices to load, unload, and transfer a workpiece between stations of a single machine or between stations of several different machines. While human intervention in these operations is conspicuous by its absence, the lack of any mechanical decision-making capacity is equally evident.

The essential difference between advanced mechanization and automation is mechanical decision-making control. Automation involves elimination of human labor, human decision-making and control. The latter characteristic is even more evident in cybernation.

It would help to think of automata or cybernetic machines as machines for decision-making and control while the linking together of machines can be regarded as operation devices. In other words, the decision-control machine dictates the operations of the power machine. The former mechanism is usually referred to as the "master," while the latter is called the "slave." Thus, while in advanced mechanization the machine operates by following the decision of a human, in automation and cybernation the machine operates by following the decisions of another machine.

Essentially, the master machine is an information-processing and

[2] For an elegant comparison of automation and mechanization see: J. Rose, *Automation: Its Anatomy and Physiology*, A Contemporary Science Paperback (London: Oliver and Boyd Ltd., 1967); S. Handel, *The Electronic Revolution* (Baltimore, Md.: Pelican, 1967); S. Beer, *Management Science* (Garden City, N.Y.: Doubleday & Co., Inc., 1968).

-transmission device in contrast to the slave, which is an energy-trans-forming and transfer instrument. Man's role in this master/slave machine interaction is to control the controller by designing into the machine properties that facilitate his role and by designing out of it whatever hinders it.

Figure 7–1 compares the role of the human operator in mechanization (including advanced mechanization) and automation (including cybernation). It can be seen that the importance of automation lies not so much in the linking together of machines, as in the ability to create feedback information for the control and comparison unit. It is the feedback loop that enables the machine to control its performance at any moment.

FIGURE 7–1
Mechanization and Automation

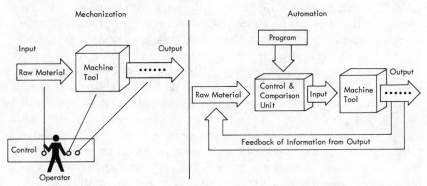

Source: Adapted from S. Handel, *The Electronic Revolution* (Baltimore, Md.: Pelican, 1967), p. 157.

INFORMATION TECHNOLOGY FUNDAMENTALS

For the contemporary student to fall into the trap of equating information technology with computers is very easy. The reason is fairly obvious. Until recently, the literature on the subject dealt almost exclusively with the capacity, speed, and accuracy of the processing, dissemination, and storage of data, since these were the obvious technical limits of the machine that somehow had to be overcome.

There are essentially four basic elements in information technology:

1. Man's Symbolic Systems
2. Hardware
3. Software
4. Metatechnology

1. Man's Symbolic Systems

Nature has equipped man with the apparatus to carry on all the symbolic communication necessary for his survival. However, this apparatus has its limits. Man has invented mechanical devices to augment his sensory and intellectual capacities. Hardware systems from the telephone to the computer extend man's cognitive apparatus just as power machines are extensions of his muscles. The computer is basically a symbol-receiving, -processing, and -communicating device. Since its symbolic apparatus differs from man's, the gap between them is bridged through software technology.

Software technology includes all the modes of translating human symbols into forms that the hardware can deal with. This conversion process consists of the following steps: human ideas are expressed in ordinary language (e.g., English); this language then undergoes various transformations: programming language (e.g., BASIC, FORTRAN, COBOL), arithmetic or algebraic, and machine language. This completes the first half of the communication cycle, from man to machine. Communication from the machine to man is just the reverse.

The last element of information technology, metatechnology, refers to the conceptualization, design, and implementation of ways of organizing man and machine into systems for the collection, storage, processing, dissemination, and use of information. It is in metatechnology that man's creativeness is put to the test. A good metatechnology is primarily a matter of the creation of new and more sensible philosophies or viewpoints of man's place in the order of things. Without it, information technology will become an anathema to many. Information metatechnology is man's last chance to defy the anti-technologists' prognostications of impending doom.

2. The Hardware

Electronic computers have indeed captured the imagination of the public. Enthusiasts call them electronic brains, intelligent boxes, and thinking machines, while the antagonists prefer to look upon them as just another but bigger calculating machine. The truth, however, is that the real power of the computer has been both underestimated and underemployed.

There is nothing mystical and mysterious about computers. Their apparent complexity and remarkable performance are based on principles that are not too difficult to understand. Indeed, as the discussion of the black box approach to complexity indicated, one does not need to comprehend perfectly the full complexity of the machine as long as one understands its role in the system.

FIGURE 7–2
The Overall Organization of a Computer System

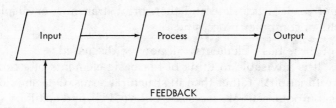

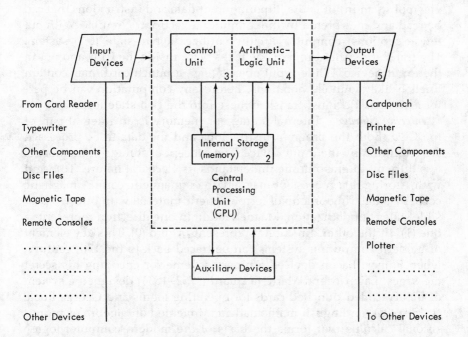

The basic elements of the electronic computer will be presented here. For a more detailed treatment, however, any of the newer standard computer textbooks should be consulted.

Disregarding for the time being the auxiliary storage devices (Figure 7–2), one can single out the five basic units in a computer system:

1. *Input* devices receive and translate data into the form required for the specific computer.

2. *Internal* storage devices store data to be operated on as well as the instructions to be performed.

3. *Control* devices regulate and coordinate all elements of the entire computer system.

4. *Arithmetic-logic* devices perform arithmetic operations and logical comparisons.

5. *Output* devices translate and then print, transmit, or display the results of computer processing.

Each of these basic elements will now be discussed.

Input. It is advisable to think of the basic elements of a computer system in functional, rather than in structural terms. One should think of inputs, outputs, central processing units (CPU), and auxiliary storage devices in terms of *what they do* rather than *what they are.* From this viewpoint, an input is the importation of data and instructions that can be read and interpreted by the computer's electronic circuitry. With but few exceptions, computer circuitry utilizes a two-state (0,1) system. The translation into binary code takes place inside the CPU and not in the input device. While input objects (e.g., a punch card) may contain English-like symbolic notations, before any computation can be performed, the CPU must translate these into the two-state form.

Internal Storage. Internal storage or memory is an integral part of the CPU. Both the program instructions and the data to be processed are fed into it by the input device. The concept of storage or memory is less difficult to grasp if one understands its electrical nature. To begin with, storing is a process of magnetizing a magnetic core. A magnetic core is a small ring or loop of ferromagnetic material with but two possible states of polarization. Magnetization in one direction represents a one (1); in the other direction it represents a zero (0). This key element of modern computing systems can be traced back to the 17th century, when Francis Bacon devised a binary scheme for encoding his secret messages. Later Joseph Marie Jacquard (1752–1834) designed a system of binary-coded punched cards for operating looms and George Boole (1815–1903), an English mathematician, invented an algebra of propositional calculus that forms the basis of the modern computer logic.[3]

The system for measuring internal storage capacity is fairly complicated. Internal storage capacity and speed are key characteristics of hardware which more or less determine the price of the CPU. Basically all storage devices are elements capable of storing one character (numerals 0–9, letters A–Z and special characters such as . , ( ) $ *). Each is made up of magnetic cores, capable of storing one bit. A bit, it will be recalled, is shorthand for *bi*nary dig*it*—the unit of measurement of information. Eight bits are used to represent one character. Units of eight bits are called *bytes*. A group of bytes must be combined to represent a computer word just as groups of bits represent a character. Word length (i.e., the number of bytes) varies with different computers. The

---

[3] For a more detailed account of the history of the binary system see: F. G. Heath, "Origins of the Binary Code," *Scientific American*, August 1972.

IBM 360 system uses an eight-bit byte and a four-byte word. A computer word in this system consists of (4 bytes) × (8 bits per byte) = 32 bits. The UNIVAC 1108, on the other hand, has a 36-bit word length. Thus, when a computer manufacturer says that the system has a 100,000-byte memory capacity, that means that the system's internal storage is 100,000 alphanumeric characters. The importance of internal storage will be seen when the subject of auxiliary storage is discussed.

Control Unit. The control unit contains the circuitry requirements to direct and integrate all units of the total computer system. It performs the same function as the feedback element of a system. In a CPU the control unit operates in two sequential cycles: fetch and execute.

In the fetch cycle, the control unit retrieves the instruction from the internal storage and places it in a temporary storage area within the control unit called the *instruction register.* In the execute cycle, the control unit separates the instruction into *operation code* and *operand address.* The operation code specifies the operation to be performed, such as read, multiply, add. The operand address provides the storage location of the data to be processed.

Arithmetic-Logic Unit. It is in the arithmetic-logic unit that the actual processing of data is done. Arithmetic operations are those restricted to numerals. Thus, multiplication is performed by storing the multiplier in one register, the multiplicand in a second register, while the product is developed in a third register. The same method is employed in other arithmetic operations.

Logical operations include comparisons, test sequences, and rearrangements of data expressed in any kind of symbols including numerals.

Although the actual performance of the four elementary arithmetic operations within a computer is rather complicated, basically it resembles the performance of the desk calculator. Thus, multiplication is performed by repeated addition, subtraction by addition of complements, and division by repeated subtractions.

Output. The output device performs the reverse function of the input: it converts the results of computer processing into a form that men can use. In some cases, however, the output is not directly intended for human consumption, but is instead transmitted either immediately or after a certain time lag to another machine as an input.

When the computer is designed to provide immediate response to the user, the output is called *on-line real-time* output. The key phrase here is "immediate response." The terms *real-time* and *on-line* are frequently misused in management information systems literature. On-line real-time systems require a large internal storage, since for fast access the programs for updating data or answering inquiries must be stored in the computer's internal memory.

Batch processing is a slower and less expensive way of obtaining an output. In batch processing the data and instructions of several jobs are accumulated and processed sequentially. Outputs are obtained in the same order and distributed to the users. Here answers to inquiries are not obtained immediately but at the end of the turnaround period which can range from an hour to a day or more.

Auxiliary Devices.   Auxiliary devices are both storage and input/output devices. These can be used to transmit data from the user to the machine, to convey the results of data processing from the machine to the user, and, finally, to supplement internal core storage. For these reasons and because of their relative economy, auxiliary devices are an indispensable part of every computer system.

The most commonly used auxiliary storage devices are magnetic discs, drums, and tapes. As far as storage is concerned, the principle is the same for all: a stored bit is represented by magnetization of a core or spot on a surface.

Magnetic drums or discs, when used as supplementary core memory, are directly linked to the CPU and controlled by it. Since auxiliary storage devices are not part of the internal storage, the time required by the control unit for the fetch or execute cycle is considerably longer. The time that elapses between the request for data (inquiry) and its delivery is called the *access time.* While the core memory access time is measured in nanoseconds (billionths of a second) or even in picoseconds (trillionths of a second), auxiliary storage access time is measured in milliseconds (thousandths of a second).

Whereas a detailed explanation of the technical properties of auxiliary devices involves a fair degree of sophistication, the basic principles behind magnetic tapes, discs, and so on, are somewhat easier to grasp. The computer's magnetic tapes and discs resemble to some extent the familiar tape recording and jukebox devices. Random access of these devices is possible, but it is not as fast as one's own brain. Magnetic tapes, for instance, are purely sequential media. Here each character (or byte) is recorded on a line across the width of the tape, successive characters of a record are recorded in sequence along the length of the tape, and successive records are arranged along the length of the tape but separated by unrecorded gaps. Of course, it is technically possible to skip over a record or a group of records to achieve some degree of random access, but in no case can the access time of a tape approach that of core memory. For example, a high-speed tape drive takes about four minutes to pass through a full reel of tape, making the average access time two minutes or more.

Magnetic drums and discs can be visualized as special arrangements of magnetic tapes. A magnetic drum would then consist of a series of tape strips (or tracks) wrapped around a cylinder, with a read-write head available for each track. Average access time, equivalent to one-half revolution of the drum, ranges from 5 to 10 milliseconds for most drums, depending upon their rate of rotation. Storage capacity, determined by the size of the drum, can be as large as several million characters or bytes.

Magnetic-disc storage is similar to drum storage, although discs provide more surface than drums. A disc-storage unit consists of a set of discs like those in a jukebox. Each disc surface is divided into a set of tracks, each wide enough to store a character across its width. Each track is divided into sectors or wedges, each sector of a track holding a fixed number of bytes, usually 100 or 200. Each sector of a track represents a separate addressable location. Average access time is about 85 milliseconds. Capacity can reach 200 million characters in the larger units.

Auxiliary devices are low-cost storage devices that in a way supplement the internal storage element of the central processing unit. However, no arithmetic-logical processing can be performed in auxiliary devices. Their use, therefore, will be determined by the problem at hand. If the problem involves the simple processing of massive data, then a small core memory could be compensated by extensive use of auxiliary devices. Auxiliary devices, as a rule, are used for storage of large amounts of data, with lower cost per item of data stored, and have slower access times than those of core storage.

Let us take a typical business situation to illustrate the above rule. A typical inventory system consists of inventory records comprising several thousand items requiring several million characters. In some cases, these records are maintained on auxiliary devices. When a change in the inventory takes place (e.g., a sale of a finished product or use of semifinished product) the computer program and the updating are read into storage. As the inventory record for a part is updated, it is pulled into core storage from auxiliary storage, processed according to the program instructions contained in the core storage, and finally returned to auxiliary storage. The access time for these two operations — that is, the pulling of instructions together with the updating of the data and the pulling of inventory records — is different. The former is accessed in nanoseconds, the latter in milliseconds.

When every item in the inventory records is to be updated — such is the case in grocery stores — the sequential device, magnetic tape, is much more practical than the magnetic disc or drum, whose positional time for reading is slower. However, not all inventory systems are like this. In most inventories only a small percentage of stock items change during each updating cycle. In this case, the computer would have to process the entire magnetic tape to update an item stored near the end of the tape. It is therefore more practical and economical to use disc storage in this case because random access is easier.

It is not difficult to imagine business situations where auxiliary devices cannot be substituted for internal storage. Engineering design, production scheduling, and automatic control of production runs require excessive and complicated arithmetic and algebraic computations. Here access time is very important, for any interruption of an assembly line for even a few minutes will prove very costly. On-line real-time output is a matter of life and death, so to speak.

Managers are often called upon to make decisions regarding the purchase, lease, or use of computer equipment on the recommendations of the experts. The more the manager understands the language of these experts the better his decisions will be. For example, the cost of supplementing core memory with auxiliary devices is not limited to the cost of tapes, discs, or drums. Normally, one or more pieces of equipment will be needed to drive these devices. In most magnetic tape applications, for instance, a single pass through a tape reel will involve either a read or a write operation—almost never both. Thus, several tape drives may be needed. But, at the same time, these same tapes, discs, and so on, and their drive mechanisms can also be used as input-output devices.

3. The Software

From the foregoing discussion of the hardware characteristics of a computer system, one might have gotten the impression that the efficiency of the system depends on these qualities. This is true as far as the central processing unit is concerned. Memory size, speed, and quality of the control unit do indeed determine its efficiency. CPU efficiency, however, seldom, if ever, creates a problem for the manager. His concern is with the system as a whole and not with a particular part, no matter how important that element might be. Obtaining the right answers to questions is a task to which all five elements of the system must contribute. As a rule, one might say that the test of the value of such a system is a test of the quality of communication between man and machine and between machine and machine. This element of information technology has been referred to as software technology.

Because the term "software" is defined as "the set of programs used with a computer to write other programs and perform standard operations," most people tend to equate software with computer-programming languages. However, such a narrow view of software is in the same class of errors as the identification of a computer system with the CPU. While computer-programming languages are part of what is termed software, they are not the whole of it. Furthermore, from the managerial viewpoint programming per se is not even the most relevant ingredient of software.

There are basically two essential elements in software technology: systems analysis and programming. Systems analysis precedes programming, although both form a continuous and recurrent cycle. In general, systems analysis determines what communication must take place between man and the machine and between one machine and another, while programming specifies the symbolic system to be used in these communication patterns. Thus, an elaborate computer program written in a sophisticated language indicating the steps of a relatively unimportant communication represents an unnecessary expense in terms of both hardware and software time and cost and managerial talent.

Systems Analysis. In the literature, the term "systems analysis" is sometimes used synonymously with the terms "the systems approach" or "systems concept" or even "management science." Although not much harm is done in equating these terms, still much confusion can be obviated if a clear definition is adopted right from the start.

Systems analysis is to be understood here as the organized step-by-step study of the detailed procedures for the collection, manipulation, and evaluation of data about an organization for the purpose not only of determining what must be done, but also of ascertaining the best way to improve the functioning of the system. There are several points worth noting in this definition. First, systems analysis is primarily concerned with the investigation of an existing information system of an organization to determine whether the system is amenable to computerization as well as to prepare the computerization process itself.

Secondly, systems analysis is primarily an analytic technique: a given system is broken down into its logical components or subsystems; subsequently, the inputs, processes, outputs, and feedbacks of each subsystem will be defined in terms of the information required to make decisions regarding the proper function of these system elements. These broad information requirements serve as guidelines for determining the sources and destinations of data transactions for each subsystem. Data must be acquired, documented, classified, processed, and evaluated before meaningful information can be obtained.

Thirdly, the outcome of a systems analysis study will result in a recommendation by the systems analysts for an improvement or replacement of the existing system. To this end, the systems analyst designs several systems charts which he turns over to a programmer who must analyze in exhaustive detail the data transactions and develop outlines of the execution procedure in the form of flow charts. These he will eventually code in some programming language. Since systems analysis is an integral part of design, it will be touched upon again in the chapter dealing with management information systems.

FIGURE 7-3
System Chart and Flow Diagram

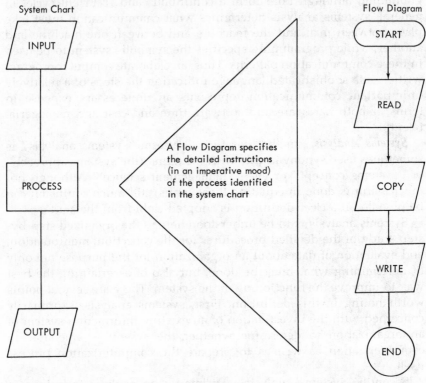

Source: Adapted from David Li, *Design and Management of Information Systems* (SRA, 1972), p. 55.

Programming. It is imperative that both facets of software—systems analysis and programming—be understood as a cyclical sequence of activities rather than as independent packages. The systems analyst's "systems chart" focuses on the inputs and the outputs of the system. His systems chart must identify programs, procedures, and data structures by name. In contrast, the programmer's "flow diagram" focuses upon the sequence of data transformations needed to produce an output data structure from an input data structure. His flow diagram must identify individual operations on portions of the data structure.

The relationship between the systems chart and the flow diagram is depicted in Figure 7–3. It should be noted that while the words inside the boxes of the systems chart are nouns, the words in the flow diagram are verbs indicating the activity character of a flow diagram.[4]

---
[4] David Li, *The Design and Management of Information Systems* (SRA, 1972).

Programming Languages. No written natural language can be directly interpreted by the central processing unit of a computer. These languages are far too ambiguous and much too complex for the simple-minded electronic circuitry. Even the universal language of mathematics, despite its inherent rigor and precision, cannot describe the full range of procedures that computer programs must include. The computer's language, the so-called machine language, is not a language in the conventional sense but, rather, the total set of operations that the computer can perform. It is more like the syntax of a natural language. The gap between ordinary language and machine language is bridged by programming languages.

A programming language is one that is used in writing the series of instructions that direct the computer to perform a specific series of operations. The machine language repertoire may contain over 100 different kinds of instructions. An instruction consists of an operation code and an operand address. Using a symbolic language, a programmer refers to the location of instructions and data by symbolic names. When the instruction is read, the computer converts the symbolic language into machine language by assigning a numerical operation code to the symbol. Symbols include both single letters and abbreviations.

The translation of symbolic language to machine language is accomplished via a computer program referred to as an *assembler.* An assembler assembles machine instructions (numerals) to replace the symbolic instructions. There is in general a one-to-one correspondence between the symbolic language source program and the machine language instructions. It is because of this one-to-one correspondence that a symbolic language is also called "machine-oriented" language.

Figure 7–4 shows the process of translation of symbolic language to machine language.

In writing assembler programs, the programmer is limited to a set of instructions peculiar to a specific machine. Every single problem must be so analyzed and programmed as to fit the particular machine

FIGURE 7–4
Translation of Symbolic to Machine Language

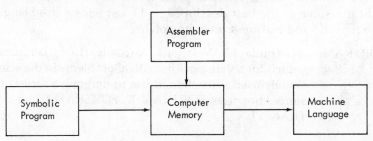

for which it is designed. For this reason, languages have been developed that permit programs to be written whose structure resembles the language of the problem at hand. These languages, "problem-oriented" languages, are unlike symbolic or machine language in that there is a one-to-many correspondence between a problem-oriented language and a machine language. In other words, while each instruction of a symbolic language is translated into *one* machine language instruction, a problem-oriented language instruction is translated into *several* machine language instructions.

Just as in the case of symbolic-to-machine language translation, so here an intermediate program is needed to translate a problem-oriented language into machine language. This intermediate program compiles a set of machine language instructions from the problem-oriented instruction. For this reason, programs that do this are called *compilers*. Figure 7–5 outlines the process of translating a problem-oriented language into machine language.

FIGURE 7–5
Problem-Oriented to Machine Language Translation

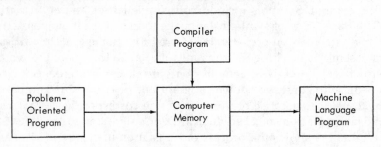

Several problem-oriented languages have been developed. All of them are designed to facilitate description of data processing or computational processes in terms of algorithmic steps. Because problem-oriented languages are detailed descriptions of step-by-step procedures, these languages are also referred to as "procedure-oriented" languages.

The following list of problem/procedure-oriented languages merely indicates some of the best developed and most widely used languages for scientific and business problem solving.

FORTRAN: (*FOR*mula *TRAN*slation). An early (1955) language designed for solving mathematical problems in the sciences. Widely used since then, it has undergone a series of revisions known as FORTRAN II, FORTRAN IV, and FORTRAN VI.

COBOL: *(CO*mmon *B*usiness *O*riented *L*anguage). Designed by a committee made up of representatives of manufacturers and users, this language is now commonly used in accounting and business applications.

ALGOL: *(ALG*ebraic-*O*riented *L*anguage). A language developed in Europe and primarily designed for solving scientific and mathematical problems. Has many things in common with FORTRAN IV.

APT: *(A*utomatic *P*rogrammed *T*ool). A special-purpose language for describing functions of numerically controlled machinery, such as milling machines.

BASIC: *(B*eginners *A*ll-purpose *S*ymbolic *I*nstruction Code). A simple language developed at Dartmouth College specifically for instructional use. It allows the user direct interaction with the computer.

PL/1:   *(P*rogram *L*anguage *1*). A general-purpose language, combining most of the features of the special languages developed for scientific, business, simulation, and command-control applications.

SNOBOL: *(S*tri*N*g *O*riented Sym*BO*lic *L*anguage). A language that can be used to manipulate information in the form of natural language; it stresses an ability to handle symbolic rather than numeric data and is used in such applications as language translation, question answering, and automatic abstracting programs.

GPSS: *(G*eneral *P*urpose *S*ystems *S*imulator). A programming language designed especially for simulation.

Figure 7–6 represents the complete cycle of man-machine-man communication.

4. Metatechnology

Metatechnology is essentially concerned with the discovery and implementation of conceptual schemes for the development of the individual elements of information technology. If progress in these individual subsystems is to be sustained, then it must be so directed and organized that the overall progress will not be impeded by any of the subsystems. Metatechnology focuses on, to borrow a phrase from Churchman, "the ethics of the whole system." It serves as an ultimate governor or controller which automatically senses and compares the contributions of the subsystems toward the progress of the whole, and

FIGURE 7-6
Overall Organization of Computer Problem-Solving-Man–Machine Interface

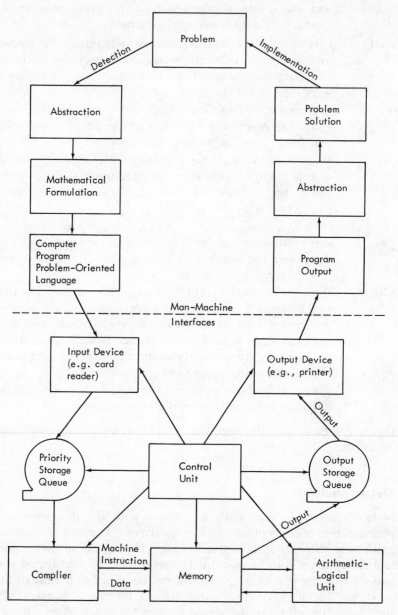

by means of negative feedback curbs any regress and accelerates progress.

The need for continuous reconceptualization of the whole and the development of new viewpoints is today more pressing than ever. Faultless hindsight can often reveal the unforeseen consequences of technological breakthroughs. While extremely helpful in assessing the once unknown, hindsight does little to develop foresight so necessary for today's manager. Optimization of subsystem operations has at times been found by hindsight to create serious bottlenecks in the overall system. Optimization of the production subsystem of an industrial enterprise, for example, can create a bottleneck in the accounting subsystem, struggling to gather the data necessary for financial control; computerization of the accounting subsystem can create a data-engulfed management, resulting in the suboptimization of the goals of the entire enterprise.

There can be little doubt that metatechnology needs reemphasizing. At times one almost wishes that progress in the other areas of information technology would come to a standstill so as to give man the opportunity to rethink and remodel the interrelationships of all the elements of information technology into better conceptual schemes. But that is unlikely to happen. Computer manufacturers are determined to supply more and more sophisticated software. Metatechnological progress must inevitably come from those who have the opportunity and foresight to see beyond the instant electronic circuitry and the DO Loop and IF Statement. But if progress does come, it ought to come from management students and practitioners.

IMPACT OF INFORMATION TECHNOLOGY

Information as a Nondepletive Resource

Man's survival depends as much upon the generation and dissemination of information as it does on the availability of material resources and the existence of free energy. Information supplies man with knowledge of the existence and the potentials of obtainable and useful resources. Thus, a strong relationship exists between information and material- and energy-resources, the latter being conditional on the former. Before any material- and energy-resource can be used to satisfy a given need or want, it must first be conceptualized and recognized as a resource. To aid in this is a function of information.

Because of this relationship between physical resources and information, there has been a marked tendency to treat information as just another resource. However, a closer look at some of the unique properties of information will convince one that equating the two is not only unwarranted; it is also misleading. It was the unique properties of information that led Daniel Bell to characterize the present society as a "postindustrial society organized around information and utilization of information in complex systems, and the use of information as a way of guiding society."[5]

Information resources, unlike any other, are not *used up* when *used.* There is no loss or diminution incurred when information is shared or transmitted from one point to another. This should indicate that informational resources are not subject to the law of conservation of energy, the reason obviously being that no physical units of mass or energy are transmitted—only symbols. Symbols lie in the public domain; their dissemination benefits all with loss to no one.

In short, information resources possess two unique properties which are not shared by other resources: (1) all other resources are dependent on them for their evaluation and utilization; and (2) unlike other resources, information is neither reduced nor lost by increased use or wider dissemination. It is nondepletive. See Figure 7–7.

Impact on the Individual and Society

A detailed treatment of the impact of information technology upon the individual and society at large may some day appear in sociological literature. Here a few basic ideas, borrowed from the informative report of the Conference Board, will be examined.[6]

In general, the effects of information technology on the individual and society can be arranged along a continuum at whose left are the favorable (i.e., survival-promoting) effects while at the right are the unfavorable (i.e., survival-impeding) effects. Initially, in the late 40s and early 50s, information technology would have been positioned at the favorable end of the continuum. Currently, expert opinion would locate it more and more toward the right (the unfavorable) end of the continuum.

In former times it was maintained that an uninformed individual is less capable of survival than one who is well informed. Therefore more information would enhance the survival capabilities of the individual and consequently of the society. Today some writers affirm that the case for information technology has been greatly overstated, and that more information creates overstimulated (a sort of information-over-loaded) individuals who are potential candidates for "future shock," —

---

[5] Daniel Bell, "Remarks of the Moderator," in *The Management of Information and Knowledge,* Committee on Science and Astronautics, U.S. House of Representatives (Washington, D.C.: U.S. Government Printing Office, 1970), p. 13.

[6] *Information Technology: Some Critical Implications for Decision Makers* (The Conference Board, Inc., 1972).

FIGURE 7-7
Information as a Nondepletive Resource

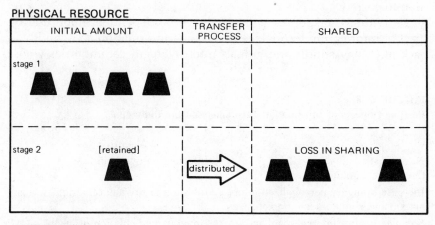

Source: Adapted from *Information Technology: Some Critical Implications for Decision Makers* (The Conference Board, 1972). Reprinted by permission of the Conference Board.

i.e., a state of minimum preparedness for dealing with survival demands of the future.

Figure 7-8 summarizes some of the possible effects of information technology on the individual. In general, one may expect the possibilities of the new information technology to present a variable distribution of negative and positive tendencies.

Given these potential positive and negative effects of information technology and given a free society, individuals will adopt certain response patterns depending upon their orientation to society. Figure 7-9 displays certain response patterns in a logical framework.

It should be pointed out that these response categories are not isolated "frozen" stages, but levels that indicate changing patterns of behavior through which individuals may move according to their orientation to society.

Individuals' responses will, of course, be dependent upon the impact of information technology on society. Information accessibility or lack of it may stratify individuals into divergent communities whose

FIGURE 7–8
Possible Impacts of Information Technology on the Individual

Positive	*Negative*
New sensory ranges available through technological means.	Information overload.
Increase in higher personalized information and communication exchanges.	Invasion of privacy of various forms.
Enhancement of interpersonal and intergroup dialogues through more sophisticated means.	Adverse manipulation of means and media to control the news and mold opinions.
More flexible and equable access to available knowledge.	Increased surveillance and monitoring of personal data "in the social interest."

Ability to use the information process to:

a. Become more aware and more self-conscious.

b. Make more free and voluntary choices.

c. Incorporate more variety into the social process.

d. Avoid needless hardship and costs of experience via simulation, i.e., social experimentation can be carried out without being tried on people.

Value enhancement
a. May free individuals to play their roles in their peculiar ways.

b. Provide for and encourage personal growth.

c. Higher levels of knowing, caring, and achieving self-determination.

Decrease in social cohesion with greater fragmentation of attitudes and motivations.

Increased discrimination via inequable access to advanced skills necessary to use information and communications effectively.

Value distortion
a. Information available in quantitative terms may be given exaggerated significance—as easier to incorporate into technology.

b. Collapse of time in information process may burden human adaptation.

c. Illusion of certainty may suppress values of voluntary action, e.g., displacement of responsibility from man to machine.

Source: Adapted from, *Information Technology: Some Critical Implications for Decision Makers* (The Conference Board, 1972). Reprinted by permission of the Conference Board.

FIGURE 7–9

Varieties of Personal Response to Increased Information

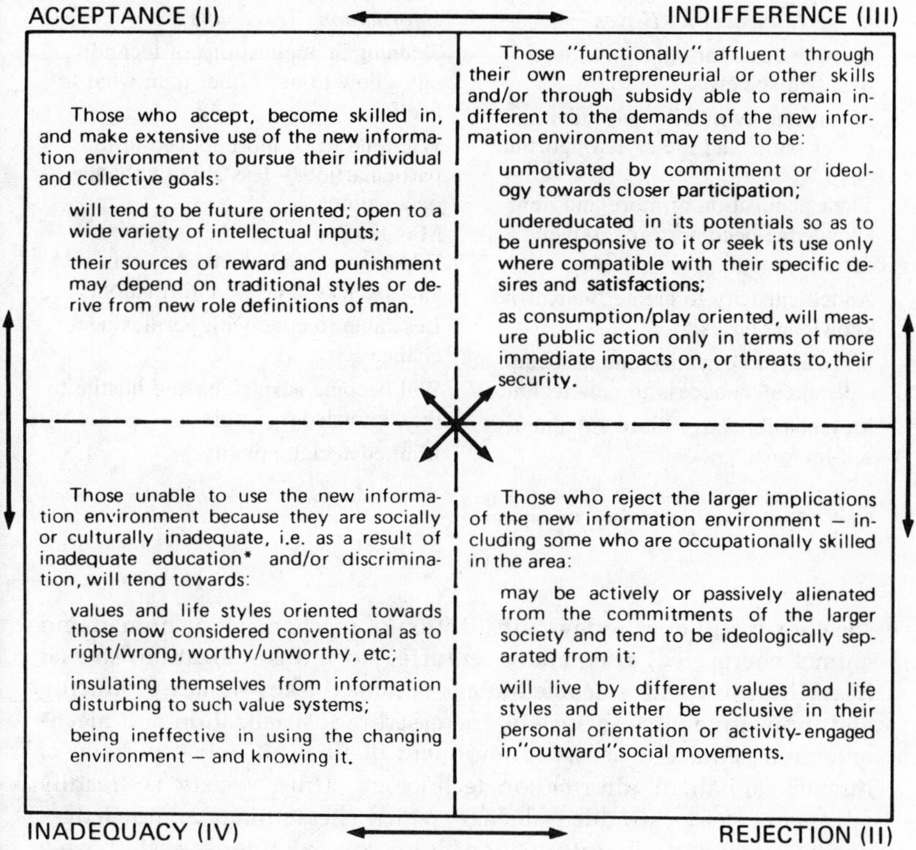

ACCEPTANCE (I) ←——————————→ INDIFFERENCE (III)

Those who accept, become skilled in, and make extensive use of the new information environment to pursue their individual and collective goals: will tend to be future oriented; open to a wide variety of intellectual inputs; their sources of reward and punishment may depend on traditional styles or derive from new role definitions of man.	Those "functionally" affluent through their own entrepreneurial or other skills and/or through subsidy able to remain indifferent to the demands of the new information environment may tend to be: unmotivated by commitment or ideology towards closer participation; undereducated in its potentials so as to be unresponsive to it or seek its use only where compatible with their specific desires and satisfactions; as consumption/play oriented, will measure public action only in terms of more immediate impacts on, or threats to, their security.
Those unable to use the new information environment because they are socially or culturally inadequate, i.e. as a result of inadequate education* and/or discrimination, will tend towards: values and life styles oriented towards those now considered conventional as to right/wrong, worthy/unworthy, etc; insulating themselves from information disturbing to such value systems; being ineffective in using the changing environment — and knowing it.	Those who reject the larger implications of the new information environment — including some who are occupationally skilled in the area: may be actively or passively alienated from the commitments of the larger society and tend to be ideologically separated from it; will live by different values and life styles and either be reclusive in their personal orientation or activity-engaged in "outward" social movements.

INADEQUACY (IV) ←——————————→ REJECTION (II)

Source: Adapted from *Information Technology: Some Critical Implications for Decision Makers* (The Conference Board, 1972). Reprinted by permission of the Conference Board.

attitudes, needs, and desires may be in conflict. Individuals or societies having access to the new technology have the same advantages as those who have access to material sources. Their attitudes will be different. Figure 7–10 summarizes some characteristics of the information "haves" and "have-nots."

Impact on Business Institutions

Any technological innovation can, if circumstances are right, alter the nature of society. For thousands of years agriculture, a technology for cultivating the land, dominated human activity, and in many

FIGURE 7–10
Implications; or "Ins" and "Outs" in Information-Rich Society

←——*Information "Haves"*————— *Information "Have-nots"*—————→

Become basis for elites in a re-stratified society.	Training in applications of technology — how to use rather than what to use for.
More socially mobile, with diverse career paths and life style opportunities.	Will tend to be more locked-in to particular jobs — less able to change occupations.
Their acquisition of more and new knowledge becomes progessively easier.	May tend to resign themselves to helplessness and alienation — will seek and use less and less information.
Added capacity to create their own knowledge bases.	Less able to cope with perplexing changes.
More able to organize and associate at a distance via access to new technics.	Will become suspicious and hostile to the "knowledge people."
May possibly have more enlightened self-interest.	Limited social mobility.

Source: Adapted from *Information Technology: Some Critical Implications for Decision Makers* (The Conference Board, 1972). Reprinted by permission of the Conference Board.

societies it still does today. Substitution of mechanical for human and animal energy led, after many centuries, to industrialization, so that heavy industry with sizeable outlays became the key wealth-producing and innovative force in society. Increased industrialization and mechanization require a heavy commitment of finances and especially of human capital, in information technology. Thus, society is steadily advancing to a postindustrial stage, which is essentially a knowledge-producing society. Its information technology will doubtlessly have a lasting impact on major social institutions, of which business enterprises are usually the first and the most affected.

Impact on the Structure of Industry

The impact of information technology upon industrial activity was the genesis and growth of an entirely new industry that became in less than a quarter of a century one of the most lucrative and fastest-growing sectors of the economy. Without displacing any of the already existing traditional industries — a characteristic of most technological developments — the new industry that was born was destined to serve the others. The ramifications of the new information industry are diagrammatically shown in Figure 7–11.

FIGURE 7-11
Information Industry and its Related Industries

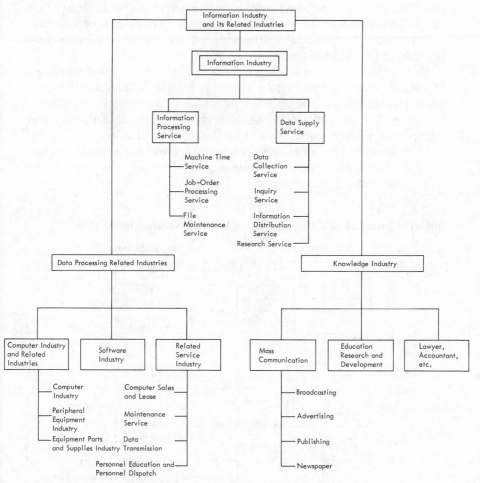

Source: Adapted from *Information Technology: Some Critical Implications for Decision Makers* (The Conference Board, 1972). Reprinted by permission of the Conference Board.

Impact on the Internal Workings of the Firm

In addition to the changes in structure, individual firms within certain industries were significantly affected by information technology. A more detailed discussion of the impacts of information technology's uses in organizations will be presented in the next chapter, when the findings of some empirical research on the subject will be reviewed.

Here we concentrate on some theoretical implications regarding the changes in the internal workings of the firm as well as its formal structure. The degree of impact is related to the percentage of activities that can be expressed in the symbolic communication system of information technology. Financial institutions are, of course, especially susceptible to this kind of communication.

The impact of information technology upon the internal workings of a manufacturing enterprise is shown in Figure 7–12. Again the same principle applies: the larger the percentage of numerical expressions, the larger the impact. Departments involved with the gathering and processing of large amounts of numerical data such as accounting, inventory control, and automated production show the greatest impact. On the other hand, strategic planning, the setting of long-range objectives, will probably never be directly affected.

FIGURE 7–12
Impact of Information Technology on the Internal Workings of the Firm

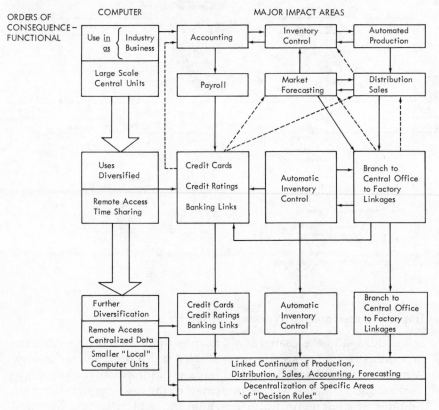

Source: Adapted from *Information Technology: Some Critical Implications for Decision Makers* (The Conference Board, 1972). Reprinted by permission of the Conference Board.

FIGURE 7–13
Information Flow and Organizational Characteristics

INFORMATION FLOW	MEDIA	FUNCTIONAL CHARACTERISTICS	PERSONNEL CHARACTERISTICS
(downward arrow)	Mainly written	Vertically oriented hierarchical bureaucracy. Organized by expertise. Written communications with "fixed" decision rules and chains of command with centralized decision points. Least horizontal communication.	People trained for highly specialized and limited functions. Little job mobility. Pyramidal authority structure with fixed procedures for access/appeal to higher levels.
(vertical + partial horizontal arrow)	Written Telephone Xerox Etc.	Horizontally organized by "function" areas. Mixture of fixed decision rules and autonomous functional rules. Shorter chains of command with more decision points.	Transitional form of organization sharing characteristics of stages 1 and 2. Mixture of line and staff functions with corresponding organizational roles well defined — but flexibly adjusted to allow for more autonomy via both formal and informal access to higher levels of decision-making. Job mobility more confined to upper level organizational tasks — other workers tend to remain tied to stated work descriptions and rankings.
(four-way cross arrow)	As above but significant introduction ot computer use at each level speeds up feedback.	Network type of organization with mission or objective foci which set flexible decision rules. Information flow includes critical man/machine interfaces (e.g. systems analysts, programmers, and comptrollers) which feedback from bottom to top. More autonomous decision making.	Skills less tied to specific sets of tasks within organization. Worker less tied to a single work situation: with developing competence and more flexible skills less attached to specific employing organization. Organizations tend to arrange work to develop capacities of people rather than use the capacities to accomplish work. Growth of serial careers — with multiple entry paths into different careers, etc.
(eight-way star arrows)	As above, plus more extended use of interactive communications modes, remote terminals, video conference techniques, etc., enabling widely distributed decision centers to interact swiftly.	More diffuse and geographically separated network type, with a high degree of adaptability and change in organizational configuration. Information and decision flows evolve in response to perceived needs rather than predefined and preset objectives or programs. Increased feedback at swifter rates enables previously autonomous decision-making to be integrated into whole system directions.	As above — mix of diverse specialties flexibly adaptive to changes in task and policy directions. The managerial executive becomes the prime interface and coodinator of "temporary"systemic clusters of specialized project groups — with multiple, mobile, and overlapping memberships. Ranking according to competence in flexible performance rather than by hierarchic position in organization.

Source: Adapted from *Information Technology: Some Critical Implications for Decision Makers* (The Conference Board, 1972). Reprinted by permission of the Conference Board.

FIGURE 7–14
Effects of Information Technology on Corporate Organization

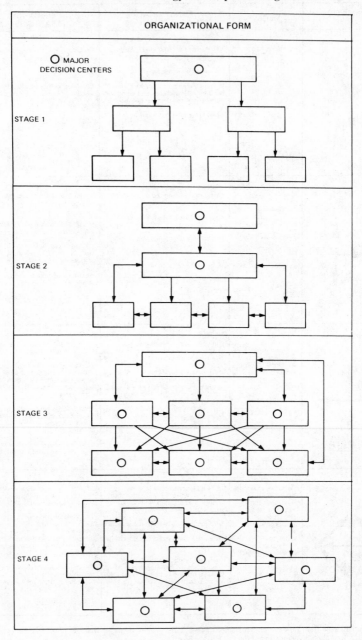

Source: Adapted from *Information Technology: Some Critical Implications for Decision Makers* (The Conference Board, 1972). Reprinted by permission of the Conference Board.

Impact on Organizational Structure

Usually organizational structures reflect the logical arrangement of their different activities. Insofar as these activities are affected by the new technology, a new logical framework must be designed to represent the changed organizational structure of the firm. Earlier organizational structures placed formal decision centers at the top of a pyramidal hierarchy. Within this framework, orders by decision makers flowed from the top downward while informative reports rose from the bottom upward. Horizontal flow of either strategic or operational information was minimal and considered inconsequential.

As more and more operations are affected by information technology, decision centers become scattered all over the organization. Now, instead of a pyramidal hierarchical structure, one finds a network of centers whose activities revolve around decision points. Information now flows from and toward all parts of the system. Although the firm's management may still exhibit its pyramidal structure to visitors and new employees, actual decision making is done in areas where data are collected and converted into information. Figure 7–13 illustrates the relationship of information flow and organizational characteristics.

The transition of traditional organizational structure into a system structure as a result of the impact of information technology is shown in Figure 7–14.

SUMMARY

This chapter attempted to explain the logic, developments, and fundamental elements of information technology. Once these are understood, one can then begin to assess the potentials, as well as the shortcomings, of the new technology. The simplicity of explanation, however, should not be interpreted as implying that the subject itself is simple.

Information technology is perhaps the most sophisticated artifact man has yet created. However, one thing is slowly becoming clear: information technology can generate raging torrents of information in which the scientist, businessman, and ordinary citizen are told to "sink or swim." There is absolutely nothing inherent in the nature of the technology itself that dictates either outcome.

Unlike many previous innovations, information technology creates a product that is consumed by man's cognitive system, the system of generating, absorbing, and disseminating information. While advancements in mechanical technology relieved man of the need of supplying muscle power, information technology actually demands greater utilization of man's cognitive capacities. Thus, the leisure created by advanced mechanization is offset by the demand to acquire and utilize the new information.

To have lasting impact, information technology must serve the needs of the user. In organizations, these needs are best assessed by those setting the goals and choosing the means. However, before the manager can know how much information hardware and software he needs, he must know what the technology can and cannot do.

> Today, information technology provides us with the capabilities for formulating and ordering our goals and priorities, from institutional to planetary, for increasing management's effectiveness, for freeing man to participate in broader and more meaningful activities, and for narrowing, if not closing, the gap between the haves and the have-nots.
>
> If we fail to recognize its potential we may drift in another direction. We may create and strengthen the power of management elites, circumscribe the freedom of man, and create a new kind of rich-poor gap between those, regardless of economic status, who know how to command the information technology and those who do not.
>
> Experts often disagree as to whether developments in information technology during the next 30 years will be evolutionary or revolutionary. Simon Ramo, in his book, *Century of Mismatch,* takes the more conservative view, but draws a compelling conclusion which this study tends to support: "By the turn of the century all we may have is a clear trend, either toward a robot society, or toward a society that uses technology to gain a higher degree of freedom for the individual. But such a trend may be set quite strongly, even irrevocably, in the near future, and there is no point in risking it. We ought to understand the alternatives now, and we ought to work to achieve the one we want."[7]

REVIEW QUESTIONS FOR CHAPTER SEVEN

1. What is to be understood under the term "information technology?"
2. What is the logic or rationale behind the development of information technology?
3. Information technology is also referred to as the "Second Industrial Revolution." Briefly compare the differences between the First and the Second Industrial Revolution.
4. What are the four basic elements of information technology?
5. An electronic data processing (EDP) system can be depicted as a simple input-process-output-feedback system. Draw such a system and identify and explain these major system parameters of an EDP.
6. What are "auxiliary devices" and what purpose do they serve in an EDP?
7. Write a job description for a systems analyst and a programmer.

---

[7] Ibid., pp. v–vi.

8. A hospital's management desires to devise a computerized plan for facility utilization of its three surgery rooms. Go through the sequence outlined in Figure 7–6 and delineate the inputs which must be provided by hospital departments such as a nursery, pharmacy, purchasing, and accounting department.

9. Your university plans to computerize its registration procedure. Determine the inputs and outputs which have to be processed by the university's computer center.

10. Easy-Going Bank and Trust Company has decided to install an EDP which will process customer checking accounts. Describe the roles that bank tellers, accountants, and the bank's management will have to play in order to get the system off the ground.

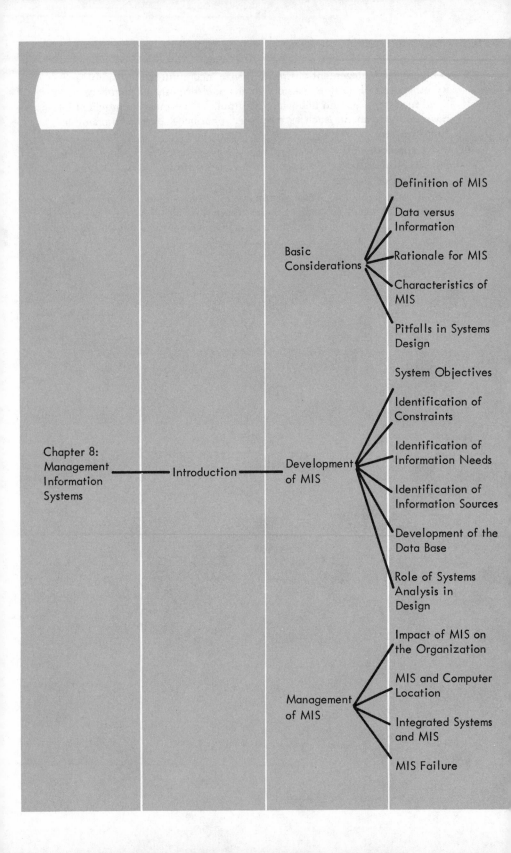

Chapter 8:
Management
Information
Systems

Introduction

Basic
Considerations

Definition of MIS

Data versus
Information

Rationale for MIS

Characteristics of
MIS

Pitfalls in Systems
Design

Development
of MIS

System Objectives

Identification of
Constraints

Identification of
Information Needs

Identification of
Information Sources

Development of the
Data Base

Role of Systems
Analysis in
Design

Management
of MIS

Impact of MIS on
the Organization

MIS and Computer
Location

Integrated Systems
and MIS

MIS Failure

Chapter Eight

Management
Information
Systems

*Leave room in the system for the Feedback of experience
to redesign the system itself.*

Max Ways

INTRODUCTION

Modern complex society presents one of the most exciting challenges
of our age—the challenge to manage those richly interacting elements
of government and industry. The problems of poverty, of pollution, of
growth, of unemployment, and of overpopulation all pose forms of
crises not adequately dealt with as yet. Like a lanky but awkward ado-
lescent, society has grown enormously; the task at hand is to provide
the proper direction, the proper regulation.

In a similar way, the task of the firm in this everchanging society is
also to provide regulation, and, as seen in the previous chapters, the
essential factor required for regulation of any system is *information*.
Throughout history both the government and the firm have been concerned with the acquisition of information for the purposes of generating change as well as understanding the rudimentary structure of
the appropriate bodies. To be sure, both entities possess vast bureauc-
racies for the collection of information, if for no other purpose than to
perpetuate the established order of things. The concept of information
underscores the notion that something of value is being communicated
to some individual or organization. Since individuals resort to multiple

191

information sources, some type of system that filters, condenses, stores, and transmits all this information must be evolved. And because these information systems are organized for the use of managers in the firm, they are appropriately labeled management information systems.

This chapter is concerned mainly with the totality of management information systems—their characteristics, their design, the possible integration of the firms' various information centers into a holistic system, and the management of such systems. But first some definitions.

BASIC CONSIDERATIONS

Definition of Management Information System

The frequency with which the term *management information system* (MIS) is employed today is matched perhaps by no other phrase in our current business vocabulary. Current computer installations all purport to be information systems, and the majority of these in turn claim to be management information systems. Connotations run from sophisticated hardware systems to routine accounting reports; from real-time systems depicting the status of space shots to basic data sources such as billings and accounts; from central data bases which collect, structure, store, and summarize information to the present inventory system in existence for the past 20 years; from the holistic "integrated" system of central management to remote terminals of specialists connected to the computer.

Despite this lack of agreement, there is, however, some commonality among the definitions, and although the following one is not distinctive in any sense, it is nevertheless functional.

A management information system is a formal system in the organization which provides management with the necessary reports to be utilized in the decision-making process.

While the system described above may also collect, store, process, structure, and retrieve data, the ultimate goal of the system is to provide information for managers to assist them in making decisions. While most MIS merely support managers in making decisions, the system may in some instances also make some of the repetitive decisions usually made at the lower levels of the organization. Inventory replenishment based on current inventory levels may be such a decision rule; production scheduling may also be done by computer; and the determination of economic order quantities is also susceptible to such a regimen. The basic distinction between lower-management decisions and those of top management is that the former are operating decisions performed on a routine basis. MIS are basically developed for providing information for planning and control purposes. Lower-level decision making can be programmed, whereas many of middle-management and upper-management decisions cannot be programmed because the inputs into the decision model come from a variety of sources both internal and external to the organization.

There is little doubt that the management information system must be designed to provide the right type of information to the manager. This implies that information must be user-oriented. Because this is so, the dimensions and the characteristics of a system must be proportionate to the user's needs. When considering user requirements, many authors feel uneasy because what users generally receive is data, not information. Such apprehension is indeed well founded, since the two must always be clearly differentiated. This discrimination is a necessary prerequisite to any intelligible discussion of management information systems.

Data versus Information

In the literature, various distinctions have been proposed, but in the final analysis they all hearken back to the original etymology of the terms used. Data, which is derived from the Latin verb, *do, dare,* meaning "to give," is most fittingly applied to the *unstructured, uninformed* facts so copiously *given out* by the computer. Information, however, is data that have form, structure, or organization. Derived from the Latin verb *informo, informare,* meaning to "give form to," the word information etymologically connotes an imposition of organization upon some indeterminate mass or substratum, the imparting of form that gives life and meaning to otherwise lifeless or irrelevant matter. It is most fittingly applied to all data that have been oriented to the user through some form of organization.

Data can be generated indefinitely; they can be stored, retrieved, updated, and again filed. Assuredly, they are a marketable commodity purchased at great costs by both the public and private sectors; however, data of themselves have no intrinsic value. Yet each year the cost for data acquisition grows on the erroneous assumption that *data are information.* The task of acquiring data presents no obstacles whatsoever, since data are generated as a by-product of every transaction or event. The real problem is data overload, for the government, the firm, as well as for the individual. Even within departments of the government there is no paucity of societal information; rather the problem is one of data overload and data organization. It has been estimated that the American economy produces one million pages of new documents every minute, of which some 250 billion pages a year must be stored.

Business firms alone store a trillion pieces of paper in 200 million file drawers, and each year they add 175 billion new pieces of paper to this enormous amount. This paper level is further raised by the outputs of educational institutions hardly able to digest their own output.

It is not so much a problem of data acquisition as of data organization; not so much of organization as of retrieval; not so much of retrieval as of proper choice; not so much of proper choice as of identification of wants; not so much of identification of wants as of identification of needs. Obviously the problem in information management is not one of gathering, organizing, storing, or retrieving data but rather one of determining the necessary information requirements for decision making.

A popular distinction among current writers restricts the label of information to *evaluated* data. Here the orientation is not so much the *function* of informed data as the explicit and specific *circumstances* surrounding the user. Accordingly, the term data is used to refer to materials that have not been evaluated for their worth to a specified individual in a particular situation. "Information" refers to inferentially intended material evaluated for a particular problem, for a specified individual, at a specific time, and for achieving a definite goal. Thus what constitutes information for one individual in a specific instance may not do so for another or even for the same individual at a different time or for a different problem. Information useful for one manager may well turn out to be totally devoid of value for another. Not only is the particular organizational level important but also the intended functional area. A production manager, for example, is typically unconcerned with sales analysis by product, territory, customer, and so on, while the person in charge of inventory control is little concerned with the conventional accounting reports that affect him only indirectly. Thus, the definition here being considered is that information concerns structured data—data selected and structured with respect to problem, user, time, and place. Figure 8–1 shows the process by which data become information.

In Figure 8–1 the internal data centers may, for illustrative purposes, be labeled as inventory, personnel, marketing research, and production scheduling. There can be little doubt that each of these functional areas is concerned with the generation of data thought to be of some worth to its respective area. However, often no analysis is made of data requirements, with the result that unstructured facts are presented in the hope that they will be utilized. It is precisely these meaningless data that have contributed to the data explosion. While the subject of data overload has been adequately treated in the literature, the magnitude of the problem is one worth noting here. Ackoff, in discussing the misinformation explosion, succinctly states:

My experience indicates that most managers receive much more data (if not information) than they can possibly absorb even if they spend all of their time trying to do so. Hence, they already suffer an information overload. They must spend a great deal of their time separating the relevant from the irrelevant and searching for the kernels in the relevant documents. For example, I have found that I receive an average of 43 hours of unsolicited reading material each week. The solicited material is half again this amount. I have seen a daily stock status report that consists of approximately 600 pages of computer print-out. The report is

FIGURE 8–1
Data Transformation with a Centralized Data Base

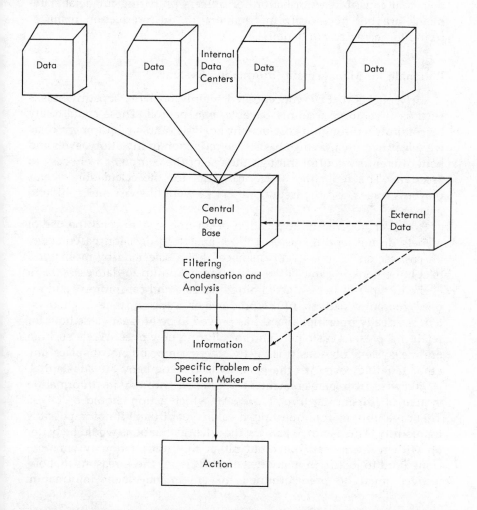

Source: Adapted from Peter P. Schoderbek, "Data, Information, and Information Theory," *The Business Quarterly*, Autumn 1971, p. 81.

circulated daily across managers' desks. I've also seen requests for major
capital expenditures that come in book size, several of which are distributed to managers each week.[1]

Thus, all information must be viewed as being imbued with relative
value. Much of the so-called information utilized in management systems today enjoys a "sacred definiteness" which in reality is subject
to wide ranges of both human and institutional errors. Valueless data
have in many instances been accepted as information simply because
of an emotional investment on the part of the practitioners who have
traditionally treated such data in their routine operations. For example,
data that constitute information for officers preparing financial statements are not necessarily information for the production manager
trying to decide on run length.

Rationale for Management Information Systems

Early efforts of EDP concentrated mainly on large repetitive tasks
such as inventory, accounts payable, and payroll. These could easily
be computed through a cost-benefit analysis, and so equipment costs
were justified on a savings basis. The reduction of inventory levels and
better inventory control through the use of the computer were easy to
ascertain. The reduction of clerical costs for the calculation of employees' wages and the issuing of checks were likewise susceptible to
measurement.

However, the recent application of on-line systems and the use of
models do not lend themselves to an analytical decision based solely
on cost factors. Suppose, for instance, that a salesman by means of a
telephone hookup can call the office and get an up-to-date assessment
of the number of items uncommitted for sale and can promise and indeed guarantee delivery of the products on a specific date. Such a system is already operational and has proved to be of great value both for
retaining present customers and for alluring new ones. While such an
on-line system obviously is costly, the system did not displace any
other tangible costs; yet the benefits to the company are substantial.

Likewise, top management is appraised through its information
system of impending trouble areas in which action should be taken.
This capability is of unquestioned value, yet difficult if not impossible
to quantify. This is not to gainsay that all organizations would be better
off with real-time capability but rather that information systems become hard to justify in many instances. It is for these reasons that objectives must be predetermined for any management information

---

[1] Russell L. Ackoff, "Management Misinformation Systems," *Management Science*
(December 1967), pp. 147–56.

system, and the expected benefits defined. Too often management information systems are not information systems at all but are merely subsumed under the all-embracing heading. Such systems distort the true meaning of MIS; the result is a proliferation of operating systems.

Characteristics of Management Information Systems

Although few information systems in existence today have all of the following desirable characteristics, this does not imply that they are all necessarily poor systems. How well a particular system operates will depend upon a host of factors, such as the type of applications, the type of equipment, the technical competence of the personnel, support by top management, and previous applications. However, the following requirements would probably find acceptance among most systems personnel.

1. That timely and accurate information be available to the decision maker, the timeliness being determined by the *user* of the information. If accurate information is not provided when needed, it may have little or no utility later. Although this point may appear obvious and commonplace, it is still a significant one, since timeliness and accuracy are the primary objectives of any information system.

2. That the system be responsive to the manager's inquiries for information. Although much of the information a manager receives may be gotten on a regular basis, the system should be able to respond to one-time requests. This may require some additional system design, since this capability calls for a different structuring of the data base. Although it is more difficult to retrieve this kind of information, still such a built-in capability may well be worth the cost.

3. That the system provide management with exception reporting. As will be noted elsewhere in this chapter, the manager is not interested in learning of the hundreds of things which are being performed satisfactorily. Rather, he wants to be informed of the factors that are or soon will be out of control. This management by exception reporting is extremely important when one considers the nearly unlimited ability of the computer to generate reports devoid of any meaningful content.

4. That the system be capable of additional integration in the future. Present systems should be compatible with future hardware and software changes that may be initiated with expanded operations. Much discussion centers about the decision whether one should design systems in piecemeal fashion or according to the holistic approach. Although this is discussed elsewhere, suffice it to say that the benefits of both methods can be had if one retains the capability of interfacing at a later date. The ability to interface is generally identified with compatibility. Compatibility means that the various models of a particular

product line utilize the same design logic: they have the identical instruction set. They will accept the very same instructions and execute them to produce identical results, although at different speeds. Programming done for a smaller model will run on a larger model but not vice versa. While compatibility usually exists with hardware, it is less true for software. Were one to use the same software package while moving up in the hardware, he would fail to utilize the larger machine's potential. This could be very costly, since cost per operation goes up if the computer is only partially utilized. Thus new programming and reprogramming must generally take place simply for the sake of economy. Compatibility is thus a valuable attribute of computers and should be carefully considered before a system is installed.

Another common feature of compatibility is modularity. Modular computers are designed to allow the user to arrange his configuration in terms of building blocks for possible expansion. In this manner, memory units, channels, input-output devices, file units, and so forth, can be added to the system one at a time. When the CPU becomes too small, additional units may be added to the system. Modularity has since become a vital part in the computer hardware business just as it plays an important part in the high fidelity sector of the economy.

5. That the system be accepted by the intended users. This point may appear to be unnecessary, but if one meets all of the previously mentioned requirements and the system is still not acceptable to the users for whatever reason, the system simply will not succeed. Many a system remains inoperative because of user rejection—this in spite of the fact that such systems have been proven to be of benefit to managers. In some instances, managers have even gone out of their way to sabotage the system.

There are other characteristics such as economy of operation, simplicity of use, and goal congruency within the organization that could be detailed, but the salient point to be made here is that if one clearly defines the objectives of the system initially, then one can judge whether the objectives are being achieved. This, then, provides a measure of system performance and a gauge of its usefulness to the organization.

Pitfalls in Systems Design

Despite the tremendous strides made in information processing, many of today's managers are in no better position for making decisions than before. As Ackoff sees it, the reason is that those who design systems work with assumptions that are unrealistic and erroneous. As a result, the system that they end up with is not a management information system but a management *mis*information system. The assumptions underlying the latter system are five in number and will here be briefly examined.

1. *Give them more.* This assumes that the manager is not getting enough information and that if he were provided with more information he would be able to perform more effectively. Ackoff makes the point that managers suffer more from an overabundance of irrelevant information. This overload only magnifies the manager's problems, since what is needed is less data of an irrelevant nature.

Ackoff proposes two processes to avert this condition: filtration and condensation. Filtration involves evaluation of data as to relevancy, while condensation involves the curtailment of redundancy in the mass of otherwise relevant information. One must be careful, however, that there be agreement on the useful limits to the processes, since indiscriminate filtration and overcondensation can lead to the other extreme — namely, paucity of the information necessary for sound decision making. Overcondensation, though probably less frequent and more serious than overabundance of data, is difficult to set objective limits to, since the degree of uncertainty surrounding the decision maker is a subjective phenomenon unknown to others and since the quantity and quality of information needed by a particular decision maker in a specialized setting and working on a definite problem are generally unknown to the nondecision maker.

The filtration process is no less problematic. Who should decide what is to be filtered out? If this filtering out is done at the lower organization levels, then the limited perspective of the men carrying out the evaluation is open to question: if done by the higher levels, then no relief from the overload is experienced. Somewhere in between overabundance and underabundance (overcondensation and undercondensation) lies the golden mean to be discovered by the top managers themselves.

2. *The manager needs the information that he wants.* This assumption probably arises in the early design stage of an MIS when the manager is asked by the designer what information he needs to make decisions. Implicit is the notion that the manager has a model of the decision-making process, which is seldom true. When managers do not possess a clear understanding of their decisions and of the requisite variables, there is a tendency to ask for "all the available information which may bear on the decision." Similarly, a systems designer who understands even less of the decision-making process, in an effort to show how effective he can be in meeting the requests of line managers, provides a veritable deluge of information which compounds the overload. Managers often do not make correct decisions because they lack sufficient information, the required analysis, or improved modes of communication. The researcher still does not know about how the

manager decides where to direct his attention. The point being made is simply that one must have an explanatory model of the decision-making process before one can specify the information requirements.

3. *Give the manager the information he needs and his decision making will improve.* Even if the MIS system can provide the manager with the relevant information, this is no guarantee that decision making will improve. Experimental results in other settings lend support to this belief that, in spite of formal models, managers tend to rely on intuition, experience, or judgment in many instances. In fact, the veritable response "my decisions are based on years of experience" have prevented many systems designers from modeling decisions. Such a response in many cases is merely a rationalization on the part of the manager either to cover up the fact that he does not have a decision model or to resist the implication that the computer can make the same decision. In still other cases, although "good" information is given to the manager, he may utilize it in an improper manner.

4. *More communication means better performance.* While the presence of an MIS may assist in providing managers with information about other divisions, departments, and managers, it is erroneous to assume that such communication is always and necessarily good. While such communication may facilitate coordination, it does not necessarily improve performance. The point is made that an organization must have appropriate measures of performance for both managers and departments so that the sharing of information does not put each in conflict or competition with one another. Thus, before permitting the free flow of information, the organization structure and performance measures must be considered.

5. *A manager does not have to understand how an information system works, but only how to use it.* This premise has been accepted both by the designers of systems and by the managers. In fact, it was sold to the managers to overcome the mysticism of the computer. Managers were assured that they have only to tell the designers what information they want and it will be provided them. Whether this attitude on the part of the designers was intended to prevent access to their domain and encroachment upon their functional specialty, or whether it was a means to prevent nonacceptance of their wares by wary managers, it was effective. The majority of managers today still cite the analogy that "my wife doesn't need to know how an automobile is made for her to operate it." The fallacy here is that if one is to judge the performance characteristics of an automobile, he must indeed have in-depth knowledge to make such an assessment. Likewise with an MIS — if managers are to be able to evaluate the system and have complete control of it, they must have more than a passing knowledge of how to use it.

THE DEVELOPMENT OF MANAGEMENT INFORMATION SYSTEMS

In a previous discussion a distinction was made between *data* and *information*. Such a distinction was necessary because many departments of an organization are concerned with the mere collection and processing of data. Such data may never become information because they do not enter into the decision-making process.

Prescinding for the moment from this distinction, and using the term in its generic sense, let us see how one goes about developing a management information system.

Even before the advent of the computer, many data centers like those in the personnel, production, sales, and quality control departments existed in organizations. These were concerned with the accumulation of records depicting the level of organizational functioning. Manual information systems thus predated computer systems. But even more basic than the existence of any information system is the consideration of the overall objectives of such a system.

System Objectives

The purpose of any information system is to provide managers with the necessary information which will reduce the range of uncertainty surrounding a decision. How much uncertainty is reduced will depend upon the particular problem at hand and the information currently available. Obviously not all the available data are relevant to the problem. This suggests that only certain information is useful and therefore this ought to be included in an information system. An information system must be designed with some particular purpose in mind; there ought to be an objective for the system. However, this determination must be made by the manager rather than by the systems designer, since usefulness, effectiveness, and overall systems value rest with the ultimate user, the manager, and not with the designer. This determination of the systems objective is quite critical, since much of the subsequent design function is geared toward it. As far as possible, specificity ought to be the rule followed in designing any system. An inventory system, for example, which will allow for a certain response rate, a minimum dollar value, a specified minimum serviceable rate, reorder points, and safety stock levels for classes of items, warehouse location, unit cost, quantity on hand, quantity on order, is much more desirable than an "all-encompassing inventory system" which fails to spell out the system objectives.

Much of what has been described previously refers to the specification of output reports to be determined by the user. While summary reports may suffice at one organizational level, detailed reports are typically required at lower levels. Finer classifications and refinements such as sales by customers, territories, income class, products, and so on may be desired. Cross classification may also become necessary, since fields containing the same data may be organized differently.

Content, format, timeliness, and accuracy are important considerations in the design of the system. File content depends on input data and the methods of aggregation employed in the system. If output requirements can be specified, then the files can be carefully structured to include only the required inputs. If output reports cannot be completely specified, then the files must contain more data to meet subsequent demands for output requirements. To be sure, the most troublesome case occurs when the manager is unable to specify his output requirements. To cope with this uncertainty, since the source cannot be structured, the data must be retained in their original form with little or no organization or condensation. The organization is thus forced to maintain voluminous files and, while maintenance costs are relatively low, costs for retrieval and structuring of data can be excessive, relative to value.

Timeliness of reports required by managers affects the file structure both in the reporting interval and the reporting delay. The reporting interval is affected by the rapidity of reporting. Thus, frequent reporting requires the separation of events into successive periods either by setting up new records or by purging the old ones or by transferring earlier reported events to less active storage. Reporting delay refers to the length of time that occurs between the event (or close of a period) and the distribution of reports. For example, in many production plants, daily updates are required for work completed each day. In some plants, the manager can ask the computer to ascertain the status of work in process on a continuous basis. Such requirements definitely affect the design of the system.

Likewise the degree of detail and of accuracy is a consideration bearing upon the design of a system. Outputs must either be in the files or derivable from them, and thus both the files and the inputs are the limiting factors.

These factors, then, comprise the objectives of a particular system.

Identification of Constraints

As it was necessary earlier to identify the boundaries of a system, so too one must identify the various constraints affecting the design of a system. How limiting these are for a particular firm will depend upon many factors.

Lack of Suitable Models. Computers are typically used for operations which are highly repetitive and, once these areas are computerized, the firm can attempt to move into more sophisticated areas. When an organization tries to model these more difficult decision areas it soon discovers that complexity is a significant barrier to model development. As pointed out earlier, modeling follows a definite pattern. It proceeds in this manner: verbal description of the problem——➤block diagram to show simplistic relationships——➤mathematization of these relationships——➤and then finally a flow chart of the model followed by experimentation or simulation. Even when one assumes little difficulty in the first two stages of the process (a verbal description of the problem and a simplified block diagram depicting the interrelationship of the parts), major problems are often encountered in the attempt to quantify or mathematize the relationships. (This point will be discussed more fully in the subsection on integrated systems.) Too often, however, the specific human values of the decision maker cannot be expressed by any quantitative measure, even granting that one can ascertain the direction and the magnitude of the interdependencies of the system.

FIGURE 8–2
Steps in Modeling

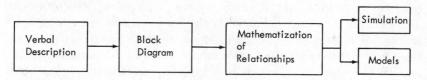

It takes no great erudition to discover that the most successful models used in industry today are still the simpler ones. As one increases the number of variables employed in the model, one increases the number of relationships and decreases the chances of successful prediction. In spite of the quantum leaps made in the past decade with respect to model application, in the real world application is still at a relatively low level. There are some notable exceptions. Some firms employ models utilizing several hundred variables. While such efforts undoubtedly pave the way for ever wider adoption, still these applications are limited to firms with sizable financial commitments to EDP.

User Constraint. Perhaps one of the most important constraints is user constraint, for the user is initially asked what decisions he makes, what areas he is responsible for, and what information he needs to make these decisions. Since it is the task of the systems designer to serve the user, it is the user who, in the last analysis, determines some of the constraints on the system. In some cases a proposal may come from the designer respecting additional data which can be provided by the system. Such data are seldom rejected by the user. The user may feel that such data may be "nice to have," but does he realize that this course of action often leads to an abundance of irrelevant data? A mediocre designer can have the computer generate literally reams of paper which overload the manager without presenting him with the information essential to decision making.

In any event, the unsuitability of reports, the irrelevance of data, and problems of a similar nature all constitute user constraints.

Cost Constraints. Cost considerations via a budget and/or cost effectiveness analysis may often dictate the configuration of the system. However, this factor should not be any different from other capital expenditures, since this, too, should show a rate of return and have a payback schedule. Nevertheless, the funds allocated to EDP are often not subjected to a rate of return but are simply allocated to the EDP department on the basis of "salesmanship."

Identification of Information Needs

The cornerstone of a management information system is the manager's informational needs. Critical is the need to understand the manager's role in the organization together with his responsibilities, to delineate the decisions managers make, and to discover the kinds and amount of information needed to arrive at these decisions. This presupposes that the executive has a model of his decision-making process. Far too often one hears of an organization's complex plan to build a computer model that will duplicate the decision models that managers carry around in their heads. In this way it is hoped the computer will behave like a rational manager when making decisions. This is sheer folly except for the most routine operations covered by simplistic decision rules. This does not mean that one should not attempt to look at the informational requirements of the manager. It does suggest that the task will most likely be a formidable one. Indeed, this modeling must precede the determination of the information requirements, since one cannot know what information to ask for unless some previous criteria exist. Neither is it an easy task to identify the informational factors inherent in decision making, since many of these are reputedly based on judgment or experience. In fact, this inquiry can be very unnerving for the individual manager because it requires him to identify the decisions to be made, to set up priorities, and to analyze his informational needs. In the process of doing so his sense of values and his disbursement of time may often be exposed for their inconsistency or lack of logic. Still some such identification of managerial needs ought to be made at each level.

Experience shows that each manager ought to verbalize his decision model in order to determine his information requirements. While he need not reduce these verbalizations to a precise mathematical form, still an ordering of the factors is a big help. If systems designers can know which variables are involved and how they interact, then they should be able to construct models with some predictive value. The more variables involved, the more complex the model and the more difficult to construct. Model complexity is a function of the number of variables involved and the identification of the interrelationship of the variables. If the variables interact in a perfectly predictable way, the model should be easy to construct. Operational models are normally confined to those in which management can control the variables. The greater the control, the more confidence the manager can have in the model. The more exogenous the factors entering into the model (this typically occurs at the higher levels of management), the more judgmental factors enter into the decision-making process.

Informational needs may be internal, environmental, competitive, economic, or situational in nature. Likewise, information may be quantitative or nonquantitative, detailed or nondetailed.

Once the informational requirements are determined, then the detailed design of the system can begin. While the methodological aspects of file structure, retrieval techniques, and so forth, occupy the attention of technicians, still basic questions as the following should concern us: How will the data be acquired? What is the source document? How is it to be stored? In what form and how often will it be presented to management? And to what level of management? Since the payoff of the information system lies in its usefulness to the manager in making decisions, these questions deserve serious consideration.

Most productive in the design of a system is the creation of a task force with representation from all users of information. The objective of this task force is to organize as much as possible the company's procedural and informational system into an MIS. This user-analyst relationship is crucial to the development of any such system. Any proposal by the analyst must be approved by the user, since the system should reflect user requirements. The user then accepts accountability for the operational aspects of the system once it is installed.

Any proposed MIS ought to include statements of (1) the particular management problems the system is expected to solve; (2) the way the problem is to be solved; (3) the value of the particular system to the organization; (4) the design, implementation, and operation costs of the system; (5) a cost/benefit analysis of the system; and (6) the control to be exercised in evaluating the real benefits of the new system. Such an approach exacts user responsibility and accountability for the system with the user.

Information requirements often cannot easily be identified. One tends to concentrate on the currently available information when, in fact, much of the information required is not available. While this generally does not hold for decisions made at the lower eschelons of management (workable formalized models already exist here), this is not the case for middle and higher managers.

Likewise, it obviously is not possible to identify *all* the information needed for *all* decisions. What is desirable, however, is to have all the information essential for major decisions and to have an effective systems design which revolves around the manager-analyst interactive process. One of the more productive ways in which to get at the essential information is to structure the questioning process. This involves setting up an inventory of the information types presently available, the suitability of such information, the frequency of usage, the significance and sources of the types, and especially the relationship of one type to another. Figure 8–3 shows how the outputs from one department can become inputs for other departments. A before-after analysis performed with such a chart can pinpoint whatever redundancy that exists.

Flow charts which trace the routing of data from origin to destination may also be utilized. These are usually arranged chronologically to show the progression of the data through the organization. Frequency, time, cost, distance, and volume can also be depicted on such charts. These charts are similar to those which trace the movement of products through a plant and derive their name from product flow or activity

FIGURE 8–3
Input-Output Chart

Outputs from Department M	Inputs to Departments				
	A	B	C	D	E
1	X		X		
2		X			
3					X
4				X	
5	X	X	X		
6			X	X	
..					

X denotes information transfer

charts. Provision should be made for new types of information as well as for external information entering into the decision-making process. Such a tops-down approach helps to ensure that the system will be user-oriented and that it will not be inhibited by the reporting formats, which typically characterize present information systems.

Someone may ask what procedures should be followed when one is intent on improving the current system rather than on setting up a brand new MIS. Does one use the same procedures? The answer is basically yes, although another consideration must be taken into account. The way to achieve a quick payoff on a system is to improve the existing one rather than to embark on an entirely new endeavor. While such an approach may be financially justifiable in the short run, in the long run the practice may be dysfunctional. At some point every system should be subject to analysis and probably altered to take account of new design principles, and so forth.

For deciding what information executives should have, one could possibly look at the decisions that executives are required to make. Some hints of the types of information that executives need can be gotten by looking at the current problems which arise. Examples of these may be: the inability of managers to explain changes in operating results; lack of knowledge about competitors' products; poor projections of growth; unexplained market factors; unexplainable returns on capital expenditures. All of the above areas deal directly with operations; other reports may actually be present but lack content or specificity, or summation. The present authors have seen all of the following defects in reports which render them less effective than they should be:

> Lack of trend analysis for analyzing the present situation
> Information gathered by differing units in the organization
> with differing results and differing analyses
> Reports too late to effect the required change
> Lack of summarization of information and too much detail
> Failure to identify causes of variance
> Lack of confidence in the data
> Lack of understanding of the data

All of the above are indicative of potential weaknesses in the present system and provide starting points for system modification.

Identification of Information Sources

How does one go about identifying the sources of information needed by managers? Much will depend on the market. While many firms rely heavily on internal sources (this would be especially true if they are operating in a mature market with little industry innovation), other firms require considerable external information (those competing

in a dynamic environment with much technological and product innovation). Even though this distinction was explicitly treated in a previous chapter, one ought to touch on it again in order to emphasize the
integrative process of systems design. Once the informational needs
are specified, a decision must then be made as to the manner in which
the information is to be obtained and from what source.

Internal records and reports form much of the internal information
sources of an organization. Production reports, sales reports, productivity reports, personnel records, accounting and financial information,
and a host of other file records, memoranda—all constitute the bulk of
this type of documentation. Furthermore, each department often maintains its own data base for specialized information banks. While such
information may be centralized, the sources still remain decentralized.

Decision models often utilize information originating outside the
organization. This is especially true for decisions emanating from the
upper levels of management which concern strategic planning and
those of long-range consequence. A change in technology or in a competitor's pricing practices, the introduction of new products, a changing
economic or political climate—all are examples of events for which
external information must be secured if a firm is to remain viable. Top-
management decisions today generally use more external data such as
industry output trends, regional production and sales classification,
industry advertising costs, and research costs. Within the firm, the
market research department and other information-acquisition centers
spend a significant portion of their time learning of sociological
changes occurring in the country, population trends, growth trends,
and so forth. Among the more notable areas in which information is
routinely sought are the government sector, the economic sector, and
the market sector.

1. *The Government Sector.* National elections can become extremely
important to a firm when, for example, one can expect price and wage
controls. Tax policies, investments credit policy, foreign exchange rates,
antitrust protection, consumer protection policies, governmental intervention in collective bargaining, the posture on conflict involving commitment of troops and/or equipment, Supreme Court decisions—all
can significantly affect planning decisions. Decisions by political bodies
of other countries such as entry into the European Common Market,
import quotas, devaluation of currency, all have a bearing on decisions
of the firm.

2. *Economic Sector.* The rising levels of income, the shifting of purchasing habits, the rise or fall of white-collar employment, and other
similar indicators of economic activity can vitally affect the firm's
operations. Many firms deem these factors of such consequence that
they employ their own economists to monitor the economic sector.

3. Market Sector. In an earlier chapter the major sectors of the environment that a firm scans in order to learn "what's going on" were discussed. A number of studies in this area have postulated that the marketing sector is accorded the most attention by executives of the firm. The importance of this sector is borne out by the results of the studies which show that more attention is devoted to this sector than to all others. The entity in the firm typically associated with this activity is the market research department. Besides monitoring the firm's own products and such factors as growth, prices, and demographic factors, this department spends much time monitoring its competitors' demands, price structures, product development, share of the market, R & D efforts, and so forth. The many lawsuits involving "industrial spying" and infringement of patents attest to the vigor with which this activity is pursued. Indeed, firms even develop complete personal profiles of the executives of competing firms in an effort to forecast the possible direction that these may go. The market research department is usually the most sophisticated and most developed sector for the gathering of environmental information deemed essential for decision making. Environmental information, in effect, constitutes an important segment of the total informational needs of the firm.

Development of the Data Base

The data base is the sum total of the data that must be obtained for use in the decision-making process. To highlight the data requirements of the system, managers at the various levels of the organization may have developed decision points in their job functions through the use of flow charts. These data inputs are typically inputs to transactions, activities, or decision rules. The data base provides a structured grouping of *data elements.* The data elements, files, and records are the component parts of a data base. Examples of data elements are "units on hand" for sale or "accounts payable." For any system there may be hundreds of classes of data elements in the data base and even more individual data elements, since there may be hundreds or thousands of customers.

A group of related data elements constitutes a *record.* An employee's record contains the data elements of his Social Security number, address, and so forth. While some files may be organized to retrieve all the data elements in a record, others may retrieve only those used most frequently.

A *file,* in turn, is a collection of records which contain common or related elements. These files, which may be in the form of punched cards, magnetic tapes, or magnetic disks, form the backbone of the data-processing system. An employee's master file is made up of tax information, year-to-date salary, credit union participation, special deductions, insurance, base rate, and so forth. The totality of all data in the organization constitutes the data base.

There has been a great deal of criticism leveled at the development of an MIS by the piecemeal approach—i.e., the development of a data center for each of the major functions and/or groups of users in an organization. The basic criticism is leveled at the duplication of input information in the files, since an individual customer may have a multiplicity of relationships in the organization. Firm A, which is a customer of Firm B, may also be a supplier to Firm B. Thus a great deal of redundancy of inputs exists in the data base. It is not uncommon to have the same information gathered by a number of different units in an organization, each for its own files. Such a practice obviously fails to recognize the interrelatedness of data and it is this which lies behind much of the current discussion of integrated systems.

A problem related to that above is the lack of commonality of data. Aside from the fact that there are legitimate reasons for developing information via the piecemeal approach, the fact remains that data development has occurred over a period of time when technological differences in documentation standards and machine capability were very great. New programming languages as well as newer equipment for which system compatability did not exist may coerce the redesign and reprogramming of older systems. In the early days of the computer, line managers knew very little about data-processing systems and so the task of designing systems rested solely on the shoulders of the systems analysts and programmers. Consequently, the design of the records and files, the development of programs and updating procedures, and so forth, were often done at the discretion and convenience of the analysts and programmers. Since the data provided by the analysts and programmers appeared either irrelevant or poorly structured to the line managers, many of the latter became by necessity involved in the structuring of the data base, so vital for decision making.

Role of Systems Analysis in Design

To evaluate an MIS one typically employs systems analysis. The prime objective of systems analysis is to learn as much as possible about a system in order to improve it (equipment, personnel, operating conditions, inputs, outputs, and so on). Systems analysis basically involves fact finding which will serve as a foundation for subsequent design or redesign of a system.

Were one to examine the following features of systems analysis and compare them with what has so far been discussed, he would note many similarities. Thus, it would be correct to assert that, if one were engaged in determining systems objectives, identifying constraints, identifying information needs and information sources, he would be performing systems analysis. Likewise, if one were to engage in the following activities he would also be performing systems analysis:

1. Fact gathering. This involves the determination of inputs and outputs, origination of documents, the maintenance of files, the flow of documents, and so forth. This step can also include the collection of sample inputs and outputs and other statistical data relating to volume of activity.

2. Organization of data. This step involves the flow charting of the documents from origination through each of the communication links into the files and to the subsequent reports. Such activity also results in the evaluation of the data-gathering process.

3. User documentation. This step involves interviews with the users of the reports for verification and possible modification. The original purposes for which the system was designed as well as the assumptions which prevailed in its implementation are critically examined.

After completing the above three steps, the analyst should be able to depict current weaknesses in the system that result from duplicate origination, redundancy in file content, meaningless documentation, unsuitable reports, and other deficiencies. The systems analyst should also include an analysis of station characteristics—i.e., description of files as to type, frequency, volume, or special requirements. Also to be noted is whether the station is an origination, storage, or satellite point in the system. Such analysis aids in the determination of possible integration or centralization.

Load analysis reports sorted by station, identification, form number, and card code all assist in the determination of work loads for each station. Document activity reports are also utilized to ascertain the flow between the stations, processing systems, and functional areas.

All the above as well as other reports provide inputs for the design of a system. Much of the above presupposes an existing system. In cases in which no system exists, it is necessary to examine the decision-making function of the manager to determine his information needs.

THE MANAGEMENT OF MANAGEMENT INFORMATION SYSTEMS

Impact of MIS on the Organization

A number of authorities have stated that with the widespread reorganization necessitated by the emergence of information technology, a large number of middle management jobs will be either eliminated or structurally altered so that all meaningful content and authority will have been designed out of the job. At any rate, more and more of the work customarily assigned to the middle management level can readily be programmed and covered by standardized procedures and decision rules.

Along with this change, the trend toward bigger and bigger middle management groups will cease. Like other middle-echelon specialists, they operate along rather narrow functional lines. Regardless of their qualifications, their role is segmented, specialized, and impersonal. These swollen middle layers of management, sometimes unfavorably referred to as "management by bureaucracy," are destined to shrink appreciably within the very near future.

Leavitt and Whistler[2] are of the opinion that centralization of management will be made easier by the new technology. The line of demarcation between top management and middle management will be more sharply drawn as time goes on, and, on the whole, middle managers will be downgraded, since more and more of their routine functions will be programmable. However, not all middle management jobs will be affected in the same way by the new technological explosion. At least two classes of middle management jobs will move upward, toward a state of "deprogrammedness." One of these will be that of the programmers, who are themselves the high priests of the new information technology. The other will be that of the research and development engineers, whose innovative abilities will become increasingly important to top management.

As for top management, the newer and more efficient information-processing techniques will free them from the more mundane executive details so that they can extend their abilities to encompass newer and broader tasks. Not only will they be freer to think, but they will be forced to do so. Since change will be the ethos of the times, one can expect a heavy turnover in top managers, who quickly burn themselves out as far as innovative ideas are concerned.

With the advent of management systems departments in numerous organizations, many of the predictions of Leavitt and Whistler have already been realized. Production problems have become increasingly routinized; simulation has proved an invaluable managerial device; tools such as PERT and similar planning and control techniques have been developed; operations researchers have been assigned important top-level roles; innovation and creativity have become increasingly vital; and the modern trend of employing models in decision making shows no signs of letting up.

Obviously not all writers concur with the view that middle manage-

---

[2] Harold J. Leavitt and Thomas L. Whistler, "Management in the 1980's," *Harvard Business Review* (November–December 1958), pp. 41–48.

ment is adversely affected by the new information technology. Burlingame[3] contends that "all managers are deeply involved in problems where judgment and human values are important elements." And more specifically, "they are concerned with situations where the decisions cannot be anticipated, the information needs predicted, or the decision elements quantified." Far from eliminating middle management, the new technology, in Burlingame's view, will provide middle management with the tools it needs to do a better job. Instead of rendering managerial ability obsolescent, it will put a high premium on native ability for exercising initiative and for shouldering judgmental responsibilities. "This effect," he believes, "should far outweigh in importance any tendency of computers to eliminate jobs where the nature of decisions is mechanical, especially when it is remembered that the growth in the complexity of business is increasing the need for effective managers."

Both of these positions seem valid. While, on the one hand, information technology has revolutionized some managerial functions, on the other hand, it has not depersonalized the functions of the middle managers. Progressive management is not so naïve as to assume that all personal needs will be adequately satisfied off the job. On the contrary, if anything, organizations have increased their efforts to cater to the needs of their managers as well as those of their rank and file. This practice, however, is in no way at variance with the demands of the new technology. Experienced managers are still the backbone of the organization and will conceivably continue to be so in the future. There is no doubt that managers have been upgraded; more information has been placed at their disposal to aid them in their decision-making function.

The following studies attempt to critically appraise the various predictions made concerning the future of middle managers. The studies by Schaul[4] and Reif[5] either reinforce one another or throw light on an aspect of the problem by introducing a necessary distinction.

Schaul interviewed some 50 middle managers and a dozen or so top managers in eight companies, and he discusses his results in terms of change in middle managers' functions and scope, decision-making authority, and status. As could be expected, change differentials were encountered in the various categories investigated.

Regarding the functions of the middle manager (planning, organiz-

---

[3] John F. Burlingame, "Information Technology and Decentralization," *Harvard Business Review* (November–December 1961), pp. 121–26.

[4] Donald R. Schaul, "What's Really Ahead for Middle Management?" *Personnel* (November–December 1964), pp. 8–16.

[5] William E. Reif, *Computer Technology and Management Organization* (Iowa City: Bureau of Business and Economic Research of the University of Iowa, 1968).

ing, staffing, directing, and controlling), less and less time was being devoted to the control function as a result of EDP and much more to the planning function. In general, none of these functions have been eliminated, nor has any specialist group arisen to appropriate any one of them. What has actually happened is that only the amount of time devoted to each has been adjusted.

The scope of the middle manager's job has been appreciably expanded. Many new activities have been assigned, while the traditional functions have become increasingly complex. The predicted reduction in the decision-making authority granted to middle managers has not been borne out by this study. The EDP system had no significant influence on their authority. What was experienced was a change in authority coincident with the increase or decrease in the number of assigned activities.

Far from being progressively eliminated, middle managers increased in number. Also, instead of the predicted lowering of status, an enhanced status was reported.

Reif's case study of three firms was meant to be representative of three major business classifications as well as of varying firm size. Furthermore, the three firms selected had utilized the computer for different lengths of time.[6]

Computers were observed to bring about a centralization of decision making in the management hierarchy: fewer decisions were being made at the lower levels and more at the higher levels, while less important decisions were left to the decentralized managers. Line-staff relationships were being redrawn, with the staff groups appropriating more decision-making authority.

Changes in the informal communication network resulted from the introduction of the computer; changes in the formal aspect had not yet been incorporated into the organizational structure.

The implications of the computer for middle management are more in line with the dire predictions reported above. However, Reif does not discriminate between the various functions reported on separately by Schaul. In both studies the control function seems to be severely restricted by EDP, since this function is most easily programmable. The susceptibility to replacement by the computer, Reif found, is tied

---

[6] In 1964 Lee studied the impact of EDP on patterns of business organization in three carefully selected firms, the same three types later reported on by Reif; viz., a public utility, a bank, and a manufacturing firm. He found that in the utility and banking firms the effect of the computer on decision making was relatively small, that an increase in the managerial personnel as a whole resulted from the introduction of EDP, and that the status of managerial jobs was generally upgraded by the elimination of clerical work and the concentration on decision making. See Hak Chong Lee, *The Impact of Electronic Data Processing upon the Patterns of Business Organization and Administration* (Albany: School of Business, State University of New York at Albany, 1965).

to the type of routine decisions for which middle managers are held responsible. The difference in the findings in the two studies, when clearly evident, may be a function of the sampling procedure used. In Reif's case study a utility company, a bank, and a large manufacturing company alone were utilized.

In his article entitled "Emerging EDP Patterns,"[7] Charles Hofer not only provides the reader with up-to-date EDP scorecards of the armchair predictors and the pavement-pounding researchers but also presents the results of his own investigation in this area. His study population consisted of two manufacturing firms, one with sales in excess of $200 million and the other with sales of approximately $8 million. While many of the studies cited in this report examined the effects of computers in but one or more functional areas, Hofer's study embraced a wide spectrum of functional areas such as marketing, finance, general management, employee relations, manufacturing, and engineering at all levels of the organization. His methodology consisted of extensive interviews with nearly 80 managers, as well as a thorough examination of supplemental company data.

Hofer's findings can be summarized according to the following five categories: EDP's effect on formal structure, on decision making, on operational planning, on budgeting, and on measurement. The three managerial levels scrutinized were those of the general managers, top functional managers, and operational managers.

As for the area of organizational structure, Hofer found no changes wrought by the computer at either the general or top functional management levels. However, where the activities involved called for the processing of a large volume of data, the operational level was affected. In this regard, some tasks were abrogated, others enhanced, and others just modified. But the principal effects were most noticeable where the processing of a large volume of data occurred.

The structure of the top levels of the organization was not visibly affected, since the specific tasks performed at these levels were also not significantly influenced by the computer. The computer does not assist managers in managing people or in understanding a subordinate's job.

As for decision making, no direct effects were observed at the general management level. However, some de facto delegation did take place at the top functional level. This delegation for the analysis and evaluation of certain classes of decisions generally involved those decisions most amenable to systems and procedures treatment. As expected, at the operational level a substantial number of decisions involving production were programmed.

---

[7] Charles W. Hofer, "Emerging EDP Pattern," *Harvard Business Review* (March–April 1970), pp. 16–170.

Again, the general management level was not affected by any changes in the area of operation planning. This may have been attributable to the fact that these managers do not get involved in the day-to-day operational planning that goes on, except for monitoring activities. Top functional managers were provided with more useful information for improving existing systems, while the more accurate information given the operational managers did assist them in improving their planning operations.

With respect to the effects of EDP on budgeting, two changes were observed. The first, occurring at the top functional level, pertained to the rapidity with which budgets could be altered to meet changing circumstances. The other related increased accuracy to a more refined cost breakdown at the operational level.

In the area of measurement, however, the impact of the computer was felt at all levels of the organization. General management was able to request additional back-up information in many areas of uncertainty. In other instances the computer allowed for in-depth penetration of problem areas by providing data hitherto unavailable. Top functional managers as well as operational managers benefited in the area of measurement by more quantitative reporting and by more objective evaluation of subordinates, replacing many qualitative subjective judgments.

Hofer's general predictions for the future are not as universal as one might expect. This is due to the nature of the diverse processes involved, the levels of the organization, the way change is implemented, and a host of other environmental variables.

The John Diebold Group, one of the largest consulting firms in computer technology, has maintained a continuing appraisal of the impact of information technology on the decision structure of European industry. A recent report[8] reviews past predictions and concludes:

ADP Impact on Organizational Structure
1. There have been major changes at the operational level, particularly in those functions which process large volumes of data. Large reductions in clerical staff have occurred, departments have been eliminated or amalgamated and remaining staff have taken on new duties.
2. At the middle management level, staffing has increased rather than the reverse. The increase has been largely in professionally trained, management science oriented personnel. They tend to be more involved with complex problems and exceptions and less with routine administration. The possible span of control has been increased.
3. There is a definite correlation between the status in the corporation accorded the ADP function and its impact on the organization. Where ADP has been embedded in finance or another functional department,

---

[8] "ADP Impact on the Organization and Decision Structure," Doc. No. E 96, Diebold Research Program—Europe, *Data Exchange* (November 1972), pp. 43–44.

the impact has been minimal. Separate organizational status has usually increased the role and impact of ADP.

4. As far as centralization or decentralization is concerned, it is apparent that data communications and ADP make either organizational approach feasible and that the choice in a given organization will depend more on history and management philosophy than on ADP.

5. Top management and physical operations such as manufacturing have been least affected organizationally by ADP because of the nature of the work involved.

6. Overall, it can be said that ADP has had an impact on the organization but it is difficult to separate and quantify that impact. So many other factors are involved, such as management philosophies, dynamic market changes, and the level of organizational sophistication already existing in a corporation. However, we believe that some past predictions underestimated the lead time required to implement commercially viable hardware and software and information systems which would have a truly revolutionary impact. ADP is a tool and its real impact is assisting and stimulating the *evolutionary* changes in organization which are occurring and will, we believe, continue.

ADP Impact on the Decision-Making Process

1. Probably the greatest impact of ADP, apart from the automation of many operational level decisions in such areas as reservation systems and inventory control, has been the increasing use of modeling techniques in support of managerial decisions. These techniques are not dependent on ADP technology, but it has encouraged and made possible their widespread use.

2. At the level of strategic decision making, past predictions were very optimistic but the least have actually been achieved. However, MIS are beginning to mature and meet some of their promises, and top management does have available much better information on which to base decisions. Again, we see the future impact of ADP as assisting an *evolutionary* process.

MIS and Computer Location

As is well known, having a computer does not automatically guarantee success, since it is only one of the many requisites for the operation of a successful electronic data processing installation. One of the significant, if not the main, determinants is the organizational climate within which it must function. A sound organization will successfully produce the conditions deemed essential to a well-managed computer department. If the EDP effort is to succeed, it must be properly organized, and the first area of concern is its location in the organization. The advice most frequently given is to:

1. Locate the activity within the accounting or financial department.
2. Establish an independent EDP activity.

Most accountants will no doubt agree that management information systems have their origin in the accounting function, since this was the only department created in the organization to generate information. Typically, every organization has some type of financial information system, and it is the oldest, the best developed, and the most important management information system in the organization. This financial information system, normally operated by the accounting department, has the responsibility of supplying management with *all* data reflecting the financial status of the business.

Since the inception of machine bookkeeping systems and the subsequent development of equipment associated with the punch card, the traditional location of the facilities for the rapid collection, processing, and analysis of data has been in the domain of the accountant. It is therefore not surprising to find the accounting function in the midst of rapid change, growth, and upheaval. Nor should it be startling to hear accountants profess themselves to be in an advantageous position for the transition to even more sophisticated information systems, since their function cuts across all areas of the organization, and, perhaps more importantly, their information flows are by this time both well developed and well received by management. Joplin notes the accountants' qualifications in this regard:

> The accountant, because of his experience with the only existing information system, has certain qualifications which should be considered when choosing the manager of information systems. Because of the length and nature of his association with top management, he is presumed to have their confidence, an important attribute in systems development. He has the knowledge of management's needs for information and the extent to which management relies upon such information for decision making. He is aware of the problems and limitations involved in supplying information for decision making.[9]

To be sure, there is more than a grain of truth in the above statement. Nor would it be difficult to make a good case for the accounting department as the skeleton for management information systems. There may indeed be substantial benefits in having a monolithic information structure in the organization. Such an arrangement, however, is not without its critics. Firmin and Linn in their article touch upon this problem:

> The seriousness of this problem is reflected in the plethora of articles in our literature about "Who'll Be in Charge?" These articles discuss the importance of the role of the information system controller, and they issue repeated warnings that the controller or management accountant will lose

[9] Bruce Joplin, "Can the Accountant Manage EDP?" *Management Accounting* (November 1967), p. 7.

ground unless he has the will and capability to broaden his horizon. There are hints that management accountants are too involved in their own discipline and too restricted in their own vision to appreciate the problems of the entire organization.[10]

Withington also notes:

The financial and accounting application will always receive favored treatment, and there may be a reluctance to start employing the computer in an operational or decision-making role which is foreign to personnel familiar only with accounting work. Therefore, a computer activity located under the controller is likely to become less and less satisfactory as the computer is put into operating areas. . . . The department responsible for meeting certain objectives is also responsible for the computer tool which helps it meet the objectives. . . . It is difficult for the computer activity to provide important services for departments other than the one in which it is located, for however well intentioned they may be, the personnel dealing with the computer system's operation and the development of its future applications are naturally going to be most responsive to those who control their promotions and salary reviews. As a result, most or all of the organization's available resources for computer applications development will be channeled into the "parent" accounting department.[11]

A recent study by Schoderbek and Babcock confirms Withington's suspicions. About 95 percent of the respondents stated that the EDP function should be independent of all operating departments, including the accounting department. What makes this even more meaningful is that 50 percent of the EDP departments were then reporting to an accounting function. Also worthy of note is the observation that 39 of the 49 EDP departments reporting to accounting or financial functions listed problems related to organizational design, accounting, and financial bias as the main shortcomings. Of the 54 firms reporting autonomous EDP departments, about two-thirds (65 percent) reported *no* significant drawbacks.[12]

Another of the findings of the above study was that the location of the computer activity is shifting away from the accounting function. Nearly 70 percent of the 109 respondents stated that the original location of EDP was within the accounting department while currently only 45 percent have this reporting arrangement. Data-processing de-

---

[10] Peter A. Firmin and James J. Linn, "Information Systems and Managerial Accounting," *The Accounting Review* (January 1969), pp. 80–81.

[11] Frederic G. Withington, *The Use of Computers in Business Organizations* (Reading, Mass.: Addison-Wesley Publishing Co., 1966), pp. 159–61.

[12] Peter P. Schoderbek and James D. Babcock, "The Proper Placement of Computers," *Business Horizons* (October 1969), pp. 35–42; also, "At Last — Management More Active in EDP," *Business Horizons* (December 1969), pp. 53–58.

TABLE 8-1
Major Computer Installations

	Percent		
Application	1966	1968	1972*
Finance and accounting	47	44	29
Production	16	19	24
Marketing	12	13	15
Distribution	11	10	12
Research, development, and engineering	8	11	13
Planning and control	6	3	7

* Estimated

Sources: James W. Taylor and Neal J. Dean, "Managing to Manage the Computer," *Harvard Business Review* (September–October 1966), p. 102; and Neal J. Dean, "The Computer Comes of Age," *Harvard Business Review* (January–February 1968), p. 89.

partments were originally responsible for the EDP function in approximately 24 percent of the companies, while currently they have responsibility in about 50 percent of the cases. The most cited reason for the change of the EDP location was the increased importance of EDP as an independent function.[13]

However, location itself may not necessarily imply leadership of the function. For example, in a recent survey conducted by the Diebold Group, Inc., of more than 2,500 executives, it was reported that managers in general are not providing the leadership in the computer area. Diebold states: "The survey indicates that *technicians*, not management, are setting goals for computers. This is one of the prime reasons why companies often fail to realize the true potential from their data processing investment."[14]

Other studies support the conclusion that accounting applications involving the computer have decreased. Two investigations by Booz, Allen, and Hamilton show that, while finance and accounting just a few years ago comprised nearly half of the computer effort, the trend is away from these applications. They estimated that three years later accounting and finance would comprise but 29 percent of the total computer effort. Table 8–1 presents some of the results of these investigations.

Irrespective of the reasons for the locational shift of EDP operations from the accounting department to an independent status, be it bias, the narrow vision of the accountant, or the diminishing scope of the

[13] Peter P. Schoderbek and James D. Babcock, "Proper Location of Computers," *Business Horizons* (December 1969), pp. 39 and 40.

[14] John Diebold, "Bad Decisions on Computer Use," *Harvard Business Review* (January–February 1969), p. 16.

accountant's function, the shift is expected to continue. The technology of the industry is simply moving at too rapid a pace for the accountant or any other functional manager to stay proficient in both functions.

Integrated Systems and MIS

For many years now the computer has been heralded as *the* vehicle for the Second Industrial Revolution. Linked inexorably to this is the idea that advances in computer and information technology necessarily lead to an all-embracing, all-purpose system best suited for managing industrial corporations. In some quarters the misconception still lingers that the primary objective of a computer system is managing information rather than providing information for management. In the former instance (and, incidentally, some of today's systems are still designed with this objective in mind) the information system is concerned with the rapid collection, processing, and display of data which often have little to do with decision making. In the latter instance, information not only is collected but also is restructured to reduce the uncertainty inherent in decision making.

It is not too difficult to see why the notion of a total system received so much attention. The notion of integrated information systems, no doubt, had its origins in integrated data-processing systems. Since raw data gathered from separate information centers could be used for more than one decision-making function, many felt that further technological advancements would lead to totally integrated systems. Within the past decade we have had the technological advances necessary for such systems, and yet it appears that we are no closer to realizing this concept than we were a decade ago. New developments in the form of visual display equipment, data entry, information retrieval, computer-assisted design, and others have now made such systems more feasible. Efforts have been made in many companies to design a total data base and then to use that base to generate the necessary information. To be sure, there has been some integration of data in many organizations, but, for the most part, this has been true for those systems utilizing the same *kinds* of related data. This similarity of inputs is one of the requirements of an integrated system; if compatibility of data does not exist, the system is not integrated.

The fact that compatibility does not exist in all the subsystems does not negate the entire concept of integration but rather coerces one to define initially the boundaries of the system. Rather than striving for one gigantic data base which embraces all informational needs, one should adopt the modular system approach, since in reality this is how progress is made, even though it may be more costly. The degree of integration is somewhat dependent upon the design of the system.

The task of designing an information system must center about the specific needs of the organization. Just as no two companies employ the same strategies or even managerial policies, neither do they have the same information needs.

One approach typically employed in the design of a management system is simply to restructure the old system. The advantage of this approach is that there is a relatively short pay-off period. Examples of this are to be found in the computerization of inventory control, ordering, and production scheduling. Its major limitation is its concern with the improvement of only a specific system which may lack integration with other systems. This approach does not provide the proper analysis for determining if the present systems information needs are adequate or whether too much information is being called for. This approach, however, is not without merit. It is of benefit for specialized functions at the lower levels of the organization. This is true because there the decisions are mostly of a routine nature and are not affected by environmental or competitive factors outside the firm. These systems tend more to be data collection centers, and decisions which emanate out of these centers are routine ones such as inventory leveling.

The second approach is one which looks at the types of information needed to make decisions. It requires that for all major decisions a decision model exist which in turn calls for an identification of the information variables. This implies that the managers can identify the important variables involved in the process, assuming at the same time that the availability of this information will lead to better decisions.

There is much weighty evidence to the contrary. There is also evidence that many managers cannot reduce the decision-making process to quantifiable expressions. This need not indicate an unwillingness on the part of the managers to cooperate but rather a genuine inability to comply, since, in order to identify the information needed, one must first have a model of how one makes decisions. Until such a model exists, one cannot specify the required information. Too often, mathematical concepts cannot capture the expressions of human values which often dictate decisions. Thus, neither simply the abundance of information nor mathematical expressions generally reflect human decisions. The lack of useful models in general obviously limits their integration into total information systems.

For the most part, the interactions of the major functions of the organization have not been duly noted in the design stage of information systems. Typically what occurs is that the organization is wrenched apart and divided into its principal function and subfunctions. This is what has accounted for the superabundance of systems—production information systems, marketing systems, and so on. To further compound the situation the above functions are further splintered into inventory systems, scheduling systems, market research systems, forecasting systems, personnel systems, and a host of others. When one attempts to integrate these many systems, all designed for varying and specific purposes, each possibly using a different mode of information, the result is inevitable. The multiplicity of subsystems which may use separate data bases suggests that different kinds of information are required, and while some information centers may be created for external reporting, others will be utilized for operating decisions, and still others for planning or forecasting purposes. There is nothing inherently wrong in developing subsystems independently of each other as long as they retain their capability for interfacing.

In discussing a problem where each system was designed by the systems group residing within the individual organization, Clinton Williams of Chrysler Corporation stated:

> Each organization took a parochial view of their own requirements without regard to the impact of these systems on the total business. This is not a unique example. I could cite many others. This problem was eventually resolved by a central staff with total corporate responsibility. Such a staff is necessary to build a MIS in a big company. Some companies have attempted to solve this problem by centralizing systems design. That satisfied the requirement of looking at the corporate view but created a more serious problem; that of not being responsive to the needs of the individual organizations. Central planning combined with decentralized systems design offers the best combination for big business organizations.[15]

Even the task of delineating an MIS for a large organization is a formidable one, much less integrating it. Says Williams:

> The job of defining an information system for a large company is an enormous undertaking. We have 160 computers installed in our company, worldwide, supporting such diverse products as cars, trucks, boats, air conditioners, chemicals, tanks and missiles, and subsidiary companies engaged in commercial credit, land development, and car leasing. The thought of putting this into an overall MIS scheme staggers the imagination.[16]

A common area of disagreement is what does and what does not constitute an integrated system. It would not be difficult to point out obvious dissimilarities of authors and practitioners alike. Some would define an integrated system as including both on-line and real-time considerations (OLRT). In William Crowley's view, if a system is integrated, it will:

---

[15] Clinton C. Williams, "Practical Problems of M.I.S. in a Business Environment," paper presented at the First Annual Meeting of the American Institute for Decision Sciences, New Orleans, La., October 30–31, 1969, pp. 3–4.

[16] Ibid., p. 4.

1. Supply historical data and analysis of that data.
2. Supply "on-line" data, that is, factual material picked right out of the system as fast as it is generated.
3. Supply data in "real-time," fast enough so that management can exercise necessary management control instantly.[17]

Others appraise the merits of real-time information systems as producers of instant and relevant information. One author states: "On-line real-time information systems are upon us. The benefits to be derived focus mainly around the "instant information" aspect. At any time we are able to query the computer and receive up-to-date information on any desired phase of the business. The information system may be limited or *total*. It may deal with financial information only, or the system may include data on all functions of the business. . . ."[18]

If integrated systems were to embrace the elements suggestive of a more sophisticated computer system, the task would be no more formidable, since an on-line real-time system of itself does not imply an integrated system. Many firms with real-time capability do not thereby claim to have an integrated system. Indeed, most of the applications of real-time systems were born of critical but very specialized problems. In a survey reported by C. Clifford Wendler in which a questionnaire was sent to 110 "knowledgable" systems people, over 92 percent of the respondents judged the following factors to be either "absolutely essential" to or "highly desirable" for the total systems concept.

1. The system provides timely and accurate management planning and control information to facilitate the attainment of the company's objectives.
2. The system generates information needed to fulfill the company's operating legal, governmental, and financial requirements in the most effective manner.
3. The various systems (or subsystems) are interlocked to attain a total system.
4. Integrated data processing techniques are employed in designing systems — data are captured or created in machine language as near to the source or origin as possible, and flow through the system automatically from that point on.

---

[17] William J. Crowley, "Can We Integrate Systems without Integrating Management?" *Journal of Data Processing* (August 1966), reprinted in *The Computer Sampler* (New York: McGraw-Hill Book Company, 1968), p. 272.

[18] James W. Pattillo, "A Study in Instant Information," *Management Accounting* (May 1969), p. 16. (Emphasis by the authors.)

5. The system is all-encompassing—the company is viewed as an integrated entity.[19]

The above points are so replete with systems jargon as to be completely meaningless. Point 1 calls for timely and accurate information. However the meaning of the word *timely* is open to a vast range of interpretations which could mean "daily" in some instances or "monthly" in others. And presumably everything that the organization does will facilitate the attainment of company objectives. Point 2 suffers from ambiguity in that "effective" can mean any level of performance which the organization is currently achieving.

In regard to point 3, it is presumed that interlocking systems utilize the same data base. This is not only impossible but also impractical in many cases. Different divisions producing completely dissimilar products have no need whatsoever for the same data base. Point 4 is not only logical but less costly than other alternatives, and yet the lack of quantification again lessens its credibility. The term "as near as possible" covers many sins of omission. The fifth point calls for an all-encompassing system, to be *viewed* as an integrated entity. At the present time there simply isn't any such thing in an organization as a total information system. The fallacy of such an approach will be presented shortly.

Some authorities, rather than demanding the presence of a real-time requirement for an integrated system, even question the legitimacy of the entire concept.[20] The salient point is that the concept of an integrated system does not depend upon the use of real-time hardware for producing instantaneous information. Many organizations do not need real-time systems and at the present time cannot justify extensive use of them; moreover, the attempt to achieve an integrated system does not cut down on the problems at all. Williams also makes the same point: "Our major batch systems have daily update. This gives us a lot of flexibility for producing needed information with high response timing. Not very many information requirements have been established that require higher response than this."[21]

One of the more formidable obstacles to an integrated management system is the unproductiveness of insight into the nonrepetitive decision-making process of the data generated by systems reports. Since

---

[19] C. Clifford Wendler, "What Are the Earmarks of Effective Total Systems?" *Systems and Procedures Journal* (July–August 1966).

[20] See for example, John Dearden, "The Myth of Real-Time Management Information," *Harvard Business Review* (May–June 1966). See also John Dearden, "M.I.S. Is a Mirage," *Harvard Business Review* (January–February 1972).

[21] Williams, "Practical Problems of M.I.S.," p. 5.

some hold that information is imperfect whenever it is unavailable, or too costly, or unproductive of knowledge, the reason why many systems do not generate insightful information may be that much of the information actually utilized in the decision-making process is information *external* to the firm. In many cases this external information dictates decisions irrespective of internal conditions. Robert Anthony succinctly states this:

> It is because of the varied and unpredictable nature of data required for strategic planning that an attempt to design an all-purpose internal information system is probably hopeless. For the same reason, the dream of some computer specialists of a gigantic bank, from which planners can obtain all the information they wish by pressing some buttons, is probably no more than a dream.[22]

It is noteworthy that typically the higher the decision making in the organization, the more judgmental the factors involved. Decisions made at the levels of middle management and above are more in response to external pressures upon the firm than to indigenous factors. Those systems which purport to employ a central data base, for the most part, do so for decisions made either at the lower management level or those which are repetitive. Daniel, in his oft-quoted article, "Management Information Crises,"[23] states that a dynamic management information system requires information of three types: environmental information, competitive information, and internal information. Even if management were able to successfully integrate the internal information, it would be impractical for a firm to attempt to synthesize a system with input data based upon continually changing political, economic, and environmental factors as well as data relating to the past, present, and probably future activities of direct and indirect competitors. Anthony makes the same point when he writes: "Strategic planning relies heavily on *external information,* that is, on data collected more from outside the company, such as market analyses, estimates of costs and other factors involved in building a plant in a new locality, technological developments and so on. . . . Strategic planning and management control activities tend to conflict with one another in some respects."[24]

Recent studies show an increasing awareness of the importance of

---

[22] Robert N. Anthony, *Planning and Control Systems: A Framework for Analysis* (Cambridge, Mass.: Graduate School of Business Administration, Harvard University, 1965), p. 45.

[23] D. Roman Daniel, "Management Information Crises," *Harvard Business Review* (September–October 1961), p. 55.

[24] Robert N. Anthony, "Framework for Analysis," *Management Services* (March–April 1964), p. 21.

---

this external information. Aguilar examined the kinds, sources, and modes of external information that executives use for strategic decision making.[25] He found that for large companies 51 percent of the information utilized for strategic decisions came from sources external to the organization.[26] Keegan examined the sources and the manner in which executives at headquarters level learn about the significant opportunities and threats to their companies.[27] He found that documents were the source of only 27 percent of the important external information received by executives. He also states: "The bulk (60 percent) of these documents are publications and information service reports from *outside* the company. Letters and reports from inside sources account for the remaining 40 percent."[28]

Those enamored of the integrated systems concept are often unaware of the heavy financial commitment required. Many managers have been sold on the notion that with a totally integrated system they would be able to query the computer and receive answers to virtually any question they desire. Even if such a system were possible, and it is not, this would require a monumental data base and a special computer access language. Such a course of action could hardly be justified in regard to the time-cost expenditure relative to the benefits received. Several years of developmental time is common just with modular systems, let alone a system purporting to include all relevant data.

Characteristically, the cost of an installation and its payoff will dictate the degree of integration that is feasible, and in many instances the computer may be cost-justified for only one or two functional areas. For these firms, then, the optimal degree of integration has been achieved at that particular point in time. On the other hand, the attempt to tie together the entire information flow would be economically unwise. The great effort expended, regardless of the hardware sophistication, would be enormously disproportionate to the benefits received.

This splintering-up approach to information systems is one which often is depicted as ignoring future requirements of the firm. On the contrary, it can be said that management seldom fully ignores future benefits but rather that it discounts the value of these benefits and

[25] Francis J. Aguilar, *Scanning the Business Environment* (New York: Macmillan Company, 1967).

[26] Ibid., p. 80.

[27] Warren J. Keegan, "Acquisition of Global Business Information," *Columbia Journal of World Business* (March–April 1968), pp. 35–41. See also Warren J. Keegan, "The Scanning of International Business Environment: A Study of the Information Acquisition Process," unpublished Ph.D. dissertation, Graduate School of Business Administration, Harvard University, June, 1967.

[28] Ibid., p. 37.

hence seems to rely upon the more verifiable short-run values.[29] The number of firms that have attempted complete integration of information (often with the computer manufacturer's warranties and vows of assistance) only to experience absolute failure is not insignificant. These firms, understandably, do not draw the wide attention to their misadventures that they deserve.

It has been suggested that the acquisition and operation of an integrated management information system be viewed in the same light as the purchase and operation of an additional amount of capital equipment. Both actions can and often do involve large amounts of financial resources. If the purchase of capital must be verified as cost-effective, an information system should also contribute its share to profit. Obviously, the verification of benefits from any integrated management information system is difficult, but certainly it is not impossible. Systematic analysis of the almost unlimited volume of output not only will expand the initial list of probable benefits but will help to ensure that the system is earning its keep.

MIS Failure

One need only attend the current seminars, conventions, or other national meetings to learn of the many advances and successes of EDP systems. Yet, in spite of the mass acceptance of EDP, many failures have been experienced along the way. This is not totally unexpected, since all technological and innovative uses of systems must have a testing period. What should be marveled at is that this industry in its short life span has so revolutionized earlier mechanical systems and arrived at its present degree of sophistication. As a result, today's managers have more data available for making informed decisions than at any other period in history. Still, the experience garnered from the nonsuccessful applications of EDP can provide the necessary discretionary guidance for avoiding the more common pitfalls.

1. Lack of Management Involvement. A McKinsey and Company study done a decade ago initially scored the point for management involvement. It found that firms with the most successful computer operations were precisely those in which top management participated. In the companies surveyed which stated outstanding economic results from computer applications, top management was simply unwilling to settle for anything less. In the less successful companies, managers tended not to become involved with computers and kept them at a

---

[29] A firm in the farm implement industry announced "that they were abandoning their efforts to integrate their international operations because of both problems encountered in the endeavor and the tremendous high cost relative to benefits expected."

"safe" distance. In these instances many managers abdicated the direction and control of this function to staff specialists. This was not so for the top computer users.

Probably the single most critical factor in the success of computer operations was the chief executive of the organization. He retained the ultimate responsibility for the success of the computer effort. The report stated that for the executive who wanted maximum results from his firm's computer effort, a minimum of five things must be accomplished.

1. He must approve the objectives, criteria, and priorities for the corporate computer effort, giving special attention to the development program.
2. He must decide on the organizational arrangements to carry out these policies and to achieve these objectives.
3. He must assign responsibility for results to the line and functional executives served by the computer system and see to it that they exercise this responsibility.
4. He must insist that detailed and thorough computer systems plans are made an integral part of operating plans and budgets.
5. He must follow through to see that planned results are achieved.

2. Resistance to Change. The lack of training and thus involvement of lower-line managers in computer efforts may create resistance to change. It is well known that people tend to resist changes that they have not helped to make; consequently, decisions made by computer personnel that affect the operating management, but which reduce their role to that of mere spectators, will not win general acceptance. The criticism frequently leveled at the computer staff—that they lack an awareness of operating realities—may carry considerable weight, thus further reducing their effectiveness for implementing change.

In many instances, the lack of management involvement can be directly traced to apprehension about the computer. If the computer is viewed as an all-powerful device understandable only to computer specialists, then predictably the operating personnel will indeed avoid it. There are far too many known instances in which operating personnel did not utilize the information presented to them simply because they lacked knowledge of how the information was derived or for what it was to be used.

This resistance to change factor, however, is not unique to computer application but emerges whenever job security is threatened. This resistance may manifest itself through a perceiving of a potential loss of status, a diminution of promotional advantage, and so forth. It is for these reasons that occasionally line personnel do not fully cooperate with systems designers. The number of managers, even of top managers, who have resisted the introduction of computers is not inconsequential.

However, the evidence is mounting that line managers today are not nearly as mesmerized by the computer as they were five years ago, and in many instances it is the operating personnel who are actually taking the initiative. This can be attributed to the increasing number of training programs offered by computer manufacturers and universities. Besides, there are commercial training programs, and in-house training programs offered by the firms themselves. Line personnel have come to realize that, if they are to be held accountable for the results of their operation, they ought to have control of their information system. Also, it is almost impossible to evaluate an information system unless one has direct knowledge of the intricacies of the system itself.

3. Lack of Qualified Personnel. The point has often been made that the requisite technical skills are in short supply, but, for those progressive organizations that have led the way in innovative applications, the successful results of the EDP efforts seem to be largely due to the initiative of the EDP manager or director. This does not contradict the earlier point of top management involvement, since the chief executive does not have the technical background for providing leadership in this area. He can, however, dictate the direction and influence the intensity of the effort but he cannot recruit the professional staff and provide the needed day-to-day leadership. If the EDP director gains the support of the functional line managers and provides the information they need, the EDP managers can engage in more innovative activities. Specialists, consultants, and other EDP-related personnel are indeed costly. The firm that makes the greater commitment to excellence in this area perhaps has the greater advantage. In any event, the level of sophistication of the EDP effort seems somewhat related to the expertise of the EDP director.

SUMMARY

Interest in systems is not merely theoretical. The businessman, especially the middle and top manager, is concerned with management information systems—systems capable of supplying the right kinds and amounts of information needed by the manager for decision making. Many existing systems are such only in name. One reason for this is the obscuring of the essential difference between *data* and *information:* the two are definitely not convertible terms. Five characteristics of management information systems were noted, chief among which is the availability of timely and accurate information for the decision maker. Also detailed were five assumptions underlying systems design. Because these assumptions are unrealistic and fallacious, what results is not a management information system but a management *mis*information system.

How does one, after avoiding the many pitfalls, design and develop a management information system? First of all, one ought to specify the systems objectives. Then one identifies the various constraints — lack of suitable models, user constraints, and cost constraints. One needs also to identify the information needs of the managers as well as the sources of information. Finally one develops a data base suitable to the company's or the organization's various functions.

The final section of the chapter discussed managing management information systems. Several factors bearing on the topic were noted: the impact of the management information system on the organization itself, the proper location of the EDP department, and the endless quest for truly integrated systems. The chapter concluded with a note on MIS failure, a point that may be of interest to groups contemplating a revamping of their management information system. Three factors singled out as contributory to this state of affairs were lack of management involvement, resistance to change, and lack of qualified personnel.

REVIEW QUESTIONS FOR CHAPTER EIGHT

1. Distinguish between data and information and discuss why such a distinction is necessary.

2. If you were the vice president of marketing of a drug company and supervised 50 district managers, what information would you want in your management information system? What environmental information would you want to include in your system and why?

3. Why is the concept of integrated information systems mere folly for most organizations? What assumption would one have to make to develop the case that such a concept is possible?

4. Why would top management have a different information system than operating management? Give examples of the types of reports required at these different levels.

5. Interview a manager in a local business firm and list the types of information that he needs to make decisions.

6. Your university gathers a goodly amount of data on students. What information system does this enter and to what purposes are the data used?

7. Trace the flow of information that is recorded on a patient when he enters a hospital?

8. Arrange for an interview with the EDP director of a local organization and develop a list of criteria utilized in the design of a particular system. Note what problems were encountered and how these might have been avoided.

9.   There is much discussion regarding the need to incorporate behavioral variables into information systems. Why is this impossible to do and what problems would be encountered even if it were possible?

10. Utilizing the university as an example of a system, note the various types of information that are required for long-range planning as well as daily operations. How do they differ?

Part Five

Application of Management Systems

I always begin with the universe: an organization of regenerative principles frequently manifest as energy systems of which all our experiences, and possible experiences, are only local instances.

R. B. Fuller

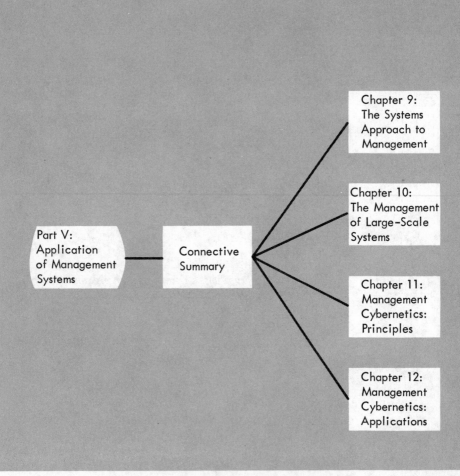

CONNECTIVE SUMMARY

The preceding four parts of this book have laid out the foundation for this last part concerning the applications of the systems approach. Part One provided us with an understanding of the interdisciplinary nature of the systems approach, as well as with a vocabulary of terms and concepts. Part Two gave us an in-depth view of the characteristics of cybernetic systems along with the tools for understanding and managing these exceedingly complex, probabilistic, and self-regulating wholes. In Part Three we saw the development of a new conceptual framework for visualizing human organizations as complex, open, and self-regulating systems. The discussion of information technology in Part Four provided the necessary means for assessing the utility of applying the systems approach to a variety of social systems such as private and public enterprise and other large-scale organizations.

The time has now come for us to consider the actual outcomes of the employment of the systems approach in the past as well as to contemplate its uses in the future.

We will begin this part on application by recapitulating the basic concepts of the application of general systems theory and cybernetics (i.e., management cybernetics), as well as by reiterating the basic steps which are involved in such an application. Our point of departure in this chapter will be the examination of some traditional "systems approaches." Chapter 10 will examine the application of the so-called network analysis to military and civilian projects. This method is variously referred to as "project management," "systems management," "large-scale systems management," and so forth. All of these terms refer to the use of network analysis as applied to the planning and controlling of complex projects.

The last two chapters represent the application of the modern systems approach to the management of complex enterprises. Since this is a novel field, we will again begin with the basic principles of management cybernetics in Chapter 11, and end with actual applications of this new hybrid to the management of systems ranging from the micro-organism, the firm, to the largest and most complex system of all, the world.

Our discussion of the application of management cybernetics to managerial problems of a firm is more detailed than other applications. The reason, of course, is fairly obvious and reflects the authors' biases as well as the fact that most progress in management cybernetics has been in that arena.

Great progress lies ahead in the area of world dynamics or in the application of management cybernetics to the field of ecology, both in the field's traditional domains and in its new sphere of global eco-systems.

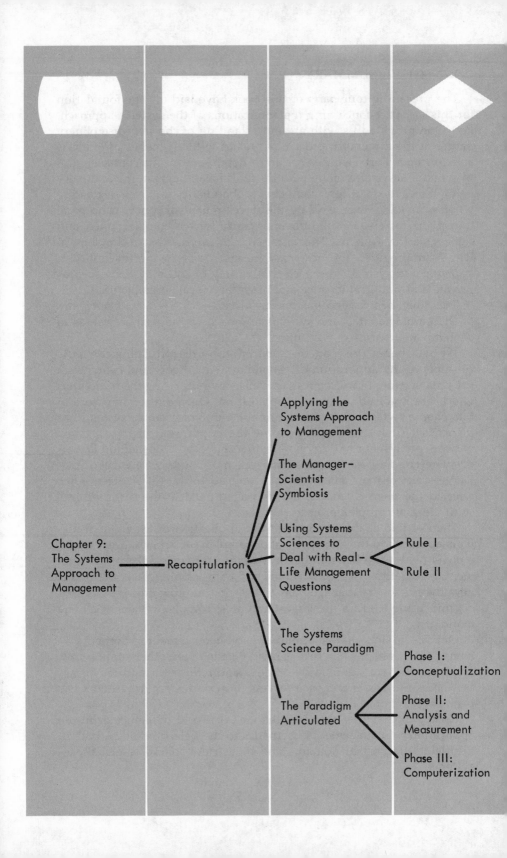

Chapter 9:
The Systems
Approach to
Management

Recapitulation

Applying the
Systems Approach
to Management

The Manager-
Scientist
Symbiosis

Using Systems
Sciences to
Deal with Real-
Life Management
Questions

Rule I

Rule II

The Systems
Science Paradigm

The Paradigm
Articulated

Phase I:
Conceptualization

Phase II:
Analysis and
Measurement

Phase III:
Computerization

Chapter Nine

The Systems Approach to Management

The gull sees furthest who flies highest!
Richard Bach, *Jonathan Livingston Seagull*

A RECAPITULATION

A basic postulate that underlies this text is that the second half of the 20th century is characterized by systems thinking—a trend that began with science and which has spread into other spheres of human activity. The study of human organizations has also been noticeably affected by this trend. Most contemporary writers in organization or management theory either implicitly or explicitly advocate a systems approach to the management of today's complex organizations. Therefore, it is not only desirable but also necessary to give the student, and through him a wider sector of the society (i.e., the practicing manager), at least some inkling of the profound conceptual changes that have been set in motion. Familiarity with certain systems concepts is indeed fundamental for the understanding of modern managerial thinking.

Systems thinking denotes alternately a technological revolution and a conceptual revolution. The latter cannot be understood without clearly tracing its origin and development. To this end Chapter One traced the origin and development of systems thinking as it evolved from the speculative ideas of the early biologists to the presently developed disciplines of general systems theory and cybernetics.

While a completely detailed historical account of systems thinking could not possibly be attempted, nevertheless, Chapters One and Three

237

include enough information to enable the inquisitive student to adequately sample this new and exciting area of intellectual activity. A brief summary of the main ideas set forth in these two chapters should prove beneficial in applying systems thinking to the study and management of organizations.

The second half of the 20th century ushered in the "age of systems." There are basically two main things associated with the age of systems: (1) the systems approach, or conceptual systems, and (2) management information systems, or applied systems. The systems approach is a philosophy or a viewpoint that conceives of an enterprise as a system — i.e., a set of *objects* with a given set of *relationships* between the objects and their *attributes*, connected or related to each other and to their *environment* in such a way as to form a *whole* or entirety.

The tremendous increase in the size and complexity of the 20th century organizations has forced students and managers into adopting a point of view that sailed along the streams of traditional analytic thinking employed by the researchers of the so-called hard or physical sciences — primarily physics and chemistry. The analytic thinker, when confronted with a complex phenomenon, attempts to understand it by breaking it into smaller and less complex parts; by studying the parts separately; and subsequently by putting his findings together to gain an understanding of the whole. This is the "age of analysis," landmarked by the works of some of the greatest 20th century philosophers such as Bergson, James, Russell, Dewey, Santayana, and Whitehead.

Systems thinking, or the systems approach, represents the "age of synthesis." Here, the researcher's approach to the understanding of complex phenomena is one of synthesizing the findings of various disciplines, with the ultimate aim of developing a method or technique which would be applicable to several seemingly different phenomena. All phenomena, whether physical or social, are treated by the systems thinker and researcher as systems. The age of systems is landmarked by the works of the late Ludwig von Bertalanffy, a biologist; the late Norbert Wiener, a mathematician; the biomathematician, Anatole Rapoport; Kenneth Boulding, a noted economist; Herbert Simon, a computer software expert; and numerous others.

APPLYING THE SYSTEMS APPROACH TO MANAGEMENT

What does this new way of thinking mean to the contemporary student of organizations? To put it differently, why should he be concerned with the systems approach? Or, assuming that he sees the need and relevance of a systems-oriented study of organizations, how does he begin to apply systems thinking to the study and management of today's exceedingly complex organizations? This present chapter will be devoted to the development of a skeletal framework for the "how-to" portion of the systems concept.

From the pragmatic point of view, the application of the systems approach to management can be conceived as consisting of the following three steps:

1. Viewing the organization as a system
2. Building a model
3. Using information technology as a tool both for model building and for experimentation with the model; i.e., simulation.

Developing a systems viewpoint of an organization is primarily a matter of the manager's adopting a new philosophy of the world, of his organization and its role within this world as well as a new viewpoint of himself and his role within this organization and this world. The manager's philosophy here advocated is, of course, systems thinking. There can be no doubt that this is a new philosophy for the practicing manager. The basic postulate of systems thinking (i.e., securing adequate knowledge of the whole relevant system before pursuing an accurate knowledge of the working of the "parts") is definitely against everything that the manager has been taught or has learned through his own personal experience.

Traditionally, organizations are departmentalized along functional lines. In business enterprises, for instance, one finds such departments as production, sales, finance, and accounting. Nonbusiness organizations follow a similar pattern. The organization or agency is divided into subagencies denoted by such names as districts, divisions, and sectors. In all these cases the individual manager or administrator perceives his own niche as the whole and consequently strives for its improvement and optimization. In reality, however, the scope of a particular manager's territory is determined by the behavior of the whole organization of which he is a part.

A systems-oriented manager is a manager of the whole. This does not imply that only organizational participants with responsibilities encompassing the entire organization can develop a systems viewpoint. Every manager can be a systems manager as long as his approach is governed by the two following principles formulated by B. Fuller:

1. I always start with the universe: An organization of regenerative principles frequently manifest as energy (and/or information) systems of which all our experiences and possible experiences, are *only local instances.*
2. Whenever I draw a circle, I immediately want to step out of it.[1]

---

[1] B. Fuller, *I Seem to be a Verb*, (New York: Bantam Books, Inc. 1970).

The manager whose style is directed by these two principles begins his investigation of the world about him not by gathering and analyzing the facts pertaining to happenings within "his" department but rather by identifying his universe—i.e., his department as it affects and is affected by its environment. This definition of the manager's department along with its environment will provisionally determine the boundary (the circle, in B. Fuller's terms) of his system. About this system the manager will want to know its inputs, processes, outputs, feedbacks, relationships, as well as their attributes. His search for these system determinants begins with the construction of a conceptual model. Thus, the model becomes the link between the real phenomenon and the manager's system. Figure 9–1 depicts the relationship between the real phenomenon (RP), the model (ML) and the system (SY).

The systems-oriented investigator who looks at phenomena from the holistic viewpoint perceives them as an orderly summary of those features of the physical and/or social world that affect his behavior. Thus, the box labeled "Real Phenomenon (RP)" represents the ob-

FIGURE 9–1
The System, the Model, and the Real Phenomenon

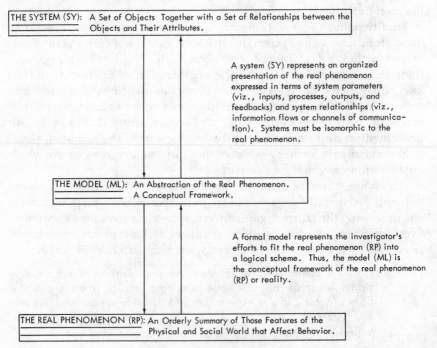

THE SYSTEM (SY): A Set of Objects Together with a Set of Relationships between the Objects and Their Attributes.

A system (SY) represents an organized presentation of the real phenomenon expressed in terms of system parameters (viz., inputs, processes, outputs, and feedbacks) and system relationships (viz., information flows or channels of communication). Systems must be isomorphic to the real phenomenon.

THE MODEL (ML): An Abstraction of the Real Phenomenon. A Conceptual Framework.

A formal model represents the investigator's efforts to fit the real phenomenon (RP) into a logical scheme. Thus, the model (ML) is the conceptual framework of the real phenomenon (RP) or reality.

THE REAL PHENOMENON (RP): An Orderly Summary of Those Features of the Physical and Social World that Affect Behavior.

Note: The Model (ML) is always "smaller" than the Real Phenomenon or the System; the System must be as complex as the Real Phenomenon. There is a *homo*morphism between the model and reality but an *iso*morphism between the system and reality.

server's interpretation of what is really out there. The contention that RP is the observer's own "reality" is, of course, supported by the fact that the observer himself at a later point in time may modify the real phenomenon to accommodate changes resulting from fresh evidence or new data. It is also supported by the fact that another observer, given the same or similar background, may very well entertain a different "picture" of the real phenomenon. In short, the real phenomenon represents the ultimate outcome of the investigator's sequence of mental activities (observation plus conceptualization) and not the outcome of the sensory system alone.

As previously stated, the systems-oriented investigator of real phenomena will treat them as systems — i.e., as orderly or organized complexities exhibiting certain characteristics such as goal-directedness, stability, ability for self-improvement or learning, openness, and so forth. In any case, were the investigator to study the real phenomenon as a system, he would soon discover its impossibility if for no other reason than its sheer complexity. Hence, he actually models the RP. The model (ML) is a representation of the RP but with much less detail than the RP itself. Again, it should be recalled that the systems thinker is most of all interested in acquiring an *adequate* knowledge of the RP. Hence the model (ML) includes only those factors or elements that are absolutely necessary for a rough description of the RP. The modeling process is not a once-for-all exercise but should be conceived as consisting of several provisional models that adequately but roughly describe the scientist's conception of the RP. In summing up the three main parts of Figure 9–1, the real world perceived as the real phenomenon (RP) is studied as a system by first being converted into a model (ML). By working between the RP and ML the systems-oriented investigator will eventually arrive at a system (SY) which will be as complex as the real phenomenon (RP) itself. This last point cannot be overstated. It should be clearly but emphatically stated that systems thinking does not advocate conceptual simplicity. The apparent simplicity involved in the modeling process is only of a temporary nature. It is used as means of comprehending the complexity inherent in the RP. The ultimate "system" which will be used to deal with the real world situation *must be as complex as the real phenomenon (SY = RP)*. That is, of course, dictated by the universal law of requisite variety: one deals with complexity through complexity.

The term "management" has often been narrowed to mean the "running" of industrial activities (primarily profit-oriented industrial concerns) where management is contrasted to "labor." In governmental or nonprofit-oriented activities such as education or health, the term "administrator" or "director" is used. Whatever the term used to describe these particular individuals, the basic activity can be easily defined as "the burden of making choices about system improvement and the responsibility of responding to the choices made in human environment in which there is bound to be opposition to what the manager has decided."[2]

How does a manager make these choices or decisions regarding the improvement of the system? What is the system that is supposed to be improved? What is an improvement? These are some of the questions that students of management have been concerned with for the last three quarters of a century. The answers given to these questions thus far could be arranged in a continuum ranging from complete knowledge of the process of managing at the one end to complete ignorance at the other end. At the complete-knowledge end of the continuum, the manager is conceived of as a rational problem-solver confronted with the identification, calculation, and evaluation of alternative solutions to a given problem for which he has only limited resources, including time. Given these constraints, the manager is supposed to maximize the degree of improvement of a given system. Thus viewed, the task of the manager is fairly simple: choose that alternative that optimizes the objective function while satisfying all the constraints.

At the complete-ignorance end of the continuum, the manager's job is enveloped in an aurora of mysticism. No one really knows how a manager makes certain decisions; not even the manager himself! And, of course, if he himself does not know how he arrived at the decision to choose alternative x_{11} and not alternative x_{22}, how can anybody else?

As in most cases the truth lies somewhere in between these two extremes. While it is certainly true that not everything that a manager does (or does not do) can be expressed in a quantitative model, it is certainly untrue that nothing can be expressed in such a form. Furthermore, conceded that an executive or decision maker is not just an arithmetic-logical machine (a computer so to speak), it certainly is untrue that nothing that a manager does can be performed by such a machine. If that is the case, if part of what a manager does can be approached scientifically through somebody else's effort, and part of what he does cannot be dealt with by anybody except himself, how then do things get done in a real organizational situation?

The answer to the above question is really simple and can be stated as follows: Every managerial (decision-making) situation can be divided into two parts: (1) preparatory decision making and (2) action decision making. Each of these is an integral part of one and the same process of management (or decision making) as performed by two different persons. What one finds in real-life managerial situations is not a fact finder or fact organizer who does not make any decisions (staff)

---

[2] C. West Churchman, *Challenge to Reason* (New York: McGraw-Hill Book Co., 1968).

and a supercalculator and chooser or judgment-passing individual (manager), but rather one finds teams of equally competent and equally contributing individuals, each performing a portion of a real, complicated activity called, for the sake of a better name, organizational decision making. Both parties are tied into a symbiotic relationship of a bipolar nature, meaning that no one can survive (perform his task) without the cooperation of the other. Both parties aim at the improvement of the same system, each preparing the ground for the execution of the other's task.

In the remainder of this chapter this symbiotic relationship between the preparatory and action decision makers will be examined. To keep in touch with conventional nomenclature let us refer to the first functionary as the scientist and to the second functionary as the manager. The question now becomes, how do these apparently different individuals communicate with each other (or if they do not communicate, why not)?

THE MANAGER-SCIENTIST SYMBIOSIS

How do managers manage? A not improbable answer seems to be that managers manage by experience. Another would suggest knowledge. But is not experience to be equated with knowledge? Still other writers would advocate intuition as a factor governing the manager's decision-making and controlling actions. Kenneth Boulding even proposes that the manager, like every other human being, bases his managerial activities upon an *image;* his subjective knowledge of what he believes to be true. The image develops as a result of all the past experiences of the possessor of the image.[3]

Whatever the experts' consensus or lack of it, one can safely argue that managers manage by experience and knowledge. While there may be considerable philosophical or epistemological controversy about the importance of knowledge as compared to experience, the truth remains that the manager uses both for decision making and control.[4]

Experience alone (i.e., the process of personally observing, encountering, or undergoing something) will prove inadequate. So too will the mere acquaintance with facts, truths, or principles (i.e., knowledge). A combination of the two, however, will provide the synergy needed for the management of today's complex organizations. For such complex systems one desires to know what the system is; what the logical

[3] K. Boulding, *The Image: Knowledge in Life and Society* (Ann Arbor, Mich.: Ann Arbor Paperbacks, The University of Michigan Press, 1969).

[4] S. Beer, *Management Science: The Business Use of Operations Research* (New York: Doubleday Science Series, Doubleday and Co., Inc., 1968).

internal relationships and the external relationships are—those with the rest of the world; and how the system is quantified. Here personal experience is supplemented and augmented by knowledge of facts, principles, and laws applicable to similar systems.

The task of science is considered to be the systematization of knowledge about the world. This systematizing involves the codification of personal experiences and knowledge of mankind as well as the organization of knowledge and experience into a form transmittable to others. This culturally transmittable organized body of knowledge and experience serves as a prototype against which new ideas may be compared and into which they may be incorporated.

As managerial problems become more complex the need for a systematized body of knowledge becomes more imperious. Thus, the scientist can be of invaluable service to the manager. There is plenty of evidence to support this assertion. Early applications of the science of mechanics to production management proved very successful as far as the technological aspects of the processing of raw material and their conversion into marketable products were concerned. Equally successful has been the application of economics to the monetary (e.g., pricing, costing, and so forth) aspects of the production and distribution of economic goods. Considerably less successful has been the attempt to utilize social science principles and postulates to deal with the so-called human side of the enterprise.

Recently attempts have been made to apply higher mathematics and sophisticated statistical methods to the management of business enterprises and most recently to the management of social nonbusiness types of institutions. The extreme enthusiasm of the so-called *quant man* has been matched with an equally strong skepticism of the practitioner. The general consensus seems to be that problems of this sort will become amenable to solution if and only if (to use the profession's lingo) managers become mathematicians or mathematicians are turned into managers. Of course, it is quite unlikely that either event will occur.

The systems approach begins with the assumption that the manager and the scientist have something in common: their viewpoint of an organization as a system. The only difference is that the manager's knowledge of that system is based upon his own experience with the system, whereas the scientist's knowledge is based upon experience with *other similar* (i.e., analogous) systems. Thus the two have different conceptual frameworks that govern their study of the system. This difference in conception is primarily the result of differing educational background and training.

Consider, for a moment, the manager who is confronted with an inventory problem.[5] He knows a lot about it; he has been with this particular job for some time, and before that he had experience with similar systems generating similar problems. If he were asked to describe the inventory problem to someone else, his description of it would be, by necessity, through use of a conceptual model. This model would represent an accurate account of the situation but it would nonetheless be incomplete and somewhat "nonscientific" to the extent that it is too person-bound. In other words, although the manager usually knows what he is dealing with, still his being so close to the real phenomenon (RP) may result in a distorted view of it. The manager's conceptual model of the real phenomenon might be unnecessarily detailed in some respects while lacking sufficient detail in other respects. In any event, the manager's too close view may interfere with his grasp of the overall problem.

Now, let us introduce the scientist. As already noted, his conceptual model of the situation will be somewhat different. He most likely has had no previous experience with the particular inventory setting that the manager is concerned with. Nevertheless, he develops a conceptual model of the situation. His modeling approach will draw heavily upon a storehouse of knowledge and scientific experience. Most likely, he will begin to quantify his crude conceptual model right away. Most verbal statements that make up the manager's conceptual model will be replaced with some kind of number system. The scientist's model will have to be tested or experimented with. Upon the satisfactory performance of the scientist's model, it is then converted into a system for dealing with the real situation.

The modeling process described in the two previous paragraphs is diagrammed in the following schematic adapted from Beer's *Management Science*.

The management scientist's modeling process actually constitutes a hierarchy of models that begins with the manager's conceptual model (CM), goes through a state in which the model has virtually no resemblance to the original (it represents the scientist's way of conceptualizing), and finally ends up as a fairly realistic model that *can* be interpreted by the manager. It is this process of creating a rigorous scientific model that can be understood and appreciated by the practicing manager that makes the manager's transformation into a scientist and the scientist's transformation into a manager a *conditio sine qua non* of successful application of the systems approach to managerial problems and opportunities.

---

[5] This discussion is based primarily upon Beer's treatment of the subject. Beer, *Management Science.*

FIGURE 9–2
The Management Scientist Modeling Process

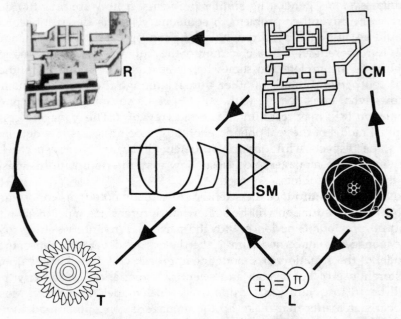

Science (S) contributes to the formation of the conceptual model and furnishes languages (L) that, together with the conceptual model, permit a scientific model (SM) of the real system to be formed. The scientific model furnishes techniques that permit the real situation (R), as well as the scientific model, to be manipulated.

Source: Adapted from Stafford Beer, *Management Science: The Business Use of Operations Research* (New York: Doubleday Science Series, Doubleday and Co., 1968), p. 69.

USING SYSTEMS SCIENCE TO DEAL WITH REAL-LIFE MANAGERIAL QUESTIONS

Rule I: Understand First, Diagnose Second, Prescribe Third

Managerial problems are to a large extent futuristic: they call for solutions whose implementation will affect future events. To the extent that the future is unpredictable, the manager must infer from incomplete information. His inferences from incomplete information will be the more realistic the more he understands the complete problem.

In discussing the differences between analytic and systems thinking, it was emphasized that the systems viewpoint advocated that the systems-oriented investigator should strive for an *adequate* knowledge of the whole relevant phenomenon rather than for an *accurate* knowledge of it. Now that we have reason to believe that managerial problems are by nature games of incomplete information, one can see the relevance of the realistic quest for adequate knowledge and the futile nature of the analytic thinker's drive for accurate knowledge.

Understanding managerial problems presupposes the realization that (*a*) life in an organic system such as a business enterprise is an ongoing process, (*b*) that one gains knowledge about the whole not by observing the parts but by observing the process of interaction among the parts and between the parts and the whole, and (*c*) that what is observed is not reality itself but the observer's conception of what is there.

Once this understanding of the whole relevant system is secured, then an understanding of a specific situation (problem and/or opportunity) is relatively easy. The diagnostic process should at least point to an array of alternative prescriptions of which the systems scientist must choose one.

It cannot be overemphasized that the use of systems science to solve managerial problems or to create managerial opportunities proceeds from understanding to prescribing and not the other way around. The most common practice of the management scientist and operations research expert usually follows the opposite direction. It is also well known that this attitude, in addition to being illogical, is also very unpredictable. Beer's aphorism seems as appropriate today as it was then:

> This warning about confusing particular solutions to stereotyped problems with a proper understanding of management science seems very necessary today. No one would confuse the pharmaceutical chemist's dispensing of a prescription with the practice of medicine. Yet there is today a widespread attempt in many industrial companies, and to some extent in government, to make use of the powerful tools of O.R. trade without undertaking the empirical science on which their application should alone be based. This is like copying out the prescription that did Mrs. Smith so much good, and hopefully applying it to oneself.[6]

Rule II: Conceptualize, Quantify, Simulate, Reconceptualize and Apply

In modeling and systematizing managerial phenomena the manager must go through a series of modeling attempts all of which are arranged in a thoroughness-abstraction hierarchy. The apex of this hierarchy is occupied by the most abstract thinking, while the basis of it houses the most detailed models of the managerial phenomena. Beer calls this hierarchical arrangement of models "cones of resolution."

---

[6] Ibid., p. 26.

FIGURE 9–3
Cones of Resolution

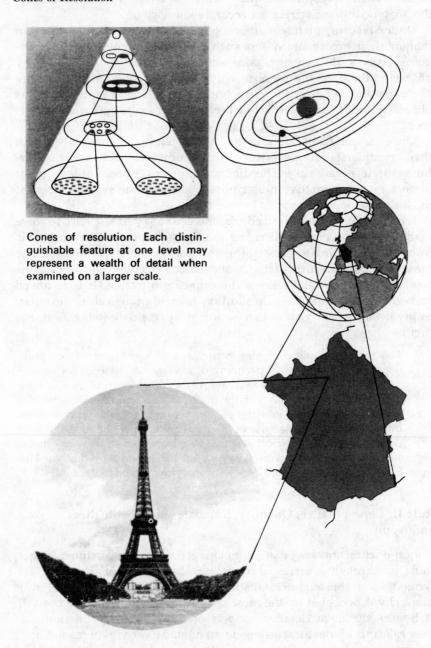

Cones of resolution. Each distinguishable feature at one level may
represent a wealth of detail when
examined on a larger scale.

Source: Adapted from Stafford Beer, *Management Science: The Business Use of Operations Research*
(New York: Doubleday Science Series, Doubleday and Company, 1968), p. 114.

Figure 9–3 shows how each level of resolution contains more and more detail. The tourist seeking to visit certain points of interest in the world may end his modeling process with a mere visit to the Eiffel Tower in Paris, France. An architect, however, might go further in his modeling process. Thus, the same objects or phenomena will occupy different levels within the same cone of resolution depending on the individual's interests.

Conceptualization of a managerial problem or opportunity begins at the top of the cone of resolution. Clearly at the top the level of abstraction is the highest and the degree of thoroughness is at a minimum. The primary concern of the systems scientist here is to comprehend the logic of the basic elements as well as the relationship among the elements—i.e., the logic of the system. Usually, the investigator would be very satisfied if he could discover a common yardstick by which he can measure the impact of one element's interaction with the other. In the business world we employ money as the common denominator of all relevant activities of the firm and its market as is shown in Figure 9–4. The inadequacy of this top view along with its definitely monetary flavor becomes clearer the more one descends the cone of resolution from the balance sheet toward the isomorphic relationships between the factory and the market.

In summary, this is then what we mean by *conceptualization*—understanding and organizing the interactions among the elements making up the phenomenon under scrutiny into a logical network of relationships in such a way as to reveal the direction of the underlying structure.

This general systems theory–like framework is then converted into a quantitative network whereby the logical relationships are assigned economic values (i.e., cost and/or benefits). In this way the original abstract arrangement of relationships becomes an econometric model (i.e., a mathematical structure of economic relationships). The systems scientist is now ready to experiment with these highly particularized econometric models. Experimentation with a model over time is referred to as simulation.

In simulating a particular model the investigator deliberately changes certain parameters of the model, certain key variables or relationships, in the hope of gaining some knowledge of the degree of sensitivity of the model to such changes. The numerous books written on the subject of simulation indicate that the process is useful, albeit not simple. However, the basic concepts of simulation which are of interest to the manager are simple. Given that every model is based upon certain assumptions regarding an uncertain future which the model is supposed to organize and eventually predict, how would the

FIGURE 9–4
Cones of Resolution for the Firm-Market Interface

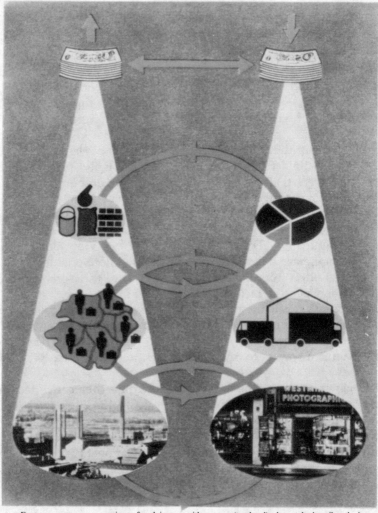

For some purposes comparison of cash income with expense (top level) adequately describes the interaction of a company with its market. For other purposes the proportion of income derived from each product
is relevant, for others the number of trade representatives, etc., is required, and so on down the cone of
resolution until we come to the actual company and market.

Source: Adapted from Stafford Beer, *Management Science: The Business Use of Operations Research*
(New York: Doubleday and Company, 1968), p. 112.

model's organizing, heuristic, and predictive power change in the event
of changes in some of the assumed conditions?

In general there are three kinds of simulation: (1) human simulation,
(2) computer simulation, and (3) man-machine simulation. The first
kind of simulation is really nothing else but the Hegelian method of
inquiry known as the dialectic method of thesis-antithesis-synthesis.

This kind of simulation can range all the way from the practicing session of a sports team, animated war battles, or managerial meetings to sophisticated "sensitivity analysis." In managerial meetings the process of simulation begins by asking the so-called what if questions to proposed plans of action. The team proposing the plan will recompute the model's most likely performance under the different conditions imposed upon it, and so on.

Computerized simulation involves essentially the same process as human simulation, the only, but big, difference being that changes in certain parameters are initiated by the computer, which in turn recomputes the most likely results of these changes. This kind of simulation, although it can be very interesting as well as very informative, is generally of scant interest and small utility to the practicing manager because of the mysticism attached to the internal workings of the machine. As a result of this romanticized attitude, management's reliance on computer simulation results is still very limited. This attitude is to some extent reinforced by the simulation expert's unwavering preoccupation with simulating more and more abstract problems which have an intrinsic interest for him but are of little practical consequence for the manager.

The third kind of simulation, man-machine simulation or business gaming, is of paramount importance to the practicing manager. The logic underlying business gaming is essentially the same as in the two previous kinds of simulation: explication and understanding of the process of problem solving via experimentation of a model of this process as well as testing the impact of possible variations in certain assumptions upon model outcome.

In gaming, the investigator takes the managerial decision function (the decision to change certain parameters) out of the computer program and restores it to the manager, while the computation of the possible results of the decision is left to the computer. The time between the change in a parameter (managerial decision) and the outcome can vary from several hours in the more traditional games to instant replay in the most advanced simulation games (on-line computer management interaction). Figure 9–5 represents a typical business game situation of a four-member four-team management game situation. The fact that the manager initiates changes rather than being forced to accept certain arbitrary and random variations in certain market or firm conditions takes a lot of the mysticism out of the computer simulation, thereby making it more realistic and believable.

In summary, Rule II again tells us that one necessarily begins with conceptualization and ends with conceptualization. As in all phenomena the manager's intellectual tasks of policy setting, decision making and control are, naturally enough, cyclical. In cyclical phe-

FIGURE 9–5
Business Gaming

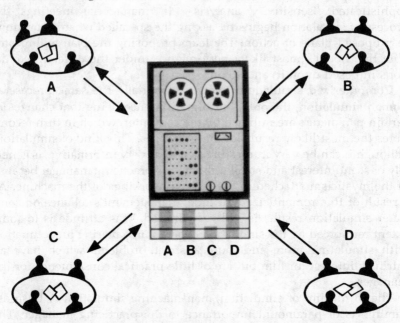

A competitive simulation situation. Four teams of "managers" operate competitive companies, trying market and production strategies, etc. The computer, furnished with a model of the complete industry and market, feeds back information to the "managers" and also keeps score.

Source: Adapted from Stafford Beer, *Management Science: The Business Use of Operations Research* (New York: Doubleday and Company, 1968), p. 86.

nomena, as Heraclitus averred centuries ago, the beginnings and the ends are the same. Again, one must not begin delving into complex managerial situations involving thousands of relationships by quantifying first, simulating second, and applying third; rather one should begin with rigorous thinking about the logical relationships among the elements of the whole, then quantify them, and so on.

Before discussing in the last section of this chapter a paradigm of systems science in action, one ought to point out again the danger of excessive and premature analysis and quantification antecedent to logical conceptualization of the problem. Managerial problems by nature involve human experience, and when one deals with human experience, what L. Mumford once said about the so-called scientific method (analytic thinking) and its ability to deal with total human experience is still relevant:

> Admittedly the sciences so created were masterly symbolic fabrications: unfortunately those who utilized these symbols implicitly believed that

they represented a high order of reality, when in fact they expressed only a higher order of abstraction. Human experience itself remained, necessarily, multi-dimensional: one axis extends horizontally through the world open to external observation, the so-called objective world, and the other axis at right angles, passes vertically through the depths and heights of the subjective world; while reality itself can only be represented by a figure composed of an indefinite number of lines drawn through both planes and intersecting at the center, the mind of a living person.[7]

THE SYSTEMS SCIENCE PARADIGM

The task of systems science, like any other science, is to develop and maintain some kind of a consensus among its practitioners regarding (1) the nature of legitimate scientific problems and (2) the methods employed for dealing with these problems. Kuhn employed the term paradigm, long familiar to students of classical languages, to connote "universally recognized scientific achievements that for a time provide model problems and solutions to a community of practitioners."[8] These paradigms, then, represent basic milestones in the development of a discipline.

Just as the invention of the telescope when combined with Newton's and Leibniz's calculus was significant for the development of classical physics, so too are the inception and development of systems science for the study of organizations. And just as F. W. Taylor's work at the beginning of this century provided a working paradigm for managers, so too does systems science in 1970s provide a paradigm for sound management of complex organizations. However, the paradigm of systems science is not a blueprint for application of systems to organizations; a rather substantial amount of exciting mop-up work must first be done in the form of matching the facts to the paradigm and in further articulation of the paradigm itself.

There are several foci for factual systemic investigation of organizations and these are not always nor need be distinct. First, there is the question of the philosophical predisposition of the systems enthusiast/practitioner: his is a world of organic-open systems. Two main processes are of paramount importance in studying organic-open systems: growth and control. Growth is a necessary condition for the survival of any system; at the same time, control (the ability of a system to sustain

---

[7] Lewis Mumford, *The Myth of the Machine: The Pentagon of Power* (New York: Harcourt, Brace and Jovanovich, Inc., 1970), p. 74.

[8] Thomas S. Kuhn, *The Structure of Scientific Revolutions*, 2d ed. (Chicago. University of Chicago Press, 1970).

a rate of growth in keeping with its capacity and the environment's tolerances) is a necessary condition for balanced growth.

Second, the organic system is investigated from the holistic viewpoint. However, operationally speaking, holism does not necessarily imply that the systems scientist must investigate everything about everything. What holism implies is that enough thought will be given to determining the critical variables influencing the growth and control patterns of an organization as well as to establishing ways of monitoring the critical parameters in the organization-environment interface. Holism means, to paraphrase Fuller, that one should begin with the universe. Again, the holistic approach does not imply that the manager should be concerned with everything that goes on within his department or division or the whole company; rather it demands that one always go one step beyond what up until now has been thought of as being satisfactory—e.g., step out of the circle that the job description has drawn.

Third, the apparently insurmountable task of holistically investigating an organization as an open-organic system under constantly changing conditions is facilitated through modeling processes. The modeling process begins with a considerably gross conceptualization of the system and ends up with a more or less precise model of an econometric nature.

Finally, the last focus of the systemic investigation has something to do with the most likely outcome of this kind of ambitious endeavor. The most likely outcome will involve an understanding of the focal situation as it relates to the rest of the organization and its environment. Once this understanding is achieved, certain quantification techniques can be utilized to calculate possible outcomes of proposed courses of action or inaction.

THE PARADIGM ARTICULATED

Phase I: Conceptualization

All too often the statement is made that "the systems approach (or the systems concept or systems in general) has not developed enough to lend itself to employment in rigorous study of organizations." This statement is, of course, true only if one perceives systems as a grab bag of unrelated clichés thrown together in a list of items under the heading of systems characteristics or systems attributes or simple buzz words. From the moment, however, that one begins to look at the systems approach as a theory or as a grown discipline with its philosophical premises and concepts (e.g., information, positive and negative feedback), hypothesized or propositional relationships among the concepts and its approach (e.g., holism, modeling) then the systems approach is quite more mature and operational than most theorists and practitioners tend to think.

Let us illustrate the point of systems operationally by examining once again the problem of assessing the relationship between the firm and its environment, using the concept of the cones of resolution discussed earlier in this chapter. To begin with, let us reiterate the steps in the systems scientist's thought process. He begins by looking at the organization as an open-organic system which is in constant interaction with its environment. His holistic approach to the study of the organization dictates that he should focus on both the organization and its environment as they interact with each other. To deal with this complexity he is forced to model this interaction. Finally, the researcher tries to understand as much of the interaction process as possible without regimenting the phenomenon to a meaningless two-member relationship.

Let us begin at the top of the cone of resolution (Figure 9–6A and B). There the organization-environment interface is pictured as just two boxes interacting with each other via two feedback loops (Feedback 1 and Feedback 2 in Figure 9–6). This, of course, is the easiest and the most economical way of gaining an understanding of what is involved. However, the informativeness of this model is exceedingly limited. Both the organization and its environment are represented by T-accounts ($) indicating the financial positions of both subsystems vis-à-vis each other.

The second level in the cone of resolution focuses on the environment in greater detail. Thus, the box "environment" is dissected into some of its most important (from the firm's viewpoint) sectors. The firm produces certain products which are sold through the market (MKT). Production is realized by combining certain factors of production (e.g., technology; sales are accomplished by competing with certain rivals (e.g., world competition) and by complying with certain regulatory agencies (e.g., Environmental Protection Agency [EPA]).

In all these interactions the firm will choose a specific relationship within a certain environmental sector (will make an offer, so to speak). The environmental sector will then indicate whether the proposed state satisfies its needs (which are, of course, organizational constraints). In cases of incongruences, situations where the proposed state does not fully satisfy the market's desires, the firm must propose another state, and so on. The point made here is that stated relationships between the two subsystems are commonly determined rather than arbitrarily chosen by either subsystem. Dominant relationships are only in the short run viable; in the long run dominance must give way to cooperation. The firm strives for a dynamic equilibrium (an equilibrium under

constant change of the rates and levels determining it) between itself
and the sectors of the external environment.

The third level in the cone of resolution involves a further elaboration of the sector of the external environment labeled MKT. Here the
systems manager can see quite a bit more about the market and its
activities. Of course, many of the boxes are still pretty much "black

FIGURE 9–6
**Modeling of the Environment-Organization Interaction System through the
Cones of Resolution Technique**

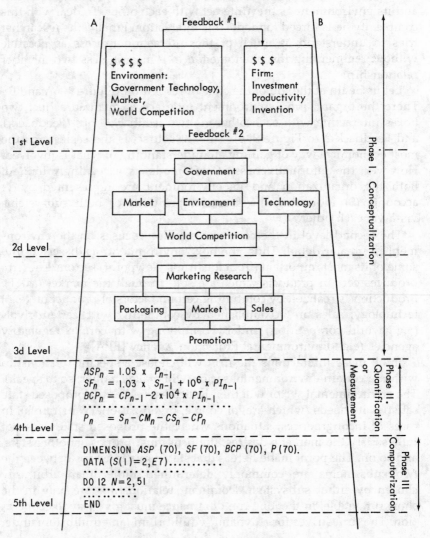

boxes" to the extent that the manager does not know everything there is to know about, let us say, the competitor's activities so that he can design an effective strategy. However, he knows enough about them to be able to identify them as well as conjecture their possible impact. Looking at the Figure 9–6 on the third level, one can perhaps identify possible decision/control points, those critical effective parameters which require extensive monitoring.

Phase II: Analysis and Measurement

So much for conceptualization. If that was all there is to systems, then, of course, no scientific status could possibly be claimed for it. While the three-level modeling process that goes on in any conceptualization of a system is necessary, it is by no means sufficient. One more detailed level of boxes and lines connecting them will do nothing more than confuse both the model maker and the user. The stage has now been set for still another modeling process to begin, although of a slightly different nature.

Quantification begins when the need for measurement of changes in the state of systems elements has arisen. Mensuration has been man's preoccupation from the beginning until now. In organizations measuring inputs (e.g., cost) and outputs (e.g., revenues) has been one of the earliest applications of science to the management of organizations known as accounting and/or finance. S. Beer seems to think that "the origins of a scientific approach to management were connected with the measurement process." One cannot but agree that measuring changes in the firm's internal and external states is as old as organizations themselves.

At the beginning of this century another quantification attempt got off the ground. This time the logic of measurement concentrated on quantifying the workers' contributions to the firm's goals. The rationale behind this was that if one knew how to measure the potential output of a worker, then, at least theoretically, one would be able to utilize that factor more effectively and efficiently. Time and motion studies were once very popular and, of course, still are important. Despite the innumerable criticisms, measuring workers' contributions did show that a better combination of man and tool can indeed increase productivity. Although only the physiological side of the human being was measured, the outcome of this measuring process did result in better tool or machine design, thereby making the exertion of human muscle energy less and less necessary. The science of biomechanics epitomizes the giant strides made since the early time and motion studies first made history.

At the beginning of the second quarter of this century the measuring efforts in organizations were concentrated on the nonphysiological aspects of the human factor of the enterprise. Configurations of monetary and nonmonetary incentives were designed to motivate workers and to some extent lower supervisory personnel so as to utilize a greater portion of their potential in achieving organizational objectives. Just as scientific management in the early 1900s assumed that healthy workers operating a better designed tool will be more productive, so did human relations assume that happy employees working within a better and happier environment will utilize their potentials more productively.

The second half of the 20th century ushered in another measurement process. This process aims at assessing the contributions of (1) managerial personnel and (2) information toward accomplishing organizational goals. The measuring process is, of course, of a slightly different nature. It aims at measuring the "measurer." For this reason the process is considerably more difficult and delicate than the previous processes. Its domain of measurement more or less encompasses the entire organization as it relates to its environment. Since the expected payoffs of this measurement process are much greater, one ought to take a closer look at this measurement process. The implications of measuring the measurer are so great and the cost of doing so so high that it behooves every modern manager to reexamine the entire measuring process.

Thus far we have been using the terms measurement and quantification more or less interchangeably. The reason for this is that measurement is frequently defined as the "assignment of numerals to elements or objects according to [certain] rules."[9] From such a definition one can easily get the impression that measuring means quantifying, as most of the literature, in what is emphatically called quant methods, seems to indicate. From this definition one might conclude that whatever cannot be quantified cannot be measured; or to carry the logic a step further, what cannot be thusly measured cannot be of any consequence for management. Practicing managers and administrators do, however, know better. They know that quantification is only one way of measuring. Another way of measuring, known as qualitative measurement does exist and is as meaningful, and under certain conditions, as useful, if not more so, than quantitative measurement.

With this in mind, it may perhaps be better to refer to the next (fourth) level of modeling in the cone of resolution depicted in Figure

---

[9] For a more detailed treatment of the subject of measurement see: C. West Churchman and P. Ratoosh (eds.), *Measurement: Definitions and Theories* (New York: J. Wiley & Sons, Inc., 1959); also C. W. Churchman, "Why Measure?" in *Management Systems*, P. P. Schoderbek (ed.), 2d ed. (New York: J. Wiley & Sons, Inc., 1971); also R. W. Shephard. "An Appraisal of Some of the Problems of Measurements in Operation Research," in P. P. Schoderbek (ed.), *Management Systems*.

9–6, as the measurement process rather than the quantification process. Operating at that particular level the systems scientist becomes a measurer. Acting in that capacity, he must decide:

1. In what language he will express his results (language).
2. To what objects and in what environments his results will apply (specification).
3. How his results can be used (standardization).
4. How one can assess the "truth" of the results and evaluate their use (accuracy and control).[10]

In Figure 9–6, fourth level, a small portion of Bonini's Model is illustrated.[11] The model indicates that the aspired profit for the period or year n (ASP_n) is equal to the actual profit of the previous period (P_{n-1}) multiplied by a factor of 1.05. Estimated sales for the same period n (SF_n) are equal to actual sales of the previous period (S_{n-1}) multiplied by a factor of 1.03 plus a pressure index which reflects the growth of the industry (PI_{n-1}) times 10^6. Budgeted production administrative expenses for period n (BCP_n) equal actual production administrative expenses for the previous period (CP_{n-1}) minus 2 multiplied by the pressure index times 10^4. Finally, actual profit for the same period n (P_n) equals actual sales (S_n) minus actual manufacturing cost (CM_n) minus actual sales administrative expenses (CS_n) minus actual production administrative expenses (CP_n).

A further elaboration on measurement and the measuring process would carry us beyond the intended scope of this work. For our purposes, it will suffice to restate that (1) the function of measurement is to develop a method for generating a class of information that will be useful in a wide variety of problems and situations. This method may involve either a qualitative assignment of objects to classes or the assignment of numbers to events and objects; in most instances, both quantitative and qualitative measuring processes will be employed; and (2) that the process of measurement is facilitated by rigorous conceptualization along the three levels within the cone of resolution and that these will definitely help or hinder the descent to the next level of computerization.

Phase III: Computerization

Ideally the ultimate product of Phase II will be a mathematical model that will be translated into a computer-consumable project. It should

---

[10] Churchman, "Why Measure?" p. 123.

[11] The Bonini Model is described in C. McMillan and R. F. Gonzalez, *Systems Analysis*, rev. ed. (Homewood, Ill.: Richard D. Irwin, Inc., 1968), pp. 424–28.

be recalled that mathematical notation is the language understood by the machine. Therefore, qualitative considerations must be inferred from the functioning and the output of the model by the human being who compares the outcome of this phase of the process with the aspirations and expectations formulated during Phase I (conceptualization). Computer simulation provides the least expensive way of performing this comparison. The outcome of this comparison, expressed in terms of simulation results, will lead to either a reconceptualization (back to Phase I) or to remeasurement and quantification (back to Phase II) or to both. In any event, the ultimate outcome of Phase III will be a computer program along the lines of the portion depicted in level 5 of Figure 9–6.

To recapitulate, it is imperative that the systems scientist proceed from conceptualization to computerization and not vice versa, as is indicated by the upside down pyramid in Figure 9–6. A better grasp of the problem along the lines of Phase I will enable the researcher to develop a better measurement method and a better computer program rather than force the problem into a preconceived computer program.

As can be seen from Figure 9–6, the systems scientist initially loses quite a bit of the real phenomenon of the environment-organization interaction system because of the abstract and aggregative nature of the manager's conceptual model. These losses, however, are rather moderate when compared to those that would be incurred were the manager to follow the "model-up" method rather than the "model-down" approach which is advocated here. The conventional modeler will begin at the base of the cone of resolution and then will proceed upward.

The typical analytic thinker's cone of resolution can be imagined as being the mirror image of the systems scientist's cone of resolution. A quant man's model will most likely begin at the fifth level of the cone of resolution depicted in Figure 9–6. In working his way up (model-up) he will incur certain losses attributable to his narrow and precise quantification of certain relationships of the organization-environment interaction system. However, unlike the outcome for the systems scientist, his losses tend to increase as he ascends his cone of resolution until eventually he crosses the boundary of reality (as perceived by the manager). From there on he loses touch with reality, eventually reaching a point of maximum irrelevance by pursuing certain solutions to problems which he alone can understand and interpret.

It is unfortunate that some, parading under the banner of operations research and management science, use their techniques as procrustean beds upon which managerial problems are amputated and distorted so that they exactly fit the sacred box of the quant man. The recently popularized discontent of members of The Institute of Management Science (TIMS) with the Institute's orientation attests to this phenomenon.[12]

SUMMARY AND CONCLUSIONS

Figure 9–7 summarizes the logic of the application of the systems approach to the study of real world phenomena. In general, the logic is the same as in Figure 9–1 at the very beginning of this chapter: a real

FIGURE 9–7
The Application of the Systems Approach to the Study of Real-World Phenomena

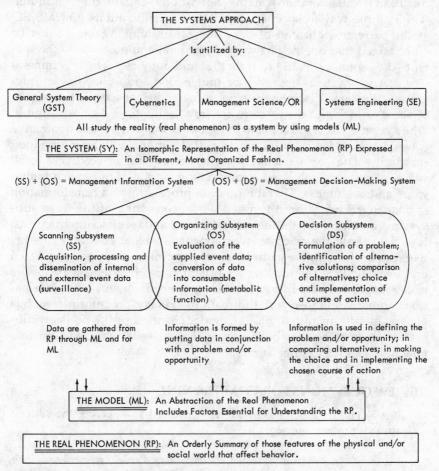

---

[12] See, for instance, D. F. Heany "Is TIMS Talking to Himself?" *Management Science,* vol. 12, no. 4 (December 1965), pp. B 146–155.

phenomenon (RP) must necessarily be studied via a model (ML) which is used in the design of a system (SY) which in turn represents organized and systematic reality. Two novelties are added in Figure 9–7: (1) the several subapproaches or subdisciplines under the name of the "systems approach" arranged in a philosophy-science or qualitative-quantitative continuum; and (2) the three subsystems of the grand system — viz., the scanning subsystem, organizing subsystem, and the decision subsystem which were originally introduced in Chapter 6.

A systems-oriented manager or student can design a system for studying an organization as it interacts with its environment by drawing from any of the four subdisciplines beginning with the general-qualitative considerations at the philosophy end of the continuum (GST) all the way to specific quantitative considerations (OR?MS?SE) at the extreme right end of the same continuum. Again, it must be emphasized that one begins with the general and proceeds to the specific — i.e., from left (GST) to right (SE) and not vice versa. It is imperative that the systems approach be understood and utilized as an integrative "linkage" discipline and not as a grab bag of specific techniques of some "quick and dirty" steps for troubleshooting. As Laszlo put it, "the system thus created [containing both quantitative and qualitative considerations] feeds on information."[13] Its inputs will be information (primarily data) of a relatively crude nature and of relatively small value; its outputs will also be information, although of a much higher value and usefulness. This information processing and transformation system functions as an integrated whole consisting of the three subsystems that interlock through feed-forward and feedback mechanisms. Raw data about the external environment as well as about the internal working of the firm are gathered by the scanning subsystem, analyzed and evaluated by the intelligence organizing subsystem, which more or less separates data into those having immediate and high information content, and those with future utility. Finally, the information thus generated is transmitted via the regular channels of the decision subsystem. It is this "metabolic power"[14] of a well-organized firm that guarantees its long-range survival.

REVIEW QUESTIONS FOR CHAPTER NINE

1. Briefly outline the three basic steps involved in applying the systems approach to management.

---

[13] E. Laszlo, *The Relevance of General Systems Theory*, Papers Presented to Ludwig von Bertalanffy on His Seventieth Birthday, G. Braziler's Series, *The International Library of Systems Theory and Philosophy* (New York: George Braziler, 1972).

[14] Stafford Beer, "Managing Complexity," in *The Management of Information and Knowledge* (Washington, D.C.: U.S. Government Printing Office, 1970), pp. 41–61.

2. What is the role of the "model" in the application of the systems approach to the study of the real world?

3. It is postulated that in managing real-life organizations the manager or administrator is assisted by the management scientist. Scientists have little knowledge of managerial situations as they unfold in real life. The manager, on the other hand, usually has little knowledge of scientific techniques. How do the two manage to solve managerial problems?

4. Discuss the two "rules" which are to be followed when using science to deal with real-life managerial questions.

5. Explain the concept of "cones of resolution" and show how it can be used in studying or examining the acquisition of a small retail shop by a big chain store operation.

6. What is the systems science paradigm?

7. Briefly outline the three basic phases of the systems paradigm.

8. An insurance company headquartered in the Eastern part of the United States is contemplating the "branching out" into a new venture. You are part of the team which is assigned to take the systems approach to this managerial problem. Briefly outline the basic steps of this approach to your team members.

9. XYZ Electric Company, faced with an increased demand for electricity, is planning the construction of a new power plant. The management desires to take the systems approach to this investment. You have been hired as a consultant for this project and asked to outline the conceptualization, analysis, and measurement phases of this problem.

10. Continue the above problem with the final phase of computerization. Name some basic management science techniques which you find can be useful in the above situation.

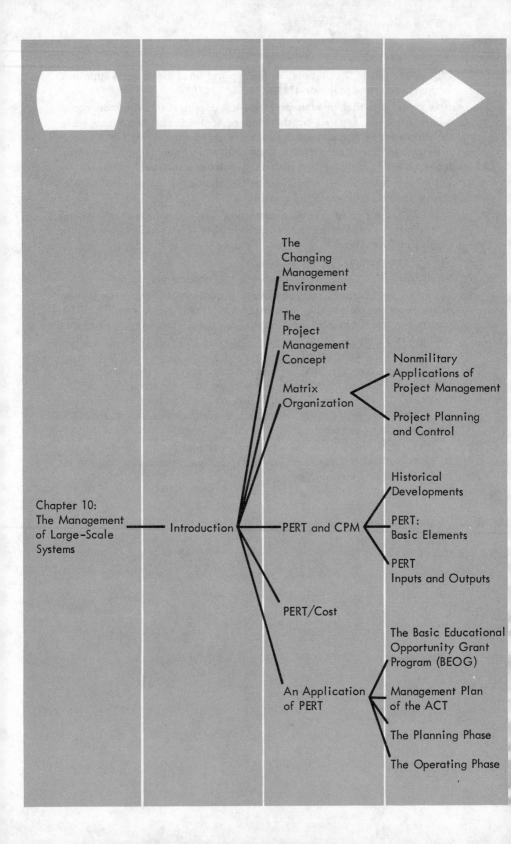

Chapter 10:
The Management
of Large-Scale
Systems

Introduction

The
Changing
Management
Environment

The
Project
Management
Concept

Matrix
Organization

Nonmilitary
Applications of
Project Management

Project Planning
and Control

PERT and CPM

Historical
Developments

PERT:
Basic Elements

PERT
Inputs and Outputs

PERT/Cost

An Application
of PERT

The Basic Educational
Opportunity Grant
Program (BEOG)

Management Plan
of the ACT

The Planning Phase

The Operating Phase

Chapter Ten

The Management of Large-Scale Systems

As a management tool Information Technology will extend management's perception of costs and benefits and affect both the concept and measurement of business profit.

George Kozmetsky

INTRODUCTION

Most students of business would agree that the basic management functions include planning, organizing, directing, and controlling. The first of these, planning, involves the setting of goals, while control relates to the measurement and correction of the deviations from the goals. Organization is concerned both with the way management structures the tasks to be performed and with the men and resources to be coordinated. The directing function deals with the form of leadership used to direct the activities of personnel.

In this chapter attention will be limited to the management of systems of special type — large scale one-time projects. Since a significant portion of the research funds expended each year are devoted to these special projects the management of these systems merits special consideration.

Of special interest here is the viewing of these projects as systems exhibiting interrelatedness and interdependencies. Some of the earlier methods for management, displaced by the current one, employed an analytical approach and ignored the interdependency aspect. As the discussion unfolds one will be able to see that some of the tools employed in the management of these systems were explicitly developed to take account of the interacting components of the systems (projects).

265

The management of these large-scale systems has been termed project management. The three specific objectives of this chapter are the following:

to acquaint the reader with the project management concept,

to present in a fairly comprehensive fashion the tools employed in project management, and

to apply the tools to a specific project for greater familiarity. However, before delving into the subject matter proper one could profitably look at the factors which precipitated the shift to project management of today.

THE CHANGING MANAGEMENT ENVIRONMENT

There is little doubt that modern industrial and governmental organizations have increased in complexity because of recent technological advances. Technology has not only ushered in new products but has created entirely new fields such as nucleonics, nuclear energy, computers, genetics, and oceanography. A phenomenon of modern society is the large-scale industrial organization engaged in equally large-scale projects and programs.

It is not surprising that much of this newly developed technology is the result of specific needs geared to and sponsored by the federal government. For example, in 1971, the federal government provided $15 billion or 53 percent of the nation's research funds.[1] A sizeable proportion of this has obviously been allocated for space exploration and defense systems. With increased technology has come increased complexity of projects, which often involve many parts of an organization and even many organizations. (The Polaris Project, part of the Fleet Ballistic Missile Program, involved over 2,000 subcontractors alone.)

Up to now organizations simply never had to deal with projects of such magnitude and complexity. Existing management technology was found to be inadequate for the management of such highly integrated projects and the demand for techniques which viewed such projects in a total systems perspective was clearly felt. The adoption of the systems approach has caused significant changes not only in basic decisioning, but also in how decisions are made and by whom. In earlier chapters it was stated that many of the major societal problems of today — pollution, transportation, the energy crisis — demand the systems approach, for it is only through such an approach that the effects of policy decisions can be realistically assessed.

Probably the first recognized application of systems on a large scale

---

[1] *National Patterns of R & D Resources — Funds and Manpower in the United States, 1953–1971* (Washington, D.C.: National Science Foundation, 1970).

was the Manhattan Project, which required the efforts of thousands of scientists. While technical personnel provided the technology, project managers were equally influential in bringing the atomic bomb project to fruition. Ever since the Manhattan Project, virtually every major weapon system has demanded the systems approach for its design, development, and production. The Department of Defense alone typically monitors over 3,000 projects at any one time. Even the layman is familiar with many of these projects — Atlas, Mercury, Minuteman, Polaris, Apollo, Nike, and the Telstar and Echo satellite communication systems. The Army, Navy, Air Force, and NASA all employ the systems approach.

THE PROJECT MANAGEMENT CONCEPT

Organizational structures are generally either of the vertical (product) or horizontal (functional) type. Organizations that are heavily technology-oriented (e.g., aerospace) typically are of the functional type. Such an organization can utilize specialty areas most effectively. Pooling specialized resources allows such resources to cross particular project lines or programs. On the other hand, some of these benefits are lost with this type of structure if there is a multiplicity of projects under development at any one time, each with its own critical time, cost, and performance elements. Since some projects are deeply rooted in a state of technology which at that point in time may not be fully developed or tested, the necessarily extremely close coordination and compliance with imposed completion dates make the management function very unwieldly.

Before the advent of project management, organizations high on technology utilized the functional structure with responsibility fragmented among engineers, top managers, marketing executives, the controller, and many manufacturing and other functional areas. With this type of structure the single critical problem was the lack of coordination and clear-cut lines of responsibility.

In project management, however, a single individual is vested with full authority and responsibility for the project. This centralization of authority allows the project manager to cross lines of authority vertically as well as horizontally. He is thus in a position to resolve technical, financial, production, or management problems.

While elements of project management existed many years ago, its first recognized application was on the Atlas Project in 1959. Because of the long lead time typically experienced in the production of an operational system and the apparent missile lag experienced by the United States, the basic objective was to develop an intercontinental ballistic missile in the shortest possible time.

Because of the success that accompanied this initial application, the government was very interested in using project management on future contracts and looked favorably on those firms that employed it. As a result, in the late 1950s and early 1960s many firms altered their organizational structure not only in response to the contractor-selection criteria used by the government but also to take advantage of the unusual benefits the new structure provided.

In its basic form the project management concept superimposes a vertical project organization on the horizontal function as shown in Figure 10–1. Basically, the only modification of the functional structure is the insertion of the triangle which represents a particular project. A single person is designated as project manager and vested with full responsibility for the completion of the project. In this way there is provided a focal point for decision making as well as for planning and control. The project manager can utilize the functional specialists in each of the several areas for carrying out the required work effort. What is unique about this concept is its dual reporting relationship—i.e., the functional managers are responsible both to their functional supervisors and to the project manager. Such a division of authority is a radical departure from accepted management principles. This arrangement appears to violate the unity of command principle which the classical organizational theorists and behavioralists have promulgated for many a year. In reality, however, the principle is not totally disregarded since project managers have authority over the functional managers only with regard to the activities to be performed but not with regard to *how* they will be performed. This latter qualification apparently safeguards this sacrosanct principle.

The authority of the project manager may also extend outside the

FIGURE 10–1

Projection Management Organizational Structure

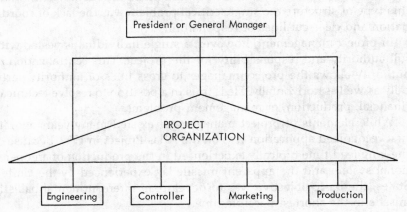

organization to those firms participating in the project as subcontractors. Since his responsibility extends to the three critical aspects of the project (schedule, cost, and technical performance), he can and does probe into the many related areas of concern. This crossing of functional lines is not without some difficulties. As noted above, support personnel are responsible to their own functional managers, who determine their promotion, pay, and evaluation. Since functional personnel return to their own home after the termination of the project, there is an obvious loyalty to functional bosses.[2]

Various terms have been used to describe the project manager concept such as program management, system management, and weapon-systems management. Although there are slight differences, the terms have a common element in that they all deal with one-time projects.

The popularity and durability of this technique is evidenced by the 37 major projects of the Army Materiel Command that in 1973 dealt with such end items as munitions, weapons, aviation, electronics, and missiles: all employ it.

MATRIX ORGANIZATION

The combination of project organization and functional organization has also been termed *matrix management*. A purely project organization obviously duplicates staff and facilities for each of the several projects. While cost and schedule performance may be better on individual projects, technology may suffer in the long run because no one group is responsible for long-range developments. On the other hand, a purely functional setup makes allowance for developing technology while sacrificing schedule performance on individual contracts. Such a setup provides for flexibility in assigning specialists to projects when and where they are needed, but overall project coordination is lacking.

What was needed then was an organizational structure that would allow technology not only to grow at a rapid pace but also to do so within specified time and cost constraints. The results of such efforts constitute what is here termed *matrix management*.

Figure 10–2 shows a matrix-type organization in which R and D, Engineering, Quality Assurance, and so forth each represents a functional unit of the organization. Personnel from each of the functional units may be assigned to specific projects for their duration. Under such an arrangement the permanent organizational structure is not altered, since some personnel concerned with pure technology may be perma-

---

[2] A thorough discussion of the project management concept appears in *System Analysis and Project Management*, by D. Cleland and W. King (New York: McGraw-Hill Book Co., 1958). See also J. Baumgartner, *Project Management* (Homewood, Ill.: Richard D. Irwin, Inc., 1963).

FIGURE 10–2
A Matrix-Type Structure

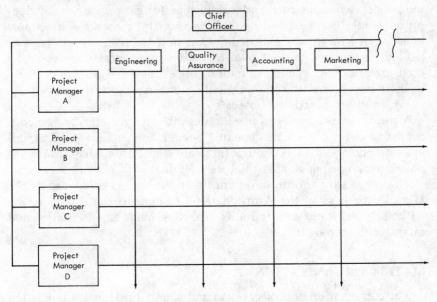

nently assigned to work within a functional area. One most often finds the matrix organization, however, on smaller projects where the functional specialists either typically serve the project and then return to their unit or serve several projects while based in the unit.

Nonmilitary Application of Project Management

Project management can be widely applicable even in the civilian area. The starting up of a new plant, the launching of a new product, the initiating of a special research project, the installation of a new computer system, model changeovers and construction projects – all are illustrative of project management applications. The staging of the Olympics, the setting up of national pavilions at world fairs, and the hundreds of other similar projects which demand close supervision and coordination among diverse personnel all make use of the project management concept or some variation of it. Many of the firms that bid on contracts issuing from governmental agencies (federal or state) often employ this same approach. It is even now speculated that newer applications will extend to urban renewal problems, health care programs, welfare systems, and to other social issues of the present. We have yet to see what the future holds in store for project management.

Project Planning and Control

In order to meet an established schedule the principal efforts of the project manager are directed toward the planning and control of the many and diverse tasks of the individual projects. To properly control, coordinate, and integrate the composite dimensions of the various projects, the project manager must know the current state of each. Here we will examine some planning and control techniques for large-scale projects. PERT (Program Evaluation and Review Technique) and CPM (Critical Path Method) both employ the network as their basic foundation. Networks enable one to view a project as a system, since they depict the interrelationships among the tasks to be performed. As with any system, a critical task is the identification of components, their interaction and interrelationship, and the feedback processes. While the network techniques presented here had their origin in large-scale research and development projects, their applicability to other endeavors is quite broad. Several of these applications will be noted later.

To appreciate the contribution of network techniques to both governmental and industrial projects, one ought to recall, at least briefly, the historical antecedents of the managerial technique that goes by the PERT acronym. New techniques, like so many discoveries and inventions, rarely spring up full blown from virgin soil. Rather they are advances in or improvements on previously known and tried techniques. PERT is no exception.

PERT AND CPM

Historical Developments

Since the end of World War II technological advances both here and abroad were both rapid and far-spreading. One sensitive area of national and international concern was the weapons and support systems of the big powers. These had become exceedingly complex. With the Cold War in progress with its threat to our national security (as apprehended by the government) a new weapons system, it was felt, had to be developed and made operational in a minimum of time.

The inadequacy experienced in managing was confined not only to defense work on weapons and support systems but also to other types of work. The type that seemed most difficult to plan and control was the special project. The more complex the project, the more difficult it was to manage. The inability of even experienced managers to identify with the aid of earlier techniques all of the tasks with their sequential relationships became increasingly evident. Often necessary tasks were initially overlooked; when they were discovered, management was forced to take corrective action that increased cost, time, or risk associated with the objective. Management also found it difficult to estimate the time required for completing a project. Often the original estimate was overly optimistic. Subsequent awareness of the actual situation led to a lengthening of the total project time or to the initiation of a crash program. Because of the difficulty of comprehending the total effect of current slippages, the decision was often made to crash the entire project. Thus, every task was accelerated rather than the few needed to bring the project back on schedule, resulting in unnecessarily higher costs. For these and other reasons the search for a more effective technique for managing large-scale systems was accelerated.

In 1957, Morgan R. Walker of E. I. DuPont de Nemours and Company and James E. Kelley, Jr., of Remington Rand introduced the network method for improving the planning, scheduling, and overall coordination of plant construction. This new method made possible the identification of job parts critical for the completion of the overall project. It became known as the Critical Path Method (CPM).

In January 1958, Willard Fazar of the Special Projects Office of the Navy Bureau of Ordnance, together with representatives of the consulting firm of Booze, Allen and Hamilton and representatives of Lockheed Missile Systems Division undertook to develop a network system for managing the Fleet Ballistic Missile (FBM) program. Since some 2,000 contractors were involved in the FBM program, a new technique was badly needed—one that would integrate the diverse efforts of the thousands of contractors so that the entire project could be effectively planned and controlled. The network system which they developed was originally called the Program Evaluation and Research Task, but this was later changed to Program Evaluation and Review Technique.

Although some differences existed in the initial development of PERT and CPM, they have in time come to be viewed as virtually identical. One is basically event-oriented, the other activity-oriented. These *are* differences, of course, but the differences are not monumental. Here attention will be confined to PERT which appears to have had wider application than CPM, perhaps not because of its inherent superiority but because of its historical development within the military arena.

PERT Network Antecedents. The PERT network approach to management planning and control is said to be but a modification and extension of the Gantt bar chart which was developed in the early 1900s. The oversimplified Gantt chart shown in Figure 10–3 lists a scheduled start, a finish date, and the time duration for each of the tasks or activities to be accomplished. The horizontal scale represents the passage of time in some time unit (a day, a week, a month, or even a year), while the vertical axis is used to represent the tasks to be performed. Such charts have been used to show progress regarding completion of tasks, production of units, deliveries, and so forth. Such charts served well for projects of limited size where the few interdependencies could be implicitly accounted for. They did present some difficulties, but before discussing these we should mention the milestone chart, which is but a slightly modified bar chart. Any comments can be applied to both.

FIGURE 10-3
Gantt Bar Chart

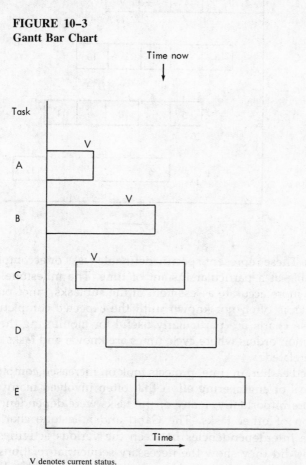

V denotes current status.

The milestone chart (Figure 10–4) is so similar to the Gantt chart that often no distinction is made between the two. Like the Gantt chart it has a time scale located along the horizontal axis while the tasks are located along its vertical axis. One significant difference, however, is that the tasks are broken down into simpler components termed

FIGURE 10–4
Milestone Chart

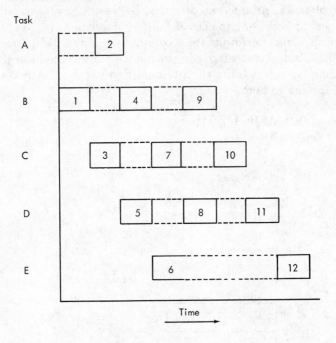

Task

A   2

B   1   4   9

C   3   7   10

D   5   8   11

E   6   12

Time

milestones. These represent specific definable tasks or accomplishments recognizable at a particular instant of time. The milestone chart can provide a more accurate assessment of the subtasks, since progress on major tasks is often not known until the expected completion of the task. These charts are particularly useful for monitoring the progress of production orders where cycle times are known and tasks are essentially nonrelated.

As noted earlier, in time, projects took on increased complexity with a great deal of engineering effort that often involved many organizations. More importantly, many of the tasks were dependent upon the completion of other tasks. The Gantt and milestone charts did not depict the interdependencies between the various activities or milestones nor did they show the necessary sequential relationships. If a progress evaluation showed that certain tasks were behind schedule, the charts did not depict the impact of this slippage on the other tasks. In many projects the slippage of some tasks will not affect the final completion date of the project because their nonaccomplishment at their scheduled time is not critical. On the other hand, delays in other tasks will cause a corresponding delay in the completion date of the entire project. Furthermore, bar charts do not permit the consideration of alternative courses of action. Once schedules are determined there is little chance to alter such schedules without undue effort. Yet even with the breakdown of tasks into subtasks (milestones), these charts do not provide sufficient detail to managers for exercising timely control and for detecting schedule slippages.

The PERT network, by portraying the sequential relationships among the significant tasks to be accomplished, overcomes many of the above deficiencies. As can be seen from Figure 10–5, it is essentially a flow diagram consisting of the activities to be accomplished if the project is to reach its objective. Because the network shows the planned sequence of tasks to be performed with their interrelations, it is the basic tool in the decision-making process. Schedules, progress reporting, and corrective action all emanate from the network. The use of a network alleviates many of the planning and scheduling problems associated with bar charts. With the earlier techniques, planning and scheduling were performed simultaneously while the available resources were utilized in a time-schedule sequence. Critical elements of the project were not readily identifiable nor was there sufficient flexibility in the scheduling of resources.

Since PERT constitutes a notable improvement over both the Gantt and the milestone chart, it is only proper that some of its components

FIGURE 10–5
PERT Network

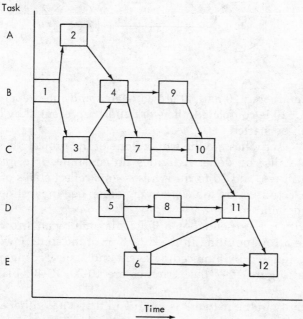

be discussed more at length. Here we shall consider PERT events, activities, the critical path, various time estimates, and slack. PERT inputs and outputs will be discussed and then a word or two on PERT/Cost. Finally a modern nonmilitary application will be used to illustrate how the network approach is used in a real life (educational) setting.

PERT: Basic Elements

Events. Events are specific definable accomplishments in a program plan, each recognizable at a particular instant in time. As such, they neither consume resources nor require time to complete. They simply reflect a *state of being;* in other words, something is developed, tested, started, accepted, completed, designed, and so forth.

Events thus represent decision points, the beginning or ending of a major phase of a job, or the transfer of responsibility either from one department to another or from one organization to another. For this reason events by themselves cannot specify all the activities connected with them. Typically they are represented on the network as circles.

An example of two related events is presented in Figure 10–6.

FIGURE 10–6
Event-Activity Relationship

Events numbered 10 and 40 are so related and interdependent that unless event 10 is completed, the work effort represented by the arrow cannot even be started.

Activities. Activities are the work efforts of the project: these are the things that must be done. Activities do consume time, money, resources, facilities, and so forth. While events reflect states of being or conditions, activities are the work being done; i.e., the testing, designing, building, and so forth.

An activity is represented on a PERT diagram by an arrow with the head of the arrow pointing in the direction of the time flow. Each activity is designated by a preceding event and a succeeding event. For example, in Figure 10–7 the activities are 10-30, 10-40, 10-20, 30-90, 40-90, and 20-90.

Each activity in the network is unique in that it has either a different predecessor, a different successor, or both. As indicated previously, the activity connecting two events cannot begin before the preceding event is completed. In those instances in which several activities lead up to a single event, all activities must be completed before the event comes into existence. For example, the activities listed in Figure 10–7 as 30-90, 40-90, and 20-90 must all be completed before event number 90 can "instantaneously" occur.

While it is sometimes convenient to number events consecutively, this practice is by no means common. It really makes little difference, since each event is a separate entity. In fact, one may even use letters as well as combinations of letters and numbers. The number of digits used for describing an event is also arbitrary. However, when one uses a "canned" PERT computer program, the specification typically calls for not more than nine digits. On most projects one would hardly experience difficulty with this size specification.

In network diagrams an event may not begin if the constraint that

FIGURE 10–7
Network Schematic

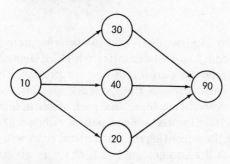

FIGURE 10–8
Merge Dummy Activities

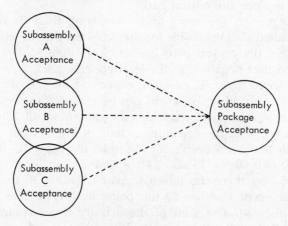

precedes it is not removed, even though the constraint does not take any time or consume any resources. Because of this a *dummy* activity, indicated on a network as a dotted line, is often used to tie the completion of several activities to the ending event or to the beginning of another event (Figure 10–8).

In other cases the dummy activity may be used between two events to indicate the completion of one activity and the beginning of another activity as shown in Figure 10–9.

FIGURE 10–9
Series Dummy Activities

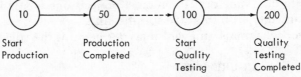

| 10 | 50 | 100 | 200 |
| Start Production | Production Completed | Start Quality Testing | Quality Testing Completed |

Critical Path. In Figure 10–10 a network is depicted within which there is one path composed of activities which, if delayed, would affect the expected completion date of the entire project. This is the most time-consuming path through the network and is designated on the diagram by the heavy black line. The path itself is termed the *critical path* and the activities located on the path are termed *critical activities.* A delay in any of the activities on this critical path will retard the final completion date of the project. Just as a chain is no stronger than its weakest link, so the earliest completion date for an ending event can be no sooner than the time required for the most time-consuming activities that lie along the critical path.

Since this path is the most time-consuming there is no leeway (technically called *slack*) possible for the activities that lie on the path. Because of this the critical path is sometimes defined as "that path having the greatest negative, or the least positive slack."

Earliest Expected Date. The earliest expected time (T_E) of an event is the earliest date on which an event can be expected to occur. The T_E value for a given *event* can be calculated by summing all *activity* times (t_e) through the most time-consuming chain of events from the beginning event to the given event. For example, in Figure 10–10, the T_Es for events 30 and 80 are 15 and 14 respectively. For each of the two events mentioned there are alternate paths leading to the event in question. For event number 30 the paths leading to this event are 0-10-30 and 0-20-30. The sum of the activity times along the path 0-10-30 is 15. This is the longer of the two paths, and therefore the T_E

of that event is assigned this value. The sum of the activity times along the path 0-20-50-80 gives a T_E of 11 which is the smallest value for event 80. The path 0-20-80 gives a T_E of 12. However, the T_E value assigned event 80 is that calculated from the activity time through path 0-40-60-80. This is the longest path! In this manner the T_E for each event throughout the network is calculated.

Latest Allowable Date. The latest allowable date (T_L) is the latest date on which an event can occur without delaying the completion of the project. The T_L value for any given event can be calculated by subtracting the sum of the activity times (t_e) from the activities on the path between the given event and the end event of the program (the latest date allowable for completing the project). For the ending event the T_L is the same as the T_E unless otherwise specified. Take Figure 10–10 in which event 100 is the ending event with a T_E and T_L of 24 weeks. We can calculate the latest allowable date (T_L) for any event by subtracting the t_es for the activity involved from the T_L of the ending event. Thus, the T_L for event 90 is 20 $(24 - 4)$; the T_L for event 70 is 19; for event 30 the T_L is 16. Since event 20 is at the juncture of four network paths, it is possible to calculate four T_Ls:

Calculated along the path 100–90–80–20, $T_L = 8$, $(24 - 16)$
Calculated along the path 100–50–20, $T_L = 12$ $(24 - 12)$
Calculated along the path 100–70–30–20, $T_L = 13$ $(24 - 11)$
Calculated along the path 100–90–80–50–20, $T_L = 9$ $(24 - 15)$

FIGURE 10–10
PERT Network Showing Critical Path and T_Es and T_Ls

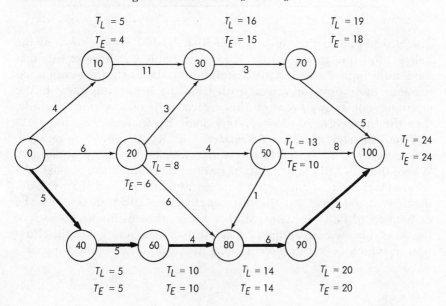

The T_L of event 20 must be the *smallest* of these values (8), since event 20 must be started 16 weeks before the project is expected to be completed. If it were started any later, there would not be enough time to complete all the activities involved.

Slack Determination. Slack (S) exists in a network system because of multiple junctions which arise when two or more activities contribute to a third. The slack for an event is but the difference between its latest allowable date and its earliest expected date $(T_L - T_E)$ and as such it may be positive, zero, or negative. To show how slack is determined, we will examine Figure 10–10 once more. The T_E, T_L, and S for each event in Figure 10–10 are shown in the following table.

TABLE 10–1
Slack Determination

Ending Event	Latest Allowable Date (T_L)	Earliest Expected Date (T_E)	Slack ($T_L - T_E$)
10	5	4	1
20	8	6	2
30	16	15	1
40	5	5	0
50	13	10	3
60	10	10	0
70	19	18	1
80	14	14	0
90	20	20	0
100	24	24	0

The critical path is the network 0-40-60-80-90-100 and is shown by the heavy solid line in the figure. It is identifiable by those events that have minimum slack. When the minimum slack activities are joined together they form the *longest* path from the beginning event to the ending event. If any event on this critical path falls behind schedule, then the final event is also expected to slip by an equal amount.

Time Estimates and Their Calculations. PERT, it should be recalled, was developed for projects for whose tasks uncertainty abounded. Where there is little uncertainty regarding the completion of tasks and where there are few interdependencies, any time-phased chart will do. In view of the great variability of engineering effort on projects, the early users of PERT initially calculated three time estimates for each of the tasks. One was an *optimistic time* estimate which reflected the situation in which everything went along just right. The second time estimate, termed the *most likely time,* was that one time estimate that would occur most frequently (modal value) were the activity to be performed repeatedly. The *pessimistic time* estimate was the maximum time that a task would require when problems were encountered. Through the use of a weighted average of the three estimates a single activity estimate (t_e) was obtained

$$t_e = \frac{a + 4m + b}{6}$$

where a = optimistic time
 m = most likely time
 b = pessimistic time.

While initially there was much debate over the merits of using three time estimates, the controversy has long since terminated and the time estimates reduced to a single one — the most likely time. Since the three estimates did not prove to be statistically more accurate despite the enormous computational work involved, by 1965 most firms had begun to settle on but a single time estimate. The time estimates may be expressed in months, weeks, tenths of weeks, or days depending on the magnitude and complexity of the project.

PERT Inputs and Outputs

Inputs. Once the network is drawn up, the events numbered, and a time estimate prepared for each activity, this information is now ready for processing. Since typical PERT networks involve thousands of events and activities, manual calculation is out of the question. The computer alone can make this formidable task of calculating T_E, T_L, and S both feasible and economical. Each activity's information will then be transferred to a punched card for inputing to the computer. Figure 10–11 shows one such card. The essential input data consist of each activity's predecessor and successor event, its activity time, and its description. This information is all that is needed for the computer to reconstruct the network and to calculate the values of T_E, T_L, slack, and so forth.

On small projects of 100 activities or less a network can be updated manually without too great difficulty. Common sense and human endurance serve as valuable guides for determining the maximum number of events and activities economically processible by manual computation. With so many "canned" PERT computer programs available today, computerization has become the accustomed way of life for many organizations.

FIGURE 10–11
PERT Input Card

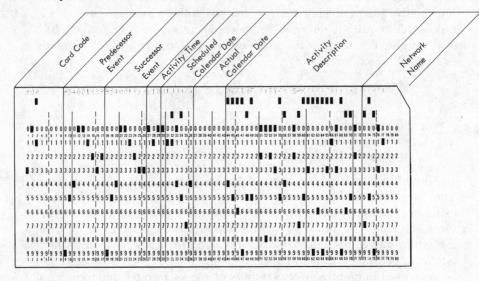

Updating Process. Since each activity is listed on a separate card, an updating of a PERT network involves the following:

Reporting of activities completed during the last report period;

Revised time estimates for activities that were not completed during the reporting cycle but which were expected to be completed;

Revised time estimates for activities expected to be completed during the next reporting cycle but which will not be completed; and

Other additions or deletions to the network.

Data Output. The basic management reports are the primary outputs of the PERT computer program. These reports will contain similar information but will be given different formats, since each format highlights a different aspect. The diverse formats not only allow for the monitoring of the key task interrelationships and interdependencies but also emphasize the many dissimilar requirements of the contract. While it is possible to generate a great variety of reports, the ones utilized by most organizations are activity and event reports.

Activity Reports. Activity reports represent work accomplished to date with regard to planned performance. While these reports will be scrutinized by managers accountable for the completion of the specific tasks, they are used at the lower work levels by department heads. These reports present enough detail to enable the manager to compare current progress against expected accomplishments. The various formats possible for activity reports are activity number sequence, expected date of completion, and slack. Activity number sequence lists all the activities in their sequential order. Expected date of completion report lists all the activities from the nearest expected date of completion to those farthest away in time. Slack report arranges all the activities in a slack sequence from the least positive to the most positive. This particular sort highlights activities deemed critical for completing the project.

Event Reports. Event reports are intended mainly for the project manager. The project manager is not concerned with the many activities leading up to any particular event but with the progress of significant happenings. Obviously he wishes to be informed of activities which are being scheduled and which may delay the project but in general he is most interested in events that reflect the completion of many activities. Event reports may be omitted if project size does not warrant them. The amount of desired control therefore can serve as a guideline in this regard.

An example of activity reports and of each of the sorts will be presented later when a specific application of PERT is discussed.

Reporting Frequency. Consideration must be given to the following factors for determining the frequency of reporting on the project:

Requirement of the contracting authority.

Duration of the project.

Magnitude and complexity of the project.

Degrees of risk and state of technology required.

Essentially, if the tasks are straightforward and the technology not too complex, a monthly reporting schedule will suffice. If a project is critical as to both time and performance, a tighter reporting period may be wanted. Biweekly reporting initially suggested by the founders of PERT no longer prevails. Since most of the value of PERT comes in the early stages of network construction in which all aspects of the project are carefully considered, many companies do update informally for internal projects and formally for outside contracts. However, if many changes in the project take place and if delays are experienced, project updating must be done more frequently.

PERT/COST

In 1962 PERT/Cost was developed to integrate the time data of the physical accomplishment with the associated financial data, and all within one systematic framework. While PERT admittedly provides the means for monitoring, coordinating, and controlling the physical aspects of a project at various operational levels, it definitely does not enable one to measure the project's financial status. Since in the scheduling process the critical path often ignored the availability of resources, PERT proved inadequate for this new job and PERT/Cost was born. (Integrating both time and cost considerations in one overall system turned out to be impractical.)

Despite the intended benefits of PERT/Cost, the technique has never really caught on. Some of the problems that arose were inherent in the system itself. Others were peculiar to the type of industry (defense) in which it was employed. Still other problems resulted from a conflict of some of its basic elements with certain organizational practices. Since the authors know of no firm employing PERT/Cost at the present time, an explication of its basic elements would have historical interest only. Rather than present a postmortem here, the authors prefer to refer the interested reader to the literature where a fuller discussion of some of the major problems that plagued this emendation of PERT can be found.[3]

AN APPLICATION OF PERT

PERT usage has been almost limitless. At one time PERT was required for all Department of Defense projects. While PERT *per se* is no longer required for government projects, *some* network technique is prescribed for most proposals. In this final section of project management we will observe how the PERT technique was utilized on one contract with the Office of Education of the Department of Health, Education, and Welfare. Although the project was not of long duration, its very complexity dictated PERT usage.

The Basic Educational Opportunity Grant Program (BEOG)[4]

The Basic Educational Opportunity Grant Program is a provision of the Education Amendments passed by Congress in 1972. Its basic aim is to provide financial assistance, based on need, to any qualified

---

[3] Peter P. Schoderbek, "Is PERT/Cost Dead?" *Management Services* (November–December 1968), pp. 43–50. Don T. DeCoster, "PERT/Cost–The Challenge," *Management Services* (July–August 1965), pp. 13–18. Peter P. Schoderbek, "PERT/Cost: Its Values and Limitations," *Management Services* (January–February 1966), pp. 29–34.

[4] The authors express their appreciation to the American College Testing Corporation and the Office of Education for their permission to reproduce portions of the BEOG project.

student in the United States attending a postsecondary institution. The basic grant allows the student to choose the school of his choice.

A significant provision of this program is the national family contribution schedule. A formula is applied to all applicants uniformly: in the past each institution made its own need analysis. This assistance program is to be augmented by the traditional student aid programs. The number of anticipated applicants attending more than 10,000 postsecondary institutions is in the millions.

The American College Testing (ACT) company, one of the largest administrators of educational testing services in the United States, was invited to submit a proposal for the design and implementation of the Basic Educational Opportunity Grant Program. Measurement Research Corporation, which is a branch of the Learning Division of the Westinghouse Corporation, was selected as a major subcontractor to American College Testing. Measurement Research Corporation is the world's largest processor of standardized educational tests and a leader in its field.

Because of the severe time constraints in the contract, close coordination was required for all phases of the project that had numerous interfaces with governmental agencies, and other subcontractors. One of the stipulations outlined in the request for a proposal was a PERT-type capability to guarantee the contracting authority that technical and schedule performance would be attained. In this setting ACT used the following PERT techniques to assist management in its planning and control function.

Management Plan of the ACT

As with other projects, ACT delineated the PERT system cycle in two distinct phases: a planning phase during which a time plan expressed in network form and a contract schedule were to be developed, and an operating phase during which contract progress was to be assessed and the impact of accomplishments on future plans forecasted. Figure 10–12 is an illustration of the overall cycle.

The Planning Phase

Establishing Objectives. This initial step in planning is concerned with the establishment of contract objectives. The objectives or goals as defined in the Basic Educational Opportunity Grant (BEOG) Program were as follows:

The Basic Educational Opportunity Grant (BEOG) Program recently

FIGURE 10–12
The PERT System Cycle

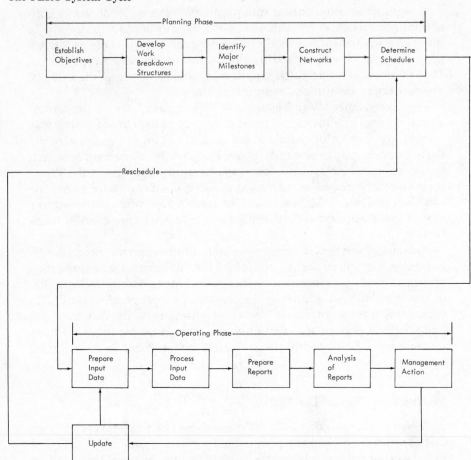

authorized by Congress represents a new concept in providing financial aid to students. The BEOG legislation provides that every eligible student be entitled to Federal assistance as a matter of right. While the student must be enrolled at a postsecondary institution, the right to aid does not depend on the location or type of institution. The Basic Grants Program is intended to guarantee a minimum amount of resources to every eligible student who wishes to attend a postsecondary institution.

Also:

The purpose of this contract is to design and implement a system to calculate and certify the expected family contribution for each student and to process student applications in accordance with specified systems procedures.

A further delineation of the objectives was articulated in the specifications of the three phases of the contract. These are:

Phase I Specifications — to assist HEW-OS with requirements definitions and systems specifications

1. To study alternative processing methods.
2. To develop system requirements definition and specification.
3. To develop the definition and documentation of data acquisition and workflow procedures.
4. To develop a system back-up and recovery procedures.

Phase II Specifications — to develop software for the basic grant program (if applicable)

1. To design the computer system.
2. To develop the computer system.
3. To test the computer system.
4. To document the computer system.

Phase III Specifications — to administer operational processes that will provide a necessary level of production support to successfully and cost-effectively deliver the contract requirements

1. To prepare a cost estimate for the production process supporting the BEOG program.
2. To prepare the cost estimate to reflect the maximum thruput cost per unit.
3. To provide a detailed work breakdown of production tasks.
4. To develop work requirement plans for data preparation, volume and processing time, space, facilities and materials, training, courier service and special requirements.

Developing Work Breakdown Structure.   The purpose of the work breakdown structure (WBS) is to provide a common framework for the accomplishment of the work to be performed. It provides a basis for uniform planning and program visibility, enables assignments of responsibilities, and delineates objectives for monitoring progress. Additionally, it establishes the basis for constructing networks at any desired level of detail by identifying the end items to be accomplished.

The work breakdown structure is developed by proceeding from the major objective of the contract to successively lower levels. A top-notch approach is used to guide planning rather than allow a detailed plan to be generated outside the common framework. While the work breakdown structure is not mandated for the development of networks, it does, nevertheless, help to clarify the contract objectives.

The following guidelines were used in developing the work breakdown structure of this contract:

Complexity and time span of the contract.

Estimated cost of the contract.

Number of subcontractors.

Contractor's organization.

HEW-OS requirements.

FIGURE 10-13
Work Breakdown Structure for BEOG Project

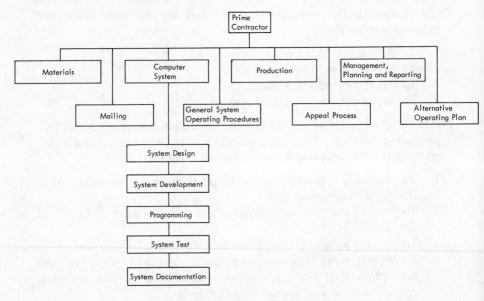

The end item subdivisions in the WBS represent (1) materials, (2) mailing, (3) computer system, (4) general operating procedures, (5) production, (6) management, planning, and reporting, (7) appeal process, and (8) an alternative operating plan that is deliverable to the government. Figure 10-13 is illustrative of the WBS concept.

Identification of the Major Milestones.   Given the above objectives it was necessary to identify the major milestones of the contract. Milestones represent the significant elements which must be closely managed, since they reflect the current status of the contract. The sequencing of these milestones provided a time-phased framework within which the network plan and subsequent schedules were developed.

The selection of the milestones to be included in the network offered management an opportunity to identify significant points in the contract that needed monitoring. In this contract emphasis was placed on the selection of milestones that were distinguishable and unambiguous points in time, coincident with the beginning or ending of a specific task or activity. Some of the major milestones in the BEOG program are illustrated in Figure 10–14.

FIGURE 10–14
Major Milestones

Milestone	Number	Event Number	Expected Completion Date	Actual Completion Date	Update Completion Date
Start Phases I and II	1	PHI000	19 March 73		
Task 1, Phase I Completed	2	PHI080	03 April 73		
Define System Inputs/Outputs	3	PHI110	16 April 73		
System Programming Finalized	4	PHII220	11 May 73		
Task 1, 2, Phase II Completed	5	PHII230	24 May 73		
End Phases I and II	6	PHII250	07 June 73		

Construction of the Network. As stated earlier the network is a flow diagram consisting of the activities and events which must be accomplished to reach the contract objectives. It is the basic tool in the contract-management decision-making process for planning the work to be performed, progress reporting, and taking corrective action. The networks for Phases I and II of this contract, though somewhat complicated, are reproduced in Figure 10–15. They are intended to illustrate the nature of a PERT network in an actual situation.

Determining Schedules. The resultant network plan with the expected time estimates provided a schedule for the work to be accomplished. The network plan was then transferred to a timetable with the current *calendar* dates which then governed the start and completion of work and the authorization of the expenditure of resources. For this particular project a six-day week was employed.

The Operating Phase

The operating phase began with the authorization of the work to be performed for that part of the contract for which a schedule had been

FIGURE 10–15
PERT Network Chart

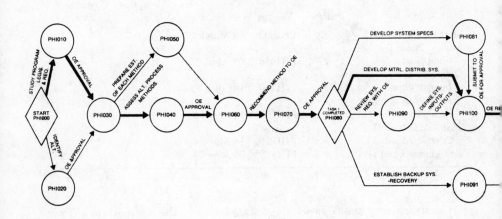

CRITICAL PATH

MILESTONES

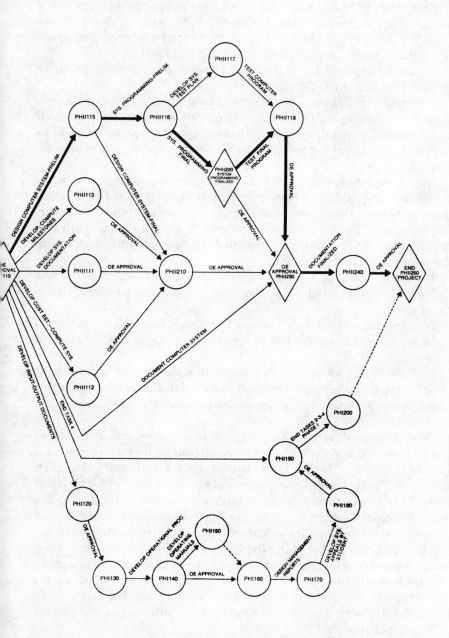

established and approved. Upon approval of the contract the scheduled tasks were performed and their completion dates noted.

Preparation of Input Data. The input data, subject to the approval of the granting authority, were presented in the activity charts. These included completed activities, changes in activity time estimates, schedule changes, and addition or deletion of activities.

Processing of Input Data. Input data were computer-processed to produce the desired PERT output reports.

Preparation of Reports. The data-processing system can easily produce a number of output reports displaying current and forecasted status needed for analysis and evaluation. These can serve as an aid in determining current problems and in assessing the impact of alternative courses of action.

Management Reporting. The basic management reports submitted to the contracting authority were the primary outputs of the PERT computer program. These contained similar information but were in different formats. The different arrangements highlight the many diverse requirements of the contract. Submitted on a semimonthly basis were the following activity reports:

1. *Activity Number Sequence Report* listed all the activities in their sequential order, sorted by predecessor and successor events (Figure 10–16).

2. *Latest Allowable Date of Completion Report* listed all the activities from the nearest expected date of completion to those farthest away in time (Figure 10–17).

3. *Slack Report* sequenced all the activities from those with the least positive slack to those with the most negative slack. This particular sort highlighted activities critical for meeting the completion date of the contract (Figure 10–18).

Gantt Chart. A Gantt chart (Figure 10–19) was also required in the request for a proposal. Its value, of course, lies in its graphical portrayal of the activities with their attendant start and completion dates. A slight modification was incorporated in the Gantt chart used. This showed the amount of slack allowed for each activity. The chart was constructed by taking the earliest completion date of each activity and extending it backward for the duration of the time estimate. Slack was calculated by taking the latest allowable date for the activity to be completed and extending a dotted line to that data. By erecting a vertical line on the Gantt chart and noting the completion of the activities one can immediately discern whether trouble exists in an activity. Of course, the same information is portrayed in the various PERT output reports.

FIGURE 10-16

Activity (Task) Status Report — Sorted by Predecessor and Successor

PROGRAM
PROJECT   19MAR73 TEST RUN   PERT1

LEVEL   DETAIL   SORTED BY PREDEC.,SUCC.

PRED. EVENT	SUCC. EVENT	CYCL. CODE	ACTIVITY DESCRIPTION	TIME ESTIMATES A	M	B	SLACK PRIMR	SECND	COMPLETION DATES EXPECTED	LATEST	SCHED. DATE	DEPT.
	PHI000	S	START				.0	.0	A19MAR73	19MAR73		
PHI000	PHI010		STUDY PROG LEGISLATION & REQ	004			.0	.0	23MAR73	23MAR73		
PHI000	PHI020		IDENTIFY ALT	003			1.0	.0	22MAR73	23MAR73		
PHI010	PHI030		OE APPROVAL	001			.0	.0	24MAR73	24MAR73		
PHI020	PHI030		OE APPROVAL	001			1.0	.0	23MAR73	24MAR73		
PHI030	PHI040		ASSESS ALT PROCESS METHODS	004			.0	.0	29MAR73	29MAR73		
PHI030	PHI050		PREPARE EST OF EACH METHOD	002			2.0	.0	27MAR73	29MAR73		
PHI040	PHI060		OE APPROVAL	001			.0	.0	30MAR73	30MAR73		
PHI050	PHI060		OE APPROVAL	001			2.0	.0	28MAR73	30MAR73		
PHI060	PHI070		RECOMMEND METHOD TO OE	002			.0	.0	02APR73	02APR73		
PHI070	PHI080		OE APPROVAL	001			.0	.0	03APR73	03APR73		
PHI080	PHI081		DEVELOP SYSTEM SPECS	004			5.0	.0	07APR73	13APR73		
PHI080	PHI090		REVIEW SYS REQ WITH OE	002			6.0	.0	05APR73	12APR73		
PHI080	PHI091		ESTABLISH BACKUP SYS-RECOVERY	006			4.0	.0	10APR73	14APR73		
PHI080	PHI100		DEVELOP MATL DISTRIBUTION SYS	010			.0	.0	14APR73	14APR73		
PHI081	PHI100		SUBMIT TO OE FOR APPROVAL	001			5.0	.0	09APR73	14APR73		
PHI090	PHI100		DEFINE SYS INPUTS-OUTPUTS	002			6.0	.0	07APR73	14APR73		
PHI091	PHI110		OE APPROVAL	001			4.0	.0	11APR73	16APR73		
PHI100	PHI110		OE REVIEW	001			.0	.0	16APR73	16APR73		
PHI110	PHI120		DEVELOP INPUT-OUTPUT DOCUMENTS	006			26.0	.0	24APR73	24MAY73		
PHI110	PHI190		END TASK 4	000			44.0	.0	16APR73	07JUN73		
PHI110	PHI1110		DUMMY	000			.0	.0	16APR73	16APR73		
PHI120	PHI130		OE APPROVAL	001			26.0	.0	25APR73	25MAY73		
PHI130	PHI140		DEVELOP OPERATIONAL PROC	007			26.0	.0	03MAY73	02JUN73		
PHI140	PHI150		DEVELOP OPERATING MANUALS	001			26.0	.0	04MAY73	04JUN73		
PHI140	PHI160		OE APPROVAL	001			26.0	.0	04MAY73	04JUN73		
PHI150	PHI160		DUMMY	000			26.0	.0	04MAY73	04JUN73		
PHI160	PHI170		DESIGN MANAGEMENT REPORTS	001			26.0	.0	05MAY73	05JUN73		
PHI170	PHI180		DEVELOP SYS APPEALS BY STUDENT	001			26.0	.0	07MAY73	06JUN73		
PHI180	PHI190		OE APPROVAL	001			26.0	.0	08MAY73	07JUN73		
PHI190	PHI200		END TASK 2-3-4, PHASE I	000			26.0	.0	08MAY73	07JUN73		
PHI200	PHII250		DUMMY	000			26.0	.0	08MAY73	07JUN73		
PHII110	PHII111		DEVELOP SYSTEM DOCUMENTATION	015			16.0	.0	04MAY73	23MAY73		
PHII110	PHII112		DEVELOP COST EST-COMPUTE SYS	017			14.0	.0	07MAY73	23MAY73		
PHII110	PHII113		DEVELOP COMPUTE MILESTONES	014			17.0	.0	03MAY73	23MAY73		
PHII110	PHII115		DESIGN COMPUTER SYSTEM-PRELIM	006			.0	.0	24APR73	24APR73		
PHII110	PHII230		DOCUMENT COMPUTER SYSTEM	030			2.0	.0	22MAY73	24MAY73		
PHII111	PHII210		OE APPROVAL	000			16.0	.0	04MAY73	23MAY73		
PHII112	PHII210		OE APPROVAL	000			14.0	.0	07MAY73	23MAY73		
PHII113	PHII210		OE APPROVAL	000			17.0	.0	03MAY73	23MAY73		
PHII115	PHII11.		SYS PROGRAMMING-PRELIM	008			.0	.0	03MAY73	03MAY73		
PHII115	PHII216		DESIGN COMPUTER SYSTEM-FINAL	005			20.0	.0	30APR73	23MAY73		
PHII116	PHII117		DEVELOP SYS TEST PLAN	005			2.0	.0	09MAY73	11MAY73		
PHII116	PHII220		SYS PROGRAMMING-FINAL	007			.0	.0	11MAY73	11MAY73		
PHII117	PHII118		TEST COMPUTER PROGRAM	008			2.0	.0	18MAY73	21MAY73		
PHII118	PHII230		OE APPROVAL	003			.0	.0	24MAY73	24MAY73		
PHII210	PHII230		OE APPROVAL	001			14.0	.0	08MAY73	24MAY73		
PHII220	PHII11.		TEST FINAL PROGRAMS	008			.0	.0	21MAY73	21MAY73		
PHII220	PHII230		OE APPROVAL	003			8.0	.0	15MAY73	24MAY73		
PHII230	PHII240		DOCUMENTATION FINALIZED	010			.0	.0	05JUN73	05JUN73		
PHII240	PHII250		OE APPROVAL	002			.0	.0	07JUN73	07JUN73		
PHII250		E	END				.0	.0	07JUN73	07JUN73		

FIGURE 10–17
Activity (Task) Status Report – Sorted by Latest Date

```
PROGRAM    ↓
PROJECT    19MAR73  TEST RUN                                    PERT1              ↓
LEVEL   DETAIL                                                        SORTED BY LATEST DATE
```

PRED. EVENT	SUCC. EVENT	CYCL. CODE	ACTIVITY DESCRIPTION	TIME ESTIMATES A	M	B	SLACK PRIMR	SECND	COMPLETION DATES EXPECTED	LATEST	SCHED. DATE	DEPT.
	PHI000	S	START				.0	.0	A19MAR73	19MAR73		
PHI000	PHI010		STUDY PROG LEGISLATION & REQ	004			.0	.0	23MAR73	23MAR73		
PHI000	PHI020		IDENTIFY ALT	003			1.0	.0	22MAR73	23MAR73		
PHI010	PHI030		OE APPROVAL	001			.0	.0	24MAR73	24MAR73		
PHI020	PHI030		OE APPROVAL	001			1.0	.0	23MAR73	24MAR73		
PHI030	PHI040		ASSESS ALT PROCESS METHODS	004			.0	.0	29MAR73	29MAR73		
PHI030	PHI050		PREPARE LST OF EACH METHOD	002			2.0	.0	27MAR73	29MAR73		
PHI040	PHI060		OE APPROVAL	001			.0	.0	30MAR73	30MAR73		
PHI050	PHI060		OE APPROVAL	001			2.0	.0	28MAR73	30MAR73		
PHI060	PHI070		RECOMMEND METHOD TO OE	002			.0	.0	02APR73	02APR73		
PHI070	PHI08		OE APPROVAL	001			.0	.0	03APR73	03APR73		
PHI080	PHI090		REVIEW SYS REQ WITH OE	002			6.0	.0	05APR73	12APR73		
PHI080	PHI081		DEVELOP SYSTEM SPECS	004			5.0	.0	07APR73	13APR73		
PHI080	PHI091		ESTABLISH BACKUP SYS-RECOVERY	006			4.0	.0	10APR73	14APR73		
PHI080	PHI100		DEVELOP MATL DISTRIBUTION SYS	010			.0	.0	14APR73	14APR73		
PHI081	PHI100		SUBMIT TO OE FOR APPROVAL	001			5.0	.0	09APR73	14APR73		
PHI090	PHI100		DEFINE SYS INPUTS-OUTPUTS	002			6.0	.0	07APR73	14APR73		
PHI091	PHI110		OE APPROVAL	001			4.0	.0	11APR73	16APR73		
PHI100	PHI110		OE REVIEW	001			.0	.0	16APR73	16APR73		
PHI110	PHI1110		DUMMY	000			.0	.0	16APR73	16APR73		
PHI1110	PHI1112		DESIGN COMPUTER SYSTEM-PRELIM	006			.0	.0	24APR73	24APR73		
PHI1113	PHI1115		SYS PROGRAMMING-PRELIM	008			.0	.0	03MAY73	03MAY73		
PHI1116	PHI1117		DEVELOP SYS TEST PLAN	005			2.0	.0	09MAY73	11MAY73		
PHI1116	PHI1220		SYS PROGRAMMING-FINAL	007			.0	.0	11MAY73	11MAY73		
PHI1117	PHI1118		TEST COMPUTER PROGRAM	008			2.0	.0	18MAY73	21MAY73		
PHI1220	PHI1118		TEST FINAL PROGRAMS	008			.0	.0	21MAY73	21MAY73		
PHI1110	PHI1111		DEVELOP SYSTEM DOCUMENTATION	015			16.0	.0	04MAY73	23MAY73		
PHI1110	PHI1112		DEVELOP COST LST-COMPUTE SYS	017			14.0	.0	07MAY73	23MAY73		
PHI1110	PHI1113		DEVELOP COMPUTE MILESTONES	014			17.0	.0	03MAY73	23MAY73		
PHI1111	PHI1210		OE APPROVAL	000			16.0	.0	04MAY73	23MAY73		
PHI1112	PHI1210		OE APPROVAL	000			14.0	.0	07MAY73	23MAY73		
PHI1113	PHI1210		OE APPROVAL	000			17.0	.0	03MAY73	23MAY73		
PHI1115	PHI1210		DESIGN COMPUTER SYSTEM-FINAL	005			20.0	.0	30APR73	23MAY73		
PHI110	PHI120		DEVELOP INPUT-OUTPUT DOCUMENTS	006			26.0	.0	24APR73	24MAY73		
PHI1110	PHI1230		DOCUMENT COMPUTER SYSTEM	030			2.0	.0	22MAY73	24MAY73		
PHI1113	PHI1230		OE APPROVAL	003			.0	.0	24MAY73	24MAY73		
PHI1210	PHI1230		OE APPROVAL	001			14.0	.0	08MAY73	24MAY73		
PHI1220	PHI1230		OE APPROVAL	003			8.0	.0	15MAY73	24MAY73		
PHI120	PHI130		OE APPROVAL	001			26.0	.0	25APR73	25MAY73		
PHI130	PHI140		DEVELOP OPERATIONAL PROC	007			26.0	.0	03MAY73	02JUN73		
PHI140	PHI150		DEVELOP OPERATING MANUALS	001			26.0	.0	04MAY73	04JUN73		
PHI140	PHI160		OE APPROVAL	001			26.0	.0	04MAY73	04JUN73		
PHI150	PHI160		DUMMY	000			26.0	.0	04MAY73	04JUN73		
PHI160	PHI170		DESIGN MANAGEMENT REPORTS	001			26.0	.0	05MAY73	05JUN73		
PHI1230	PHI1240		DOCUMENTATION FINALIZED	010			.0	.0	05JUN73	05JUN73		
PHI170	PHI180		DEVELOP SYS APPEALS BY STUDENT	001			26.0	.0	07MAY73	06JUN73		
PHI1250		E	END				.0	.0	07JUN73	07JUN73		
PHI110	PHI190		END TASK 4	000			44.0	.0	16APR73	07JUN73		
PHI180	PHI190		OE APPROVAL	001			26.0	.0	08MAY73	07JUN73		
PHI190	PHI200		END TASK 2-3-4, PHASE I	000			26.0	.0	08MAY73	07JUN73		
PHI200	PHI1250		DUMMY	000			26.0	.0	08MAY73	07JUN73		
PHI1240	PHI1250		OE APPROVAL	002			.0	.0	07JUN73	07JUN73		

FIGURE 10–18

Activity (Task) Status Report — Sorted by Primary Slack

PROGRAM
PROJECT   ↓   19MAR73 T_ST RUN   PERT1

LEVEL   DETAIL   ↓ SORTED BY PRIMARY SLACK

PRED. EVENT	SUCC. EVENT	CYCLE CODE	ACTIVITY DESCRIPTION	TIME ESTIMATES A M B	SLACK PRIMR SECND	COMPLETION DATES EXPECTED LATEST	SCHED. DATE	DEPT.
	PHI00J	S	START		.0 .0	A19MAR73 19MAR73		
PHI000	PHI01J		STUDY PROG LEGISLATION & REQ	004	.0 .0	23MAR73 23MAR73		
PHI010	PHI03J		OE APPROVAL	001	.0 .0	24MAR73 24MAR73		
PHI030	PHI04.		ASSESS ALT PROCESS METHODS	004	.0 .0	29MAR73 29MAR73		
PHI040	PHI06J		OE APPROVAL	001	.0 .0	30MAR73 30MAR73		
PHI060	PHI07J		RECOMMEND METHOD TO OE	002	.0 .0	02APR73 02APR73		
PHI070	PHI08J		OE APPROVAL	001	.0 .0	03APR73 03APR73		
PHI080	PHI10.		DEVELOP MATL DISTRIBUTION SYS	010	.0 .0	14APR73 14APR73		
PHI110	PHII11J		DUMMY	000	.0 .0	16APR73 16APR73		
PHI100	PHI11J		OE REVIEW	001	.0 .0	16APR73 16APR73		
PHII110	PHII115		DESIGN COMPUTER SYSTEM-PRELIM	006	.0 .0	24APR73 24APR73		
PHII115	PHI11J		SYS PROGRAMMING-PRELIM	008	.0 .0	03MAY73 03MAY73		
PHII116	PHI22.		SYS PROGRAMMING-FINAL	007	.0 .0	11MAY73 11MAY73		
PHII220	PHII11J		TEST FINAL PROGRAMS	008	.0 .0	21MAY73 21MAY73		
PHII118	PHI23J		OE APPROVAL	003	.0 .0	24MAY73 24MAY73		
PHII230	PHII24.		DOCUMENTATION FINALIZED	010	.0 .0	05JUN73 05JUN73		
PHII250		E	END		.0 .0	07JUN73 07JUN73		
PHII240	PHII25J		OE APPROVAL	002	.0 .0	07JUN73 07JUN73		
PHI000	PHI02J		IDENTIFY ALT	003	1.0 .0	22MAR73 23MAR73		
PHI020	PHI03.		OE APPROVAL	001	1.0 .0	23MAR73 24MAR73		
PHI050	PHI05U		PREPARE LST OF EACH METHOD	002	2.0 .0	27MAR73 29MAR73		
PHI050	PHI06U		OE APPROVAL	001	2.0 .0	28MAR73 30MAR73		
PHII116	PHII117		DEVELOP SYS TEST PLAN	005	2.0 .0	09MAY73 11MAY73		
PHII117	PHII118		TEST COMPUTER PROGRAM	008	2.0 .0	18MAY73 21MAY73		
PHII110	PHII23C		DOCUMENT COMPUTER SYSTEM	030	2.0 .0	22MAY73 24MAY73		
PHI060	PHI091		ESTABLISH BACKUP SYS-RECOVERY	006	4.0 .0	10APR73 14APR73		
PHI091	PHI11J		OE APPROVAL	001	4.0 .0	11APR73 16APR73		
PHI080	PHI081		DEVELOP SYSTEM SPECS	004	5.0 .0	07APR73 13APR73		
PHI061	PHI100		SUBMIT TO OE FOR APPROVAL	001	5.0 .0	09APR73 14APR73		
PHI080	PHI09U		REVIEW SYS REQ WITH OE	002	6.0 .0	05APR73 12APR73		
PHI090	PHI10J		DEFINE SYS INPUTS-OUTPUTS	002	6.0 .0	07APR73 14APR73		
PHII220	PHII23J		OE APPROVAL	003	8.0 .0	15MAY73 24MAY73		
PHII112	PHI121U		OE APPROVAL	001	14.0 .0	07MAY73 23MAY73		
PHII110	PHII112		DEVELOP COST EST-COMPUTE SYS	017	14.0 .0	07MAY73 23MAY73		
PHII210	PHII23J		OE APPROVAL	001	14.0 .0	08MAY73 23MAY73		
PHII110	PHII11J		DEVELOP SYSTEM DOCUMENTATION	015	16.0 .0	04MAY73 23MAY73		
PHII111	PHI121U		OE APPROVAL	000	16.0 .0	04MAY73 23MAY73		
PHII113	PHI121U		OE APPROVAL	000	17.0 .0	03MAY73 23MAY73		
PHII110	PHII11J		DEVELOP COMPUTE MILESTONES	014	17.0 .0	03MAY73 23MAY73		
PHII115	PHI121U		DESIGN COMPUTER SYSTEM-FINAL	005	20.0 .0	30APR73 23MAY73		
PHI110	PHI12J		DEVELOP INPUT-OUTPUT DOCUMENTS	006	26.0 .0	24APR73 24MAY73		
PHI120	PHI13J		OE APPROVAL	001	26.0 .0	25APR73 25MAY73		
PHI130	PHI14J		DEVELOP OPERATIONAL PROC	007	26.0 .0	03MAY73 02JUN73		
PHI150	PHI160		DUMMY	000	26.0 .0	04MAY73 04JUN73		
PHI140	PHI16U		OE APPROVAL	001	26.0 .0	04MAY73 04JUN73		
PHI140	PHI15C		DEVELOP OPERATING MANUALS	001	26.0 .0	04MAY73 04JUN73		
PHI160	PHI17J		DESIGN MANAGEMENT REPORTS	001	26.0 .0	05MAY73 05JUN73		
PHI170	PHI180		DEVELOP SYS APPEALS BY STUDENT	001	26.0 .0	07MAY73 06JUN73		
PHI200	PHII25J		DUMMY	000	26.0 .0	08MAY73 07JUN73		
PHI190	PHI20C		END TASK 2-3-4, PHASE I	000	26.0 .0	08MAY73 07JUN73		
PHI180	PHI190		OE APPROVAL	001	26.0 .0	08MAY73 07JUN73		
PHI110	PHI190		END TASK 4	000	44.0 .0	16APR73 07JUN73		

FIGURE 10-19
Gantt Chart

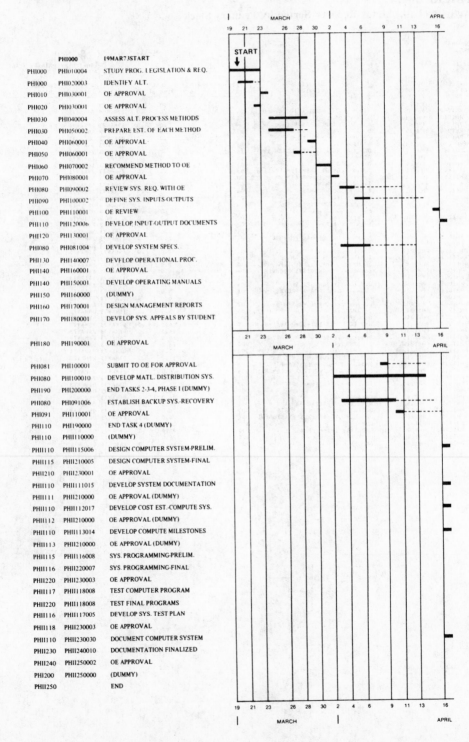

	PHI000	19MAR73START
PHI000	PHI010004	STUDY PROG. LEGISLATION & REQ.
PHI000	PHI020003	IDENTIFY ALT.
PHI010	PHI030001	OE APPROVAL
PHI020	PHI030001	OE APPROVAL
PHI030	PHI040004	ASSESS ALT. PROCESS METHODS
PHI030	PHI050002	PREPARE EST. OF EACH METHOD
PHI040	PHI060001	OE APPROVAL
PHI050	PHI060001	OE APPROVAL
PHI060	PHI070002	RECOMMEND METHOD TO OE
PHI070	PHI080001	OE APPROVAL
PHI080	PHI090002	REVIEW SYS. REQ. WITH OE
PHI090	PHI100002	DEFINE SYS. INPUTS-OUTPUTS
PHI100	PHI110001	OE REVIEW
PHI110	PHI120006	DEVELOP INPUT-OUTPUT DOCUMENTS
PHI120	PHI130001	OE APPROVAL
PHI080	PHI081004	DEVELOP SYSTEM SPECS.
PHI130	PHI140007	DEVELOP OPERATIONAL PROC.
PHI140	PHI160001	OE APPROVAL
PHI140	PHI150001	DEVELOP OPERATING MANUALS
PHI150	PHI160000	(DUMMY)
PHI160	PHI170001	DESIGN MANAGEMENT REPORTS
PHI170	PHI180001	DEVELOP SYS. APPEALS BY STUDENT
PHI180	PHI190001	OE APPROVAL
PHI081	PHI100001	SUBMIT TO OE FOR APPROVAL
PHI080	PHI100010	DEVELOP MATL. DISTRIBUTION SYS.
PHI190	PHI200000	END TASKS 2-3-4, PHASE I (DUMMY)
PHI080	PHI091006	ESTABLISH BACKUP SYS.-RECOVERY
PHI091	PHI110001	OE APPROVAL
PHI110	PHI190000	END TASK 4 (DUMMY)
PHI110	PHI1110000	(DUMMY)
PHI1110	PHI1115006	DESIGN COMPUTER SYSTEM-PRELIM.
PHI1115	PHI1210005	DESIGN COMPUTER SYSTEM-FINAL
PHI1210	PHI1230001	OE APPROVAL
PHI1110	PHI1111015	DEVELOP SYSTEM DOCUMENTATION
PHI1111	PHI1210000	OE APPROVAL (DUMMY)
PHI1110	PHI1112017	DEVELOP COST EST.-COMPUTE SYS.
PHI1112	PHI1210000	OE APPROVAL (DUMMY)
PHI1110	PHI1113014	DEVELOP COMPUTE MILESTONES
PHI1113	PHI1210000	OE APPROVAL (DUMMY)
PHI1115	PHI1116008	SYS. PROGRAMMING-PRELIM.
PHI1116	PHI1220007	SYS. PROGRAMMING-FINAL
PHI1220	PHI1230003	OE APPROVAL
PHI1117	PHI1118008	TEST COMPUTER PROGRAM
PHI1220	PHI1118008	TEST FINAL PROGRAMS
PHI1116	PHI1117005	DEVELOP SYS. TEST PLAN
PHI1118	PHI1230003	OE APPROVAL
PHI1110	PHI1230030	DOCUMENT COMPUTER SYSTEM
PHI1230	PHI1240010	DOCUMENTATION FINALIZED
PHI1240	PHI1250002	OE APPROVAL
PHI1200	PHI1250000	(DUMMY)
PHI1250		END

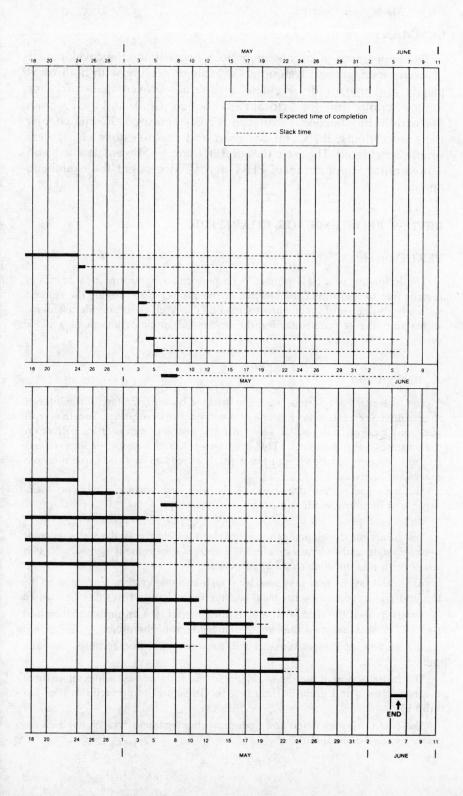

MAY JUNE

18 20 24 26 28 1 3 5 8 10 12 15 17 19 22 24 26 29 31 2 5 7 9 11

Expected time of completion
Slack time

18 20 24 26 28 1 3 5 8 10 12 15 17 19 22 24 26 29 31 2 5 7 9

MAY JUNE

END

18 20 24 26 28 1 3 5 8 10 12 15 17 19 22 24 26 29 31 2 5 7 9 11

MAY JUNE

SUMMARY

The appearance of large-scale projects necessitated the development of newer management techniques capable of dealing with both complexity and magnitude. Previous techniques proved inadequate for the job. In due time the Critical Path Method (CPM) and the Program Evaluation and Review Technique (PERT) appeared. In this chapter their antecedents, the Gantt bar chart and the milestone chart, were briefly considered. The essentials of PERT and CPM were detailed, and a nonmilitary application of PERT and CPM brought the chapter to an end.

REVIEW PROBLEMS FOR CHAPTER 10

PERT Project

The following is a class project to be performed by groups of students of four or five to a group. The objective of the project is to provide the student with a high degree of involvement with PERT as well as to provide useful insights into the requirements for the successful implementation of a PERT system.

A. Background

The National Park Service has been directed by the Secretary of the Interior to establish a recreation area in one of the larger Western States. Together with state and local officials, N.P.S. selected a site at the mouth of Bildon River on the shore of Lake Boondock. This area was selected because of its excellent fishing, camping, and swimming potential, as well as its close proximity to a main U.S. highway.

The area has certain drawbacks. It is wild and uncleared with no roads capable of handling modern traffic.

The funds for development were obtained jointly through an arrangement between state and federal agencies. The federal government agreed to match whatever funds the state could appropriate.

An initial layout was prepared by a well-known Western landscape architect and was approved by the state and the federal authorities (See Exhibit 1).

To insure that the recreation area is constructed in a minimum of time and cost, N.P.S. was assigned the overall management of the project.

A Park Service Project Manager will have complete line authority over the project.

The initial action of the Project Manager and his staff was to negotiate three contracts: roads and grounds, building facilities, and camp facilities (See Exhibit 2).

The financial, personnel, and operations management will be handled directly by the National Park Service.

Following negotiation of the contracts, a meeting was held between the project manager and his three contractors to establish policy for the overall management of the development effort. All parties agreed that a system was required to insure detailed planning and control of time and cost, since highly skilled craftsmen will be required and would be paid whether they work or not.

B. Requirements

Prepare a final oral and written report on the project. The report should include discussion of the following items. These are the minimum basic reports or charts for an adequate presentation.

Management summary network of not less than ten events or more than 20 events.

Networks charts for each of the three major contractors. These networks should show interfaces with the other contractors.

A total detailed network on not less than 75 activities which incorporates the directed requirements.

Gantt charts for each of the contractors.

Computer output reports.

EXHIBIT 1
Bildon River Project* (proposed vacation area)

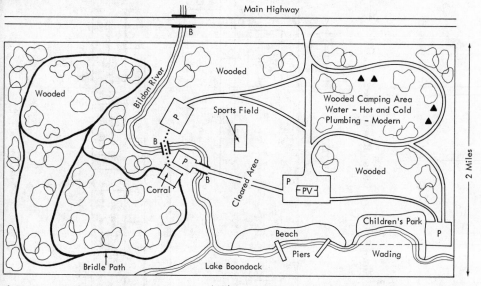

Key: B — Bridge
     P — Parking
     PV — Pavilion

All roads macadamized.

* Not drawn to scale.

EXHIBIT 2
Work Breakdown Structure, Bildon River Project

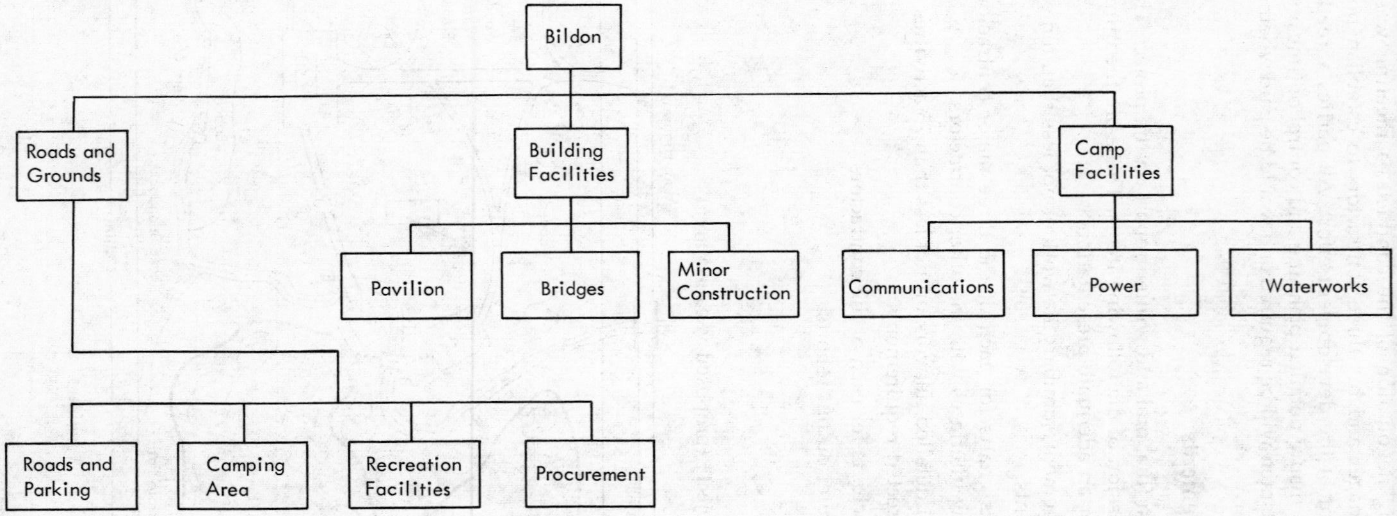

EXHIBIT 3

Project Assumptions

    Specifications and plans have been approved

    River width is 12 feet wide

    Power and telephone available at the main road

    Modern plumbing in the pavilion

    No sewer line at the main highway

    Project must be completed in maximum of 23 weeks.

Other Notes

    You may alter the location of facilities if desired

    You may build any type bridge desired so long as it can carry two-way traffic

    It is not necessary to show detailed layout of the pavilion or other facilities

    Do not go into details. If you do, the project will overwhelm you and much time will be wasted.

Note: Special thanks to William Shallman and Donald Brewer of the Army Management Engineering Training Agency (AMETA) for their permission to use the above project.

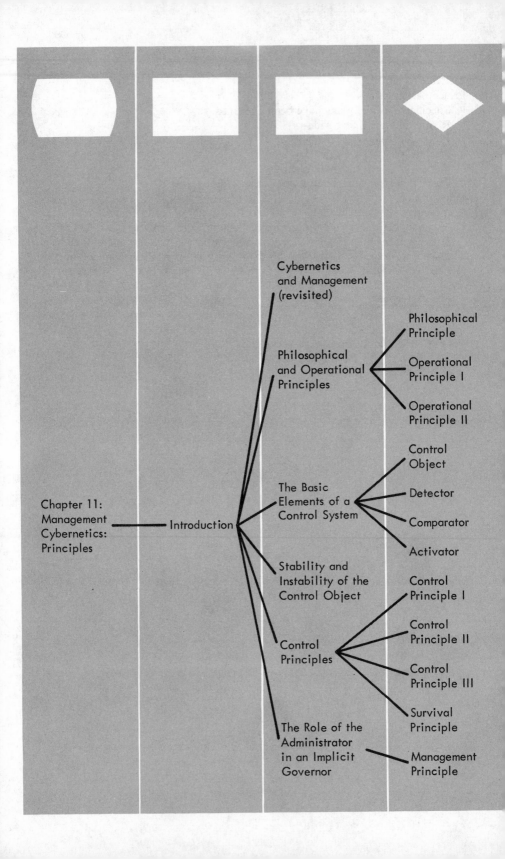

Chapter 11: Management Cybernetics: Principles

Introduction

Cybernetics and Management (revisited)

Philosophical and Operational Principles

Philosophical Principle

Operational Principle I

Operational Principle II

The Basic Elements of a Control System

Control Object

Detector

Comparator

Activator

Stability and Instability of the Control Object

Control Principles

Control Principle I

Control Principle II

Control Principle III

Survival Principle

The Role of the Administrator in an Implicit Governor

Management Principle

Chapter Eleven

Management Cybernetics: Principles

For if cybernetics is the science of control, and if
management might be described as the profession of control,
there ought to be a topic called management cybernetics —
and indeed there is. It is the activity that applies the findings
of fundamental cybernetics to the domain of
management control.

Stafford Beer

INTRODUCTION

The purpose of these last two chapters is twofold: (1) to highlight the basic principles of cybernetics and to relate them to the management of enterprises; and (2) to provide a brief description of several real-life situations in which the science of cybernetics is being applied as a managerial conceptual and practical framework.

To accomplish these two aims, this chapter begins with a recapitulation of the subject of cybernetics as it relates to the art and science of management. This brief recapitulation is to be achieved by organizing into a coherent framework the rules, principles, and guidelines presented thus far in the book.

Two main points will emerge from this recapitulation: (1) a cybernetic view of an enterprise; and (2) a cybernetic view of management. These two points taken together provide enough insight into the governing principles of cybernetics per se and of management cybernetics. Both subjects have often been misinterpreted, and to some extent misunderstood, as a result of the mysticism unnecessarily attached to them, and of the highly mathematical treatment previously accorded the subject of cybernetics,[1] and of the erroneous association of cybernetics with science fiction literature.

The next and last chapter is devoted to applications of management cybernetics (MC) to the management of modern enterprises. The description of applications of management cybernetics begins with micro and unfolds into macro considerations. Thus, the workings of MC are first examined in connection with the management of the operations of a micro enterprise such as a manufacturing firm. This description is then extended to the management of the firm as a whole under the subheading "The Whole Enterprise Control System."

Having thus described the application of MC for the micro enterprise (firm), the discussion continues with the management of a larger enterprise, such as a city. Under the heading, "Management Cybernetics and City Government," cybernetic principles employed to assist the management of a city government are highlighted. MC is next applied to political life in general. In the final section is to be found an application of cybernetic principles to ecosystems under the heading "Management Cybernetics and Global System Equilibrium."

CYBERNETICS AND MANAGEMENT (REVISITED)

Hitherto in this book numerous attempts have been made to relate the subject of cybernetics to the process of managing modern enterprises. These were primarily illustrative examples of possibilities for diffusing cybernetic thinking into the management process. As such, these examples were primarily intended to give certain cybernetic principles a more or less commonsensical explanation by associating them with conventional administrative principles and practices. Now, however, the time has come to discuss, more at length, the fundamentals of this hybrid discipline, management cybernetics.

All too infrequently the literature draws a dichotomy between general systems theory (GST) and cybernetics rather than viewing them as being integrated.[2] The integration of GST and cybernetics was, of

---

[1] See for example, N. Wiener, *Cybernetics* (Cambridge, Mass.: MIT Press, 1961); J. Klir and M. Valachi, *Cybernetic Modeling* (Princeton, N.J.: Van Nostrand, 1967); O. Lange, *Wholes and Parts: A General Theory of System Behavior* (Oxford, England: Pergamon Press, 1965).

[2] L. von Bertalanffy, "General Systems Theory—A Critical Review," in W. Buckley, *Modern Systems Research for the Behavioral Scientist* (Chicago: Aldine Publishing Co., 1967), pp. 16–17.

course, implicit in many writings, including Bertalanffy's, as he clearly indicated:

> . . . It appears that in development and evolution dynamic interaction (open system) precedes mechanization (structured arrangements particularly of a feedback nature). In a similar way, G.S.T. can logically be considered the more general theory; it includes systems with feedback [Cybernetics] constraints as a special case, but this assertion would not be true vice versa. It need not be emphasized that this statement is a program for future systematization and integration of G.S.T. rather than a theory presently achieved.[3]

Pursuant to this goal of "future systematization and integration," management cybernetics is proposed as the unifying framework or theory which integrates GST and cybernetics into a coherent scheme for dealing with control and communication, as well as with the evolution of complex dynamic open systems.

PHILOSOPHICAL AND OPERATIONAL PRINCIPLES

What follows immediately is, in part, a recapitulation of the material on modern systems thinking (GST, cybernetics, operations research/ management science, and systems engineering), but it is a necessary one if a successful transplant of the ideas developed in the preceding chapters is to occur.

The basic principle governing management cybernetics concerns the modern systems man's philosophy, which has been repeatedly referred to in this book as the "four pillars of systems thinking," namely, *organicism, holism, modeling,* and *understanding.* The researcher, manager, or administrator who wishes to use management cybernetics when dealing with the complexity of modern enterprises must adhere to this philosophical conviction. Adherence to this philosophy, of course, requires complete intellectual commitment. Anything short of full commitment can bring on catastrophic results, caused by the improper applications of the so-called systems approach, so vividly dramatized in the numerous criticisms of narrow quantitative models of a small portion of an organic system while lip service is paid to the "whole."[4] Pseudosystems work is, of course, worse than any conventional technique for dealing with complexity.

This leads us to the formulation of the following two principles of management cybernetics.

---

[3] Ibid., p. 19.

[4] C. West Churchman, *The Systems Approach* (New York: Delacorte Press, 1968); Ida Hoos, *Systems Analysis in Public Policy: A Critique* (Berkeley, Calif.: University of California Press, 1972).

Philosophical Principle

Cybernetically, enterprises/organizations are viewed as organisms (organicism); organisms are studied as wholes (holism); holistic studies of complex organic wholes can only be approached via a multistage modeling process (modeling); modeling of complex organic wholes can *only* lead to an appreciation or understanding of their structure, function, and evolution (understanding).

Corollary. Attempts to deal with the study of modern enterprises in a detailed, precision-oriented fashion (analytic thinking) will lead to damaging fragmentation of generic relationships that tie subwholes into coherent and cohesive entities; this fragmentation, in turn, gains precise knowledge about the *parts* at the expense of ignoring the basic principles governing the structure, function, and evolution of the *whole.*

Operational Principle I

Cybernetics, the science of control and communication, conceives of organic complex wholes (enterprises) as *purposeful control systems that feed on transmission of information (communication).*

The nature of this principle has also been repeatedly discussed. It will be recalled that teleology, purposefulness, or its cybernetic equivalent, goal-directedness, was listed as one of the basic properties of systems behavior, along with hierarchical order and adaptation and learning. Systems that exhibit goal-directedness are classified as purposeful and are characterized by the presence of a third-order feedback. Third-order feedback systems exercise governing or organizing function over first- and second-order feedback systems, which are, of course, inherent in every complex enterprise. These ideas about the orders of feedback were discussed at some length in Chapters 3 and 5.

The term "control" has been used frequently in this book. The time has now come to clear up any ambiguity in the reader's mind concerning this notion.

It must be emphasized at the outset that the conventional interpretation of the term is both inadequate and misleading. Stafford Beer pointed out this inadequacy over 15 years ago!

> The biggest lesson of all, however, is the interpretation that is placed on the notion of control itself. The fact is that our whole concept of control is naive, primitive and ridden with an almost retributive idea of causality. Control to most people (and what a reflection this is upon a sophisticated society!) is crude process of coercion. A traffic policeman, for example, is alleged to be "in control." He is, in fact, trying to determine a

critical decision-making point on much too little information, by a fundamentally bullying approach (because it is backed by legal sanctions).[5]

It is fairly obvious from the above quotation that the popular association of control with coercion is definitely anticybernetic. The cybernetic notion of control is that it is an integral, natural attribute of a system's behavior; i.e., it is an *implicit control.*

Operational Principle II

Cybernetically, *control is that function of the system via which a critical variable of system behavior is held at a desirable level by a self-regulating mechanism* (homeostatic control, implicit control, or control from within).

How does the implicit controller manage to perform such an important function? This apparently formidable task is in actuality very simple. The reason it is simple is that it is nature's own way of doing things, although it really took nature several billion years to perfect it.

The workings of a human body's homeostatic (implicit) controllers have long been understood.[6] The investigation of the same implicit controls for the achievement of a balanced growth of natural systems has been the concern of physical scientists for centuries. Contemporary ecosystem theory or ecology has given us new insights into the ingenious simplicity of natural control processes.[7] In man-made systems, servomechanical control principles have been extensively used for guaranteeing system control. The Watt governor and the common household thermostat are familiar examples of man-made implicit controllers.[8]

All these implicit controllers, whether natural or man-made, operate in accordance with certain basic principles which govern the behavior of their basic elements.

---

[5] S. Beer, *Cybernetics and Management* (New York: John Wiley & Sons, Inc., 1958), p. 21.

[6] See, for example, the classic work of Walter Cannon, *The Wisdom of the Body* (New York: W. W. Norton and Company, Inc., 1939).

[7] Of particular interest, we believe, are the following works: Eugene Odum, "Ecosystem Theory for Man," in J. A. Wiens (ed.), *Ecosystem Structure and Function* (Corvallis, Ore.: Oregon State University Press, 1971), pp. 11–23; E. P. Odum, *Fundamentals of Ecology,* 3d ed. (Philadelphia: W. B. Saunders, 1971); H. T. Odum, "Biological Circuits and the Marine Systems of Texas," in T. A. Olson and F. J. Burgess (eds.), *Pollution and Marine Ecology* (New York: John Wiley & Sons, Inc., 1967), pp. 99–157; H. T. Odum, *Environment, Power and Society* (New York: John Wiley & Sons, Inc., 1970).

[8] Description of man-made homeostats can be found in every standard text in servomechanics. For our readers, we suggest Stafford Beer, *Management Science: The Business Use of Operations Research* (New York: Doubleday and Co., Inc., 1968).

THE BASIC ELEMENTS OF A CONTROL SYSTEM

Implicit controllers are subsystems whose main function is to keep some behavioral variables of the focal or operating system within predetermined limits. They consist of four basic elements which are themselves subsystems. These basic elements are:[9]

1. A control object or the variable to be controlled;
2. A detector or scanning subsystem;
3. A comparator; and
4. An activator or action-taking subsystem.

These four basic subsystems of the control system, along with their functional interrelationships and with their relationship with the operating system, are depicted in Figure 11–1.

As can be seen from this figure, neither the subsystems of the control system nor their relationships to each other and to the operating system are completely foreign to the reader of this book. They have often been dealt with in previous chapters (see, for instance, Figure 6–3). The novelty here is that, while in other diagrams the control function was just shown as a feedback line going from the output of the system back to its input, the present diagram (Figure 11–1) represents a complete anatomy of that particular function. In addition, it can readily be seen that, as far as the control system is concerned the source of disturbance is the operating system itself whose behavior it is supposed to be regulating.

Since the reader is more or less familiar with the meaning of these four basic elements of a control system, a brief discussion of them should suffice.

Control Object

A control object is the variable of the system's behavior chosen for monitoring and control. The choice of the control object is the most important consideration in studying and designing a control system. Variations in the states of the control object—i.e., its behavior—become the stimuli which trigger the functioning of the control system. Without these variations, the system has no reason for existence. Since in reality there is never a perfect match between desired and actual outcome, variations will always exist; ergo, the need for control.

From the foregoing it should be clear that great care and much

---

[9] A much more detailed discussion of the basic elements of a control system can be found in R. A. Johnson, F. E. Kast and J. E. Rosenzweig, *The Theory and Management of Systems*, 2d ed. (New York: McGraw-Hill, 1967).

thought ought to be given to "what must be controlled." Let it suffice to point out the rather obvious observation that the control object must be chosen from the system's output variables. Well-balanced quantitative and qualitative attributes of the system's output should provide the best choice of control variables. Focusing on controlling system output does not necessarily imply an ex-post account of system behavior. Feedback control systems can just as easily function as anticipatory mechanisms as well as post-factor devices.

FIGURE 11–1
The Major Elements of a Control System

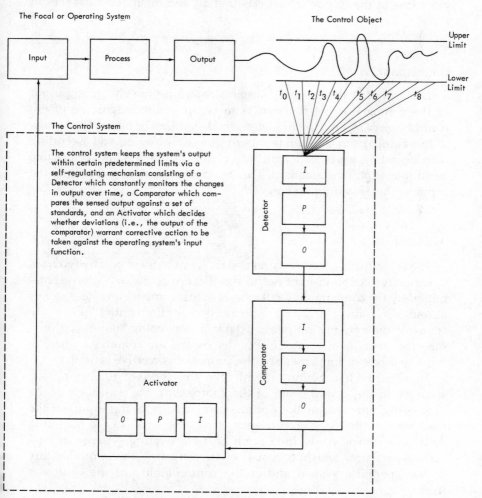

Detector

The structure and function of a detector or scanning system has been the subject of an entire previous chapter. There the entire organization was conceptualized as a scanning system. The only point that must be repeated is that scanning systems feed on information. Again, as in the case of the control object, the detector operates on the principle of selective acquisition, evaluation, and transmission of information. As such, a detector system is another name for a management information system (MIS). Therefore, the ideas expressed in the chapter on MIS are relevant here as well.

Frequency, capacity, efficiency, accuracy, and cost of detector devices are some of the important aspects that an administrator must reckon with.

Comparator

The output of the scanning system constitutes the energizing input of the comparator. Its function is to compare the magnitude of the control object against the predetermined standard or norm. The results of this comparison are then tabulated in a chronological and ascending or descending order of magnitude of the difference between actual performance and the standard. This protocol of deviations becomes the input to the activating system.

Activator

The activator is a true decision maker. It evaluates alternative courses of corrective action in light of the significance of the deviations transmitted by the comparator. On the basis of this comparison, the system's output is classified as being "in control or out of control." Once the status of the system's output is determined as being "out of control," then the benefits of bringing it under control are compared with the estimated cost of implementing the proposed corrective action(s).

These corrective actions might take the form of examining the accuracy of the detector and of the comparator, the feasibility of the goal being pursued, or the optimal combination of the inputs of the focal system, that is, the efficiency of the "process" of the operating system. In other words, the output of the activating system can be a corrective action which is aimed at investigating the controllability of the operating system and/or the controllability of the controller itself.

STABILITY AND INSTABILITY OF THE CONTROL OBJECT

The state of the control object's behavior can take either of the two following forms: (1) it can be stable, or (2) it can be unstable. Both of these states are necessary for system survival. While stability is the ultimate long-run goal of the system, short-run and periodic instability is necessary for system adaptation and learning. The system, in other words, pursues a long-run stability via short-run changes in its behavior manifested in its output's deviations from a standard.

Let us briefly explore the nature of stability and instability, as well as some of the reasons for instability. In general terms, stability is defined as the tendency for a system to return to its original position after a disturbance is removed. In our systems nomenclature, stability is the state of the system's control object which exhibits at time t_1 a return to the initial state t_0 after an input disturbance has been removed. Were the system's control object not able to return or recover the initial state, then the system's behavior would exhibit instability. The input disturbance may be initiated by the feedback loop or it may be a direct input from the system's environment. The particular behavior pattern that the system will exhibit is dependent on the quality of the feedback control system (detector, comparator, and activator) in terms of sensitivity and accuracy of the detector and comparator as well as the time required to transmit the error message from the detector to the activator. Oversensitive and very swift feedback control systems may contribute as much to instability as do inert and sluggish ones.

Time delay is the most important factor for instability of social systems such as business enterprises and governments. Although the application of information technology (MIS, EDP) has made considerable progress toward accelerating the transmission of information from the detector to the activator, as well as expediting the comparison and evaluation of information inside the comparator, still the impact of the corrective action upon the control object's behavior is felt after a considerable time lag.

Continuous oscillations of the kind exhibited in Figure 11–2 are the results of two characteristics of feedback systems: (1) the time delays in response at some frequency add up to half a period of oscillation, and (2) the feedback effect is sufficiently large at this frequency.[10]

Figure 11–2 demonstrates the behavior of the system's output which is controlled by a feedback system characterized by a one-half cycle time delay.[11] When a time delay of that magnitude exists, the impact of

---

[10] Arnold Tustin, "Feedback," in *Automatic Control* (New York: Simon and Schuster, Inc., 1955).

[11] Johnson et al, *Theory and Management of System*, 2d ed., p. 89.

FIGURE 11-2
Control Object's Behavior

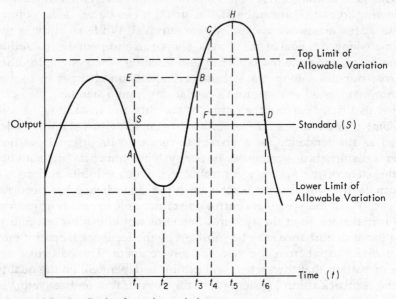

A = Point where direction of output is recognized
B = Corrective input is added
C = Error is noted—directive to remove resources
D = Resources are removed
EB and FD = Information time lag

Source: Adapted from R. A. Johnson, F. E. Kast, and J. E. Rosenzweig. *The Theory and Management of Systems*, 2d ed. (New York: McGraw-Hill, 1967), p. 89.

the corrective action designed to counteract the deviation as sensed, compared, and communicated comes at a time when this deviation is of a considerably different magnitude although it has the same direction. This causes the system to overcorrect. In Figure 11-2 a deviation of the magnitude equal to SA is detected at time t_1. At time t_2, new inputs are added to bring the output back to the standard (S). The impact of this corrective action upon the system's output is not felt until t_3. By that time the actual system's output is at point C; i.e., after a time lag equal to $t_4 - t_3$. The detector senses this new deviation and initiates new corrective action. Because of the one-half cycle time lag, the actual system's output has oscillated above the upper limits at the point H. New corrective action initiated, aimed at bringing the output back to the standard, will be felt at time t_6.

This basic principle of time-delay and its impact upon the system's control behavior is illustrated very clearly in Tustin's diagrams, Figure 11-3.

It might seem from the above brief description of the function of the

FIGURE 11–3
Oscillations in Feedback Systems

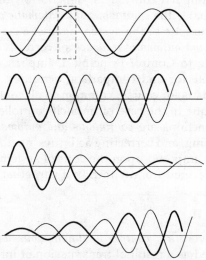

Oscillation is inherent in all feedback systems. The drawing at top shows that when a regular oscillation is introduced into the input of a system (*lighter line*), it is followed somewhat later by a corresponding variation in the output of the system. The dotted rectangle indicates the lag that will prevail between equivalent phases of the input and the output curves. In the three drawings below, the input is assumed to be a feedback from the output. The first of the three shows a state of stable oscillation, which results when the feedback signal (*thinner line*) is opposite in phase to the disturbance of a system and calls for corrective action equal in amplitude. The oscillation is damped and may be made to disappear when, as in the next drawing, the feedback is less than the output. Unstable oscillation is caused by a feedback signal that induces corrective action greater than the error and thus amplifies the original disturbance.

Source: Adapted from Arnold Tustin, "Feedback," in a *Scientific American* book, *Automatic Control* (New York: Simon and Schuster, 1955), pp. 20–21.

basic elements of the control system that the task is a formidable one when measured in terms of cost and/or time. In conventional control systems, this might indeed be the case. In cybernetic control systems, however, this is definitely not true. The reason, of course, is that this task is performed as part of the normal operation of the system and requires no extra effort.

CONTROL PRINCIPLES

The principles governing control functions in a cybernetic system are universal and simple. These principles formulated by Beer read as follows:

Control Principle I

Governors, or implicit controllers, depend for their success on two vital tricks. The first is the *continuous and automatic comparison* of some behavioral characteristic of the system against a standard. The second is the *continuous and automatic feedback* of corrective action.[12]

Thus, according to Control Principle I, implicit controllers are engaged in both detector and comparison activities, as well as in corrective action. This is, of course, common to all control systems. However, what is unique in the case of implicit controllers is the prerequisite that these functions are *continuous and automatic*. That is to say, detecting, comparing, and correcting activities are not initiated periodically, nor are they imposed upon the control system from outside, but they are rather executed from within in a perpetual manner.

Control Principle II

In implicit governors, *control is synonymous with communication*. Control is achieved as a result of transmission of information. Thus, to be in control is to communicate. Or, in Norbert Wiener's original words, "Control . . . is nothing but the sending of messages which effectively change the behavior of the recipient. . . ."[13]

This is indeed the most basic and universal principle of cybernetics. The realization that control and communication are two sides of the same coin motivated Wiener to use them as the subtitle of his classic pioneering work on *Cybernetics*. This is where one reads: *to control is to communicate and vice versa!*

It is evident from the above principles of control (I and II) that the system whose behavior is subject to this type of control becomes literally a slave to its *own* purpose. Since every deviation from standard behavior is autonomously and automatically communicated (through the sequential activities of the detector, comparator, and activator) the more frequently "out-of-control situations" occur, the more frequently communication takes place and consequently the more corrective action is taken. It is for this slave type of function that implicit controllers are also referred to in the literature as "servomechanisms" ("servo" means "slave").

This observation allows us to formulate another basic principle of cybernetic control, originally conceived by S. Beer:

---

[12] S. Beer, *Management Science: The Business Use of Operations Research* (New York: Doubleday and Co., 1968), p. 147.

[13] N. Wiener, *The Human Use of Human Beings* (New York: Avon Book Division of Hearst Corp., 1967), p. 45.

Control Principle III

In implicit controllers, variables are brought back into control *in the act* of and *by the act* of going out of control.[14]

This principle follows directly from our explanation of the basic structure of the control system. It will be recalled that what triggers the detector subsystem is the existence and magnitude of the deviation (d) between the goal and actual performance; i.e., the output of the operating or focal system (cf. Chapter 6). It follows that the more frequently deviations occur, the more frequent the communication between the detector and the comparator (control phase I). In addition, the more frequent and more substantial the magnitude of the deviation, the more likely it is that corrective action will be initiated and executed (control phase II).

From the foregoing discussion of the basic principles of cybernetic control, the following question is inescapable: given the unique nature of a cybernetic control system, what kinds of demands do these control principles impose upon goal-directed systems?

The most important demand facing the system is that it be an adaptive learning system. In other words, the function of the implicit controller demands that the operating system eventually learn that being in control is as necessary a condition for its survival as its growth capabilities. Thus, another basic principle of cybernetics:

Survival Principle

In organic wholes, *growth and control are the two sides of the same coin. A system's growth is checked and facilitated by control.*

This is neither a contradiction nor a paradox. It is indeed an axiom! Control prevents growth tendencies from becoming exponential, thereby running the risk of reaching limits imposed from without—a coercive and insidious control. Implicit control is, in fact, as natural as nature itself!

THE ROLE OF THE ADMINISTRATOR IN AN IMPLICIT GOVERNOR

The basic conviction underlying the ideas developed in the previous chapters of this work is that an administrator or manager is essentially a decision maker. Most modern literature in organization or management theory would agree with this contention.[15]

---

[14] Beer, *Management Science*, p. 147.

[15] R. M. Cyert and J. G. March, *A Behavioral Theory of the Firm* (Englewood Cliffs, N.J.:

The foregoing development of the basic principles of cybernetic control, along with their accompanying discussions should have provided an obvious hint that in cybernetics, decision-making and control are two very closely related, if not identical, activities. This allows us to formulate the first management principle:

Management Principle

Cybernetically, *decision making and control are similar if not identical managerial activities.* Both activities are initiated and maintained through communication.

The relationship between information acquisition, evaluation and dissemination (communication), was explained in Chapter 6 of this book. The discussion of the basic principles of cybernetics has also emphasized the dependence of control on communication. It thus appears that the common denominator between the two basic managerial activities of decision and control is communication. This is, of course, the reason an entire chapter has been devoted to this subject.

The cybernetic framework or way of looking at decision, control and communication is of tremendous importance for the administrator of modern enterprises, for the precise reason that enterprises consist of human beings who are by definition communicative. In communicating, humans decide; in deciding, they communicate; in communicating, they control; in controlling, they communicate; . . . and the cycle goes on as long as the systems enterprises remain what they are, namely living entities.

SUMMARY

In summary, the brief recapitulation of the basic principles of cybernetics and management points out two main viewpoints of management cybernetics:

Viewpoint I.   Enterprises, if viewed as wholes, can most effectively be studied as control systems of the type referred to here as implicit controllers or governors. As such, the basic goal of the enterprise—survival in an everchanging environment—is achieved through a balanced

Prentice-Hall, 1963); J. G. March and H. A. Simon, *Organizations* (New York: John Wiley & Sons, Inc., 1958); J. G. March, *Handbook of Organizations* (Chicago: Rand McNally, 1965); H. A. Simon, *Administrative Behavior* (New York: Macmillan, 1959); F. E. Kast and J. E. Rosenzweig, *Contingency Views of Organization and Management* (SRA, Inc., 1973).

growth; i.e., an increasing complexity of the enterprise's structure and function which is governed by an
implicit controller.

Viewpoint II. Managers and administrators in a cybernetic enterprise are implicit controllers. Their main functions of
decision and control are carried out via the process of
communication. *Decisioning, controlling, and communicating constitute the three basic characteristics of the administrator* as a whole man and, through him, the three
basic pillars of the enterprise.

Management cybernetics represents a hybrid discipline which may
be defined as the application of general systems theory and cybernetics to the management of complex enterprises of all types.

Figure 11–4 presents the basic relationships between the goal of an
enterprise and management cybernetics. Enterprises, like all organic
wholes, have one goal: survival. To survive they must grow and evolve.
Evolutionary growth requires that increases in the system's complexity
contribute to the system's ability to survive. That is to say, growth
processes must be true improvements of systems' performance to cope
with the everincreasing and continuously changing environmental
complexity. Progress toward that goal is facilitated by implicit governors which prevent exponential growth and decay processes.

General systems theory studies the general evolutionary growth
pattern and principles of biological systems. Cybernetics is concerned
with the function of implicit controllers in both natural and man-made
systems. Finally, management cybernetics, by combining both GST and
cybernetics, determines the growth and evolutionary requirements of
enterprises and designs the requisite implicit controllers.

These then, are the basic philosophical, operational, and managerial
principles governing the application of management cybernetics to the
management of modern complex enterprises. The following accounts
of applications of management cybernetics are not intended to be either
exhaustive or exclusive. Rather, they constitute a cursory treatment of
the huge ground swell begun by the pioneering works of von Bertalanffy, Wiener, Beer, and others, which is slowly but steadily affecting
managerial philosophies and day-to-day practices.

FIGURE 11–4

FIGURE 11–4
Management Cybernetics: Anatomy and Meaning

The Goal of an Enterprise

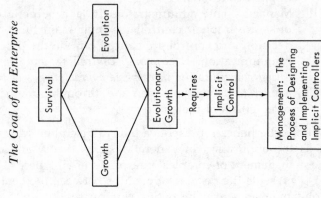

The Goal-Reaching Processes

Growth: Successive positive feedbacks. Unchecked feedbacks lead to counterevolutionary growth.
Examples: Exponential growths of weight, population, pollution, epidemics, and so forth.

Control: Successive negative feedbacks. Too frequent and lightly triggered feedbacks (i.e., small deviations) stifle learning processes.
Examples: bottlenecks in an overcontrolled economy, tightship outfits, shy individuals because of disciplinary parents, Clockwork Orange societal controls, etc.

Survival: Successive +FBs and −FBs.
Examples: Balanced populations of species, economies, adaptable individuals, organizations, nations, successful rehabilitation programs, and so forth.

The Methods of Understanding

GST: Organizations are open systems. Evolutionary survival is guaranteed by homeostasis; in biological organizations (systems), homeostasis is an inborn characteristic.

Cybernetics: Organizations as open systems must possess implicit governors.

Management cybernetics: In enterprises implicit governors are not inborn but must be rationally designed by the manager.

REVIEW QUESTIONS FOR CHAPTER ELEVEN

1. Most businessmen have never heard of cybernetics, yet in spite of this operate their businesses in an efficient manner. Taking a retail store in your town, show how they do employ cybernetic principles, although they are not aware of it. This calls for an identification of the components of the system and a discussion of the functions of the components.

2. The federal government funds many programs dealing with health care, alcoholism, drugs, urban development, and so on. In spite of these many programs critics argue that they are not doing the job. Cybernetically speaking, what components are missing or ill-defined in such systems so that it may be difficult to determine whether the job is being done?

3. The text states the need for continuous and automatic feedback in a cybernetic system. What is meant by continuous? Is this a time dimension or can it be something else? Show how it can be both and show how the time dimension can vary with the process of controlling the system.

4. The text discusses the need for a firm to be an adaptive learning system. Is it possible for a firm to reject this principle without negative consequences' being incurred?

5. The text states that a "system's growth is checked and facilitated by control." Is it possible to grow too fast? Give a particular instance of this situation and show how control could have been helpful.

6. The authors state that decision making and control are similar if not identical managerial activities. Make the case that these are not similar through the use of examples. After you have done so then refute the case that you have made.

7. Show how the accounting department acts as an implicit controller for the organization. Show the major elements of the control system as their functions.

8. In a university setting, give examples of situations in which it is comparatively easy to establish control systems, or very difficult to do so. Which parts of college or university might understand cybernetics more easily than others? How does measurement enter into such a system?

9. Most of the material in this chapter deals with people-control systems. Give some examples of mechanical control systems and identify the principles involved.

10. Taking the city in which your college or university is located, identify the major control systems employed.

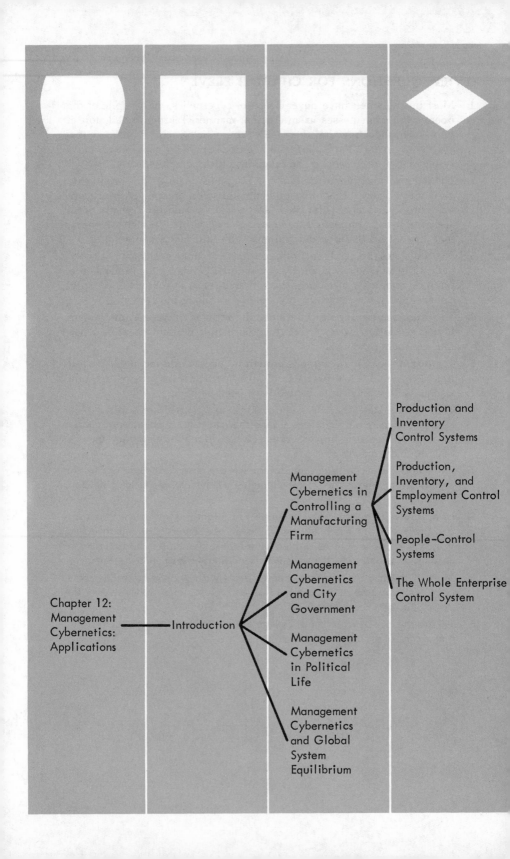

Chapter 12:
Management
Cybernetics:
Applications

Introduction

Management
Cybernetics in
Controlling a
Manufacturing
Firm

Management
Cybernetics
and City
Government

Management
Cybernetics
in Political
Life

Management
Cybernetics
and Global
System
Equilibrium

Production and
Inventory
Control Systems

Production,
Inventory, and
Employment Control
Systems

People-Control
Systems

The Whole Enterprise
Control System

Chapter Twelve

Management Cybernetics: Applications

Because wealth is energy, people everywhere will be rich in power when integrated world-wide industrial networks are established. A world-wide electrical energy network linking the day and night hemisphere would result in staggering economic gains. As Vladivostok sleeps, Con Edisonnovitch's current would be channeled to California. Industry works best as a world system. Newly emerging nations must realize that their independence depends on their participation in the world industrialization.

R. Buckminster Fuller,
I Seem to be a Verb

INTRODUCTION

In this last chapter we shall attempt to report briefly on some actually operating control systems which function in accordance with the basic principles of cybernetics explained in the previous chapter. To preserve authenticity, and to avoid as much as possible the imposition on the reader of our own biased interpretation, we have chosen to freely present the crux of these systems in the language of the original designer. We hope that we are making a contribution to the field in terms of our organization of these applications of management cybernetics into a comprehensible and logical framework. Beyond that, the glory of originality belongs to those individuals whose works we quote.

We will therefore begin with the illustration of the application of management cybernetics to specific operations of the firm. However, it must be kept in mind that the difficulty of designing implicit controllers increases as one goes from a production and inventory control system to a system for the whole enterprise. The reason for this difficulty is that our understanding of the detailed structure and function of the enterprise itself and of the environment, as well as our understanding of the interactions between the two, is still very meager.[1]

Even though the task of designing implicit controllers for larger systems seems exceedingly difficult, it is by no means impossible. Moreover, the benefits to be derived from the dependable operation of such control devices far outweigh their costs. In addition, one has little choice in deciding whether or not to design such systems. One must in varied degrees abide by the laws governing natural behavior. This is especially true today, when the physical or natural environment seems to more or less dictate the systems' design.

MANAGEMENT CYBERNETICS IN CONTROLLING A MANUFACTURING FIRM

It must be pointed out at the outset that the greatest successes are to be found in the management of the operations rather than in the management of the so-called human side of the enterprise. The reason for this is that operations deal primarily with material flows and processes and, to a large extent, are quantifiable and deterministic. The human element is less important, and decision making is more routinized. In cybernetic terms, there is less "variety" in such systems and therefore less variety is needed for controlling such systems. Because of this the field of cybernetics has been dominated by the engineer. Only recently has cybernetic control of organizational activities been applied to the human element. Systems of this type will be discussed in the section entitled People-Control Systems.

Production and Inventory Control Systems

In his application of cybernetics to the manufacturing process, Beer has utilized the familiar block diagram below to depict a production control system.[2] The input (θ_i) into the production system is the raw material, and its output (θ_o) is the product. Work flows through the

---

[1] S. Beer, *Decision and Control* (New York: John Wiley & Sons, Inc., 1967), pp. 301–2.

[2] S. Beer, *Cybernetics and Management* (New York: John Wiley & Sons, Inc., 1959), pp. 171–72.

system at the rate of (μ). θ_L represents the arrival of new orders. There are two control loops, as can readily be seen. The major loop of $\theta_i - \theta_o$ is the difference between the desired state and the actual state, in other words, the error (ϵ). This error is fed back into the system through the operator (K_2), who makes the appropriate rate of adjustment with respect to the goal. The planned level (η) becomes the actual rate (μ) by operator (K_4). However, the actual rate of production is also modified by previous rates and new orders (θ_L). The new orders constitute an input affecting the planned rate via operator (K_3). This load directly affects the output and its action is represented by operator (K_1). This order level represents the second control loop.

FIGURE 12–1
Production Control System as a Cybernetic System

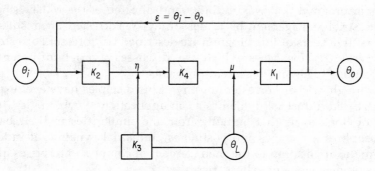

The output of this system may be described mathematically:

$$\theta_o = K_1(\mu - \theta_L) \tag{1}$$

The output of the system (θ_o) is the rate of production as affected by the order load (θ_L), the influence being represented by the primary operator K_1. The error feedback (ϵ) is, of course, the difference between output and input. The actual production rate is being determined by the planned rate (η) through its operator K_4. The planned rate (η) itself is influenced by the error (ϵ) with its operator (K_2) and the order load (θ_L) with its operator K_3.

These relationships are presented in equations 2, 3, and 4.

$$\epsilon = \theta_i - \theta_0 \tag{2}$$
$$\mu = K_4\eta \tag{3}$$
$$\eta = K_2\epsilon + K_3\theta_L \tag{4}$$

The four equations effectively define the system. With the treatment of the variables as functions of time, it is possible to complete the analysis. For the interested reader, the complete mathematical analysis may be followed in Beer[3] and especially in Simon's original work,[4] where the mathematical relationships are worked out in greater detail.

Production, Inventory, and Employment Control Systems

A more encompassing cybernetic production-inventory-employment control system comes from attempts to apply servomechanic principles to industrial management. These attempts are known as industrial dynamics or, as most recently rechristened, systems dynamics. Industrial dynamics is defined as "a philosophy which asserts that organizations are most effectively viewed (and managed) from this control system perspective. It is also a methodology for designing organizational policy." The two-pronged approach is the result of a research program that was initiated and directed at the M.I.T. School of Industrial Management by Professor Jay W. Forrester. The results of the first five years of this program are described in Professor Forrester's book, *Industrial Dynamics,* which also discusses a variety of potential applications to key management problems.[5]

Although neither Forrester nor any of his disciples have ever explicitly labeled industrial dynamics as an application of cybernetics, it is very obvious to us that, judging from the similarities in their basic philosophies, principles and postulates, industrial dynamics is indeed cybernetic application to the management of complex enterprises, that is to say, management cybernetics.

For this reason, and because industrial dynamics has recently been widely popularized because of Forrester's association with a remarkable study on world dynamics and world equilibrium, the applications of industrial dynamics will be given preferential treatment. It is hoped that this rudimentary exposure will reverse the unfortunate conventional association of industrial dynamics with strictly inventory management. Its potential application to social systems is far-reaching indeed. Figure 12–2 depicts an enterprise as a control system. Although in this diagram some slight differences exist from the cybernetic one presented previously, the two diagrams are essentially the same.

The decision-making process is a response to the gap in organizational objectives and actual results. In cybernetic terms, it was the

---

[3] Ibid., p. 172.

[4] H. A. Simon, *Models of Man* (New York: John Wiley & Sons), 1957.

[5] E. B. Roberts, "Industrial Dynamics and the Design of Management Control Systems," in P. P. Schoderbek (ed.), *Management Systems,* 2d ed. (New York: John Wiley & Sons, Inc., 1971), p. 416.

measurement process (detector) which detected not only the existence of a gap but also its direction and magnitude. The difference between real and apparent achievements in Figure 12–2 may or may not be significant, depending on the communications channels and the particular management information system available and/or desired by the firm. For example, for the sake of simplicity or timeliness of information, one may sacrifice accuracy. Thus, a difference may exist between real and apparent achievements.

As can be seen in Figure 12–2, a decision transformation process exists which transforms decisions into results (inputs into outputs) through a complex process which involves the organizational structure, production function, market factors, and so forth.

Attention must also be given to the multiple feedback loops which exist in both the total system (the organization) and its subsystems. These loops are iterative in that the cycle is a never-ending one in which goals, decisions, measurement, and evaluation are the prime elements.

FIGURE 12–2
Control System Structure of Organization

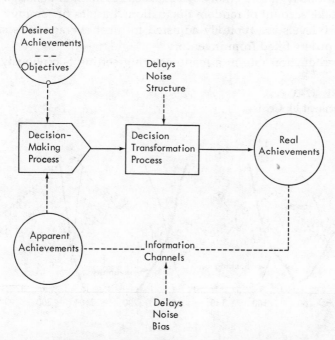

Source: Adapted from E. Roberts. "Industrial Dynamics and the Design of Management Control Systems," in P. P. Schoderbek (ed.), *Management Systems*, 2d ed. (New York: John Wiley & Sons, Inc., 1971). Reprinted with permission of Professor Roberts.

This being the basic viewpoint of an enterprise advocated by industrial dynamics, let us begin with the description of a control system for production-inventory-employment management.

The application of industrial dynamics which began with what appeared to be a simple inventory problem proved, upon an in-depth analysis, to be much more encompassing. In graphing the major variables associated with inventory (Figure 12–3) one can readily see that wide oscillations occur in inventory and employment, and back orders which appear extreme when compared to the oscillations of incoming orders. Attempts by management to control each of these problem areas as they arose were largely futile. As soon as the inventory problem was solved an employment problem arose; as soon as the employment problem was solved another problem arose. This analytical approach was soon abandoned in favor of the systems approach, which focused on the dynamic integrative behavior of the four variables. The problem was now being approached from a different perspective. The question when reformulated simply asked: Could the oscillations in inventory and production rates come from the *interactions* of management's internal policies with its customers rather than from the firm's internal policies?

Traditionally, an inventory system seeks to maintain enough inventory to take account of random fluctuations in sales (incoming orders). Inventory levels are typically adjusted to meet average sales and the ratio of orders filled from inventory.

The production rate in a manufacturing setting is obviously deter-

FIGURE 12–3
Management by Crisis

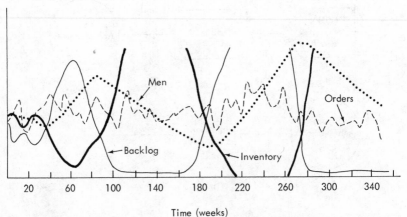

Time (weeks)

Source: Adapted from E. Roberts, "Industrial Dynamics and the Design of Management Control Systems," in P. P. Schoderbek (ed.), *Management Systems,* 2d ed. (New York: John Wiley & Sons, Inc., 1971). Reprinted by permission of Professor Roberts.

mined by the number of people employed. Finished products either are shipped directly to the customers or are put into inventory, depending on the backlog of orders. Clearly one controls both the size of the inventory and the backlog size by the number of employees in production.

What happens, of course, is that when orders increase, management hires new employees in order to increase production so as to meet the demand and to build up some inventory. As soon as inventory is built up and the demand is met, employees are laid off. The decline in the rate of employment causes the production level to drop, a fact that leads to delivery delays, backlog order accumulations, and, ultimately, of demands for new hiring of employees. The cycle repeats itself indefinitely.

FIGURE 12–4
Effects of Management Control Systems

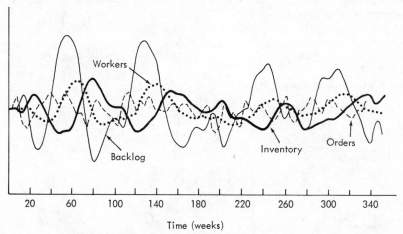

Time (weeks)

Source: Adapted from E. Roberts, "Industrial Dynamics and the Design of Management Control Systems," in P. P. Schoderbek (ed.), *Management Systems*, 2d ed. (New York: John Wiley & Sons, Inc., 1971). Reprinted by permission of Professor Roberts.

Figure 12–4 shows the behavior of the four basic variables of the system after conventional control methods were installed. Although the firm has obviously benefited from the operation of the control system by dampening oscillations and preventing some overshootings (except in the case of backlog around the 50th and 130th week), the basic dynamic behavioral patterns observed in Figure 12–3 are still present.

This, then, was the behavior of the basic control variables at the beginning of the industrial dynamics program which lasted for several years. One of the things the researchers did was not only to duplicate the manufacturer's organization (shown in Figure 12–5) but also to add the customer subsystem. As shown in the figure, incoming orders come through the engineering deck which has responsibility for developing specifications for the products. These orders were for component products which would be used in the customer's own manufacturing process. The orders are prepared and released to production as dictated by the customer delivery dates.

Figure 12–5 is a graphic model that represents the decision policies in the system and how they are related to one another. The model is closed except for the customer system. The presence of the customer system uncovered the existence of a feedback loop previously ignored. It was found that the delays which are quoted for delivery to customers affect new ordering rates, that is to say, that if longer delays are being quoted to customers, they simply order earlier to compensate for such delays. This in turn affects the delivery delays. This loop obviously amplifies the problems of the company which manifest themselves through the wide oscillations in inventories, backlog of orders, employment, and profits.

FIGURE 12–5
Company-Customers Systems

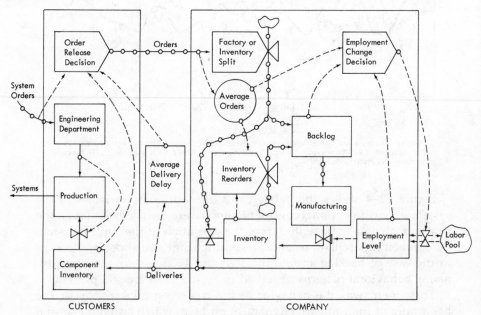

Source: Adapted from E. Roberts, "Industrial Dynamics and the Design of Management Control Systems," in P. P. Schoderbek (ed.), *Management Systems*, 2d ed. (New York: John Wiley & Sons, Inc., 1971), p. 424. Reprinted by permission of Professor Roberts.

Of the 140 variables initially identified in this model 30 of them represented decision points, and of these, five were singled out as critical. These were:

1. The purchasing rate by the customers as it depends on design releases from their engineering department, their inventory policies, and their response to changing delivery-delay quotations coming from the supplier.
2. The factors that determine the fraction of the incoming-order flow which the factory can fill from inventory.
3. The basis for quoting delivery delays from the supplier to the customer.
4. The basis for inventory reorder from the stock room to the production department.
5. The policies governing changes in employment and work week which control the production rate.[6]

As discussed earlier, the time history of the dynamic model using conventional methods to reduce the oscillations was successful; however, Figure 12–6 shows the behavior of the system model when it was set up with the policies that the researchers believed to be governing

FIGURE 12–6
Effects of Industrial Dynamics Policies

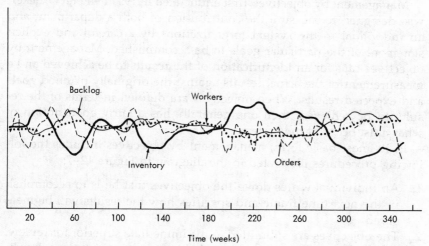

Time (weeks)

Source: Adapted from E. Roberts, "Industrial Dynamics and the Design of Management Control Systems," in P. P. Schoderbek (ed.), *Management Systems*, 2d ed. (New York: John Wiley & Sons, Inc., 1971), p. 425. Reprinted by permission of Professor Roberts.

---

[6] Jay Forrester, "Managerial Decision Making," in Martin Greenberger (ed.), *Computers and the World of the Future* (Cambridge, Mass.: MIT Press, 1962), p. 59.

the system. An examination of this latter figure will show that while inventories still fluctuate employment oscillations are about one-third as great. Oscillations in backlog of orders as well as delivery delays are also dampened.

The purpose of introducing the student to the basic concepts underlying industrial dynamics is to show an application of the systems approach. This particular example highlights the inadequacy of the analytical method and stresses the importance of viewing the organization as a system composed of many interacting subsystems whose performance affect total systems performance.

People-Control Systems

Thus far we have discussed applications of management cybernetics to the management of a production-control system (Beer), and employment system (ID). Although all of these activities are performed by people, we deliberately left out the management of the human element. If one were to ask: How do managers control their subordinates, as well as each other? the answer is through implicit controllers. The concept that resembles most closely our implicit governor control system is that of management by objectives, or, for short, MBO. This technique, introduced nearly two decades ago, has been found to be a powerful management control tool.

Management by objectives, first enunciated in 1954 by Peter Drucker, was designed to measure the contribution of both a department and an individual to the system (organization) by a careful and explicit statement of the particular goals to be accomplished. Management by objectives calls for an identification of the results to be achieved and a measurement of the actual results against the originally planned goals and expected results. When objectives are defined in terms of the results to be achieved, then one generally has a fairly good notion of what must occur in the system.

In its most basic form, management by objectives includes the following procedures illustrated in the diagram in Figure 12–7:

1. An individual writes down the objectives that he is to accomplish in the next time frame and specifies how the results will be measured.
2. The objectives are submitted to his immediate superior for review. Out of this review comes a set of objectives to which the subordinate commits himself.
3. Evaluation of performance is carried out in light of the previously agreed-upon objectives. Modifications in the individual's behavior may occur because of variances between the results achieved and the results expected.

FIGURE 12–7
The Management by Objectives (MBO) System

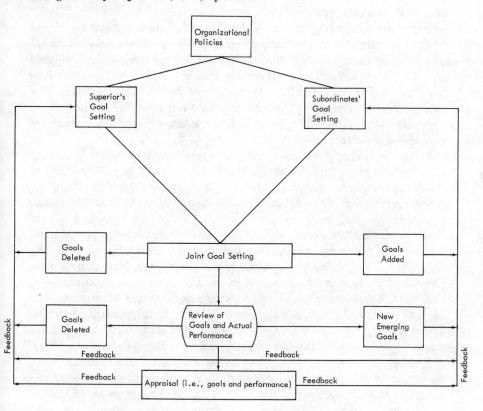

One can easily detect the similarities of goal setting in management by objectives with goal striving or goal achievement in cybernetics. Drucker himself stated that the main advantage of management by objectives is that it allows the manager to exercise *self-control,* which is nothing else than the "self-regulating device" in cybernetics. The negative feedback concept is thus employed, since actual performance is compared with the goal and corrective action is taken as a result of any "error" detected.[7]

Any system which can reflect upon its goals, search its memory for past behavior, and change its goals is a third-order feedback system. Such is the management by objectives system. In the MBO system, deviation from a goal forces one to reexamine the reasons why the goal

---

[7] A similar diagram is presented in G. I. Morrisey, *Management by Objectives and Results* (Reading, Mass.: Addison-Wesley Publishing Co., 1970), pp. 98 and 140, under the title "The MOR Process."

was not attained. This reexamination could lead to the formulation of a new goal which could then result with a change in the environment or a change in the system.

From the foregoing, it should be clear that management by objectives is applicable in situations other than those of business organizations. When one considers that all living systems are teleological (goal striving), then it isn't too surprising that management by objectives has such wide applicability. Management by objectives, instead of being a revolutionary tool, can be regarded rather as the application of well-tested principles of cybernetics to the management of human resources of an enterprise.

Where multiple goals exist, each of these may be treated as having its own receptors and feedback loops. The system may even be expanded to include multiple actions and receptors. This can help account for the fact that certain actions may result, not from deviation from a single goal, but from deviation from more than one goal. Figure 12–8 illustrates this point. While the actual condition of a single goal, say x_3, may never be maximized, the overall system objectives (x_1, x_2, x_3, x_4, etc.) may indeed be. Management, of course, has many objectives, and these must be handled by means other than maximization. One cannot maximize profit and let customer services deteriorate, nor can one disregard union-management relations in the pursuit of profits. There are approximately five key areas with organizational objectives: market standing, profitability, research and development, productivity, and financial resources. In each of these areas there are distinct objectives, and any attempt to optimize one alone will ensure the suboptimization of the others. In practice, a firm accepts less than optimization and seeks an "acceptable" level of performance in each major area. In reality, the number of independent organizational goals is rather small.

FIGURE 12–8
Multiple Goal, Multiple Detector and Multiple Action System

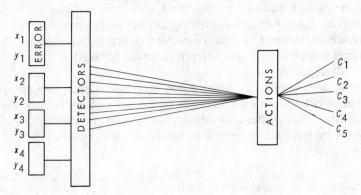

These are usually spelled out in terms of acceptable and/or desired levels of performance rather than the optimal.

Although the benefits of management by objectives as applied in industry today reach far beyond the expectations of its early practitioners, the system has been and will continue to be a basic control tool. As noted in the previous inventory, production, and employment system, the iterative cycle is goal formulation, feedback, measurement, corrective action, goal formulation, and so on. The point being made here is simply that the vastly popular management by objectives system which is utilized throughout the world today is really a simple cybernetic system. Recognition is made of the fact that goals in this system are not self-maintaining ones but rather third-order feedbacks.

The Whole Enterprise Control System

The cybernetic model for a control system for the whole enterprise is a rather complex one. This complexity, however, is dictated by the very nature of the enterprise as an organic system that functions within a dynamic complex environment. To the extent that real life enterprises are complex (in the sense of organized complexities, as the term was explained in Chapter 5), and to the extent that the environment is in reality also complex (another organized complexity), the law of requisite variety, explained in Chapter 3, dictates that a control system designed to guarantee survival of an enterprise must also be complex.

In addition to complexity, a second characteristic of a cybernetic whole enterprise control system is that it should be an open system. This means that the system must encompass both the enterprise per se and its environment. This, of course, follows from the basic contention that the enterprise and its environment are the two halves of the same whole, as was explained in Chapter 6. Thus, it follows that this kind of control system focuses on the interaction between these two subentities and not on the entities themselves, as most conventional financial controls do.

The third characteristic of a cybernetic whole enterprise control system is that it must be a self-regulating or closed-loop system. This is, of course, a basic prerequisite for all implicit controllers. Self-regulation does not necessarily mean that all control activities will be completely autonomous from other managerial activities; rather it implies that the control system constitutes the suprasystem of a multiplicity of quasi-independent control domains, and is, therefore, itself quasi-independent. Self-regulation implies autonomy within a structure.[8]

---

[8] This is the same autonomy that "profit centers" enjoy. Within certain broad market and financial policies set by the headquarters, the centers can operate as they see fit.

The necessity of structure constitutes the fourth characteristic of a cybernetic whole enterprise control system. The nature of this structure has been characterized throughout this book as being that of a hierarchical order, that is to say, a hierarchy of feedback control systems of different orders constituting a hierarchy of homeostats. As was explained in Chapter 5, this hierarchy is not imposed upon the control system, but is implicitly informed. In other words, the functions of the local control subsystems (production, inventory, employments, accounting, etc., control systems) are mediated centrally to avoid suboptimizations. Suboptimizations occur, it will be recalled, when certain improvements in the operating system's performance do not contribute toward increases in the performance of the whole system.

A model for a whole enterprise control system which encompasses all four basic prerequisites of an implicit controller—namely, complexity, openness, self-regulation, and hierarchical structure—is provided by Beer. His main premise is that the model provides a "newly oriented insight, an enriched vocabulary, a way of thinking that rises above the platitudes of orthodox management training. *Cybernetics is about control, which is the profession of management*" [emphasis added].[9]

Figure 12–9 below depicts the model in its entirety. The figure clearly illustrates the well-known fact that the modeling begins with the world situation and ends with certain concrete managerial policies designed to cope with this world situation. For the enterprise, this world situation can be bisected into two semi-independent areas: the internal world situation (i.e., the enterprise itself, its resources, strengths and weaknesses) and the external world of the enterprise environment (i.e., that portion of the world situation which has a direct bearing upon the enterprise, but which is beyond its immediate control).

The cybernetic control system for the whole enterprise is a series of nine homeostats or implicit controllers ($H_1 - H_9$). The full description of the nature of these homeostats and the relationships among them can be found in Beer's ingenious work. Here we confine ourselves to a very brief description. The first homeostat (H_1) connects the internal and external world. H_2 is the basic control center for the enterprise. It controls the moment-to-moment state of affairs. The two *Gestalt* memory homeostats H_3 supply historical data to H_2. H_4 prepares for immediate action in the real world. While H_2 is a scanning system which receives information, evaluates it, and records it in the two *Gestalt* memories, H_4 is a parallel activity which prepares to transmit instructions on the basis of its knowledge. Because these instructions will lead to specific actions in the real world of immense variety, to comply with the universal law of requisite variety a variety generator is needed to

[9] Beer, *Decision and Control*, p. 398.

regain variety for the control system. H_5 undertakes this proliferation of variety by acting as an optimizing device.

The output of H_5 represents a detailed program of activity which feeds back to the real world situation. H_6 is an operational homeostat. It interlocks H_1 and H_2 in both world situation halves: H_6 for the external and H_6 for the internal. These six homeostats guarantee control of the immediate (short-run) operations of the enterprise by constant comparison and adjustment of the state of the enterprise-environment

FIGURE 12–9
A Control System for the Whole Enterprise

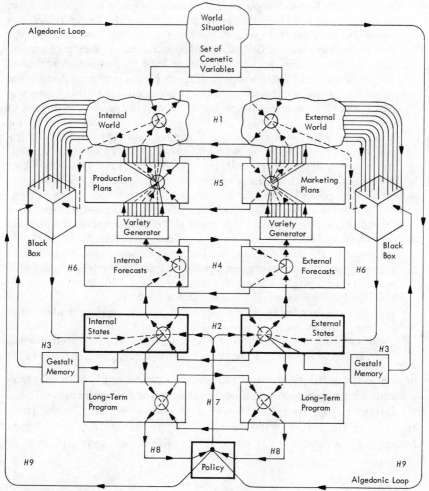

Source: Adapted from S. Beer, *Decisions and Control* (New York: John Wiley & Sons, Inc., 1967). Reprinted by permission of John Wiley & Sons, Inc.

relationship through flows of information. This information undergoes first a variety reduction in the two black boxes and subsequently a variety amplification in the homeostats H_4 and H_5.

Short-run operations are not the only concern of management, however; management must also guarantee long-run survival of the enterprise. Homeostats H_6, H_7, H_8, and H_9 are designed to provide management with the information necessary for future assessment and prognostications. The operation of these four homeostats is described by Beer as follows:

> If we return to the central core of the control, the homeostat H_2, we may see that it sponsors a further activity at the bottom half of the diagram. Just as H_2 can be used to generate immediate forecasts of events, so it can be used to examine long-range prognostications. The homeostat H_7 is a continued management exercise, not operated in real time, in which the internal and external situations are balanced for many years to come. The representative points are determined, at least in part, by extrapolated outputs from H_2. The adjustment of these points is determined by the operation of the homeostat H_7 itself. It is the output of this homeostat which largely determines the policy of the enterprise—as can be seen in the drawing. Now this policy is at once fed back to the homeostat H_2, for it is the long-range intention of the management which conditions the way in which the present state of affairs (H_6) is to be conducted. The ability of the system to cope with this fact is recognized in the two-part homeostat H_8 which produces an interaction between what is now going on and the policy of the future.
>
> . . . The system so far is robust and not easily upset. Yet if there is real trouble, its very robustness will make it a poor adaptation machine. So the whole system is enclosed by an algedonic loop, which can effectively short-circuit the total machinery. This will guarantee speedy reaction to pleasure and pain (cashing-in on the market and crisis respectively, without damaging the routine, self-organizing, self-regulating control procedures). This loop is the channel defining the final homeostat, H_9, by which the policy being promulgated through the system is allowed to impinge directly on the world situation through other channels such as an announcement to the Press about the intentions of the enterprise and the reaction of the world situation is directed straight back to the policy.[10]

These are, then, the theoretical foundations of the application of management cybernetics to the management of the whole enterprise. These foundations appear complex. However, in reality they are simple. Their simplicity is the result of the operation of the inexorable law of "continuous and automatic feedback control" which has been explained in Chapter 11.

---

[10] Ibid., p. 400.

MANAGEMENT CYBERNETICS AND CITY GOVERNMENT

By way of analogy, the following figures illustrate feedback control systems of governmental operations.[11]

Here the government of New York City is depicted in terms of the block diagram. While one may say that such treatment is too vague to be beneficial, an examination of the feedback process throws some light on the usefulness of the cybernetic treatment. Each of the city offices in effect has objectives, and both the desirability and attainment of these objectives are modified by the feedback. The following constitute sources of feedback for a city government:

1. Direct observation by the mayor.
2. Information provided by subordinates.
3. The press.
4. Public officials.

FIGURE 12–10
City Government as a Cybernetic System

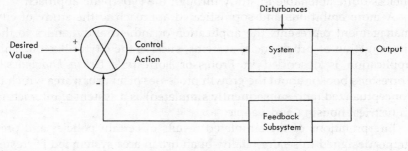

A Conventional Feedback Control System Diagram

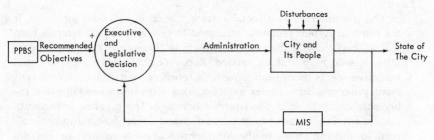

Source: Adapted from E. S. Savas, "City Hall and Cybernetics," in Edmond M. Dewan (ed.), *Cybernetics and the Management of Large Systems*, American Society for Cybernetics (New York: Spartan Books, 1969), pp. 134–35.

---

[11] E. S. Savas, "City Halls and Cybernetics," in E. M. Dewan (ed.), *Cybernetics and the Management of Large Systems*, Second Annual Symposium of American Society of Cybernetics (New York: Spartan Books, 1969), pp. 134–35.

5. Public at large.
   a. Vocal individuals.
   b. Special interest groups.
   c. Elections.
   d. Civil disorders.[12]

All of the above are signals, and as such constitute information, even though it may be biased and distorted. When the voters turn down a bond issue, this is feedback; when riots occur in the streets, this is feedback; when officials are defeated in elections, this is feedback. It may be true that a system such as that of a city government lacks quantification because of its extreme complexity. Still, this is the situation as it exists, and it is precisely this type of problem that can benefit from the cybernetic treatment. City government with its numerous vested interests, at times conflicting objectives, subjected to predictable and unpredictable constraints, reveals its full complexity when studied in such a manner. Regardless of the fact that such an operation defies complete identification of its numerous interrelationships, it is nevertheless quite amenable to study through the cybernetic approach.

A more ambitious and sophisticated approach to the study of city management represents the application of industrial dynamics to the investigation of urban social systems, such as the city of Boston. This application is described in Professor Forrester's *Urban Dynamics*.[13] Forrester's book is about the growth processes of an urban area which is conceptualized (and subsequently simulated) as a system of interacting industries, housing and people.

Interpretations of the simulated results of certain policies and programs designed to secure vitality of an urban area system led Forrester to the following conclusions:

> The city emerges as a social system that creates its own problems. If the internal system remains structured to generate blight, external help will probably fail. If the internal system is changed in the proper way, little outside help will be needed. Recovery through changed internal incentive seems more promising than recovery by direct-action government programs. In complex systems, long-term improvement often inherently conflicts with short-term advantage. The greatest uncertainty for the city is whether or not education and urban leadership can succeed in shifting stress to the long-term actions necessary for internal revitalization and away from efforts for quick results that eventually make conditions worse. Political pressure from outside to help the city

---

[12] Ibid., pp. 137–38. See also, *Bulletin*, TIMS, vol. 2, no. 4, August 1972.
[13] Jay Forrester, *Urban Dynamics* (Cambridge, Mass.: MIT Press, 1969), p. 1.

emphasize long-term, self-regulating recovery may be far more important than financial assistance.[14]

The relevance of urban dynamics cannot really be overemphasized. It is precisely even more pertinent today, when even the most conservative television commentators advocate limiting the growth of cities as a solution to every urban problem (transportation, crime, poverty, and so forth). It therefore behooves the student and manager of organizations to take a closer look at this significant contribution to the contemporary struggle for the survival of our decaying cities.

MANAGEMENT CYBERNETICS IN POLITICAL LIFE

One cannot but marvel at the fascinating progress which has been made toward the development of a theory of politics, both national and international. Most experts in the field concede that the intellectual models set forth by Karl Deutsch and David Easton represent key milestones in these developments.

Karl Deutsch's *The Nerves of Government: Models of Political Communication and Control* is indeed a classic example of cybernetic application to the study and management of government. He states in the preface:

> In the main, these pages offer notions, propositions, and models from the philosophy of science, and specifically from the theory of communication and control—often called by Norbert Wiener's term "cybernetics" —in the hope that these may prove relevant to the study of politics, and suggestive and useful in the eventual development of a body of political theory that will be more adequate—or less inadequate—to the problems of the later decades of the twentieth century.

Here again the basic principle of cybernetics, that communication and control are the two sides of the same coin, is evident in the following statement by Deutsch.

> This book concerns itself less with the bones and muscles of the body of politic than with its nerves—its channels of communication . . . [it] suggests that it might be profitable to look upon government somewhat less as a problem of power and somewhat more as a problem of steering; and it tries to show that steering is decisively a matter of communication.[15]

After presenting an historical account of some conventional models for society and politics, the basics of cybernetics, and the role of com-

---

[14] Ibid., p. 9.
[15] Karl Deutsch, *The Nerves of Government* (New York: The Free Press, 1966).

munication models and political decision systems, Deutsch offers a crude model of control and communication in foreign policy decisions. The model represents a complete account of the main information flows which are necessary in making foreign policy decisions. Suffice it to say that information-scanning systems, both foreign and domestic, constitute inputs to the system and that internal and external policies constitute the major outputs of the system. Of particular interest is the transformation process. Deutsch here shows the integration of all three types of feedback systems, which were discussed earlier—first-, second-, and third-order systems. The interested student is referred to the original source for a fuller explanation of this example.

Another application of cybernetics to the political arena is depicted in Figure 12–11 below. Here the demands of the system provide the basic ingredients for the inputs.[16] Responsible authorities (officials) are constantly converting the demands of the populace (raw materials) into some form of suitable output.

In this third-order feedback system, with the feedback loop running from the outputs to the total environment, communications are represented by solid lines connecting the environment with the political system; the arrows indicate the direction of the flow of information in the system; the broken lines indicate that the environments are changing as a result of the outputs.

In the box labeled "political system," one notes that the authorities acquire information about the consequences of previous actions. This information is then taken into account in the formulation of new goals. In reality, there is a constant interchange of information between the officials and the total environment. Indeed, without it no system could survive. Every system must be able to adapt to the threats and opportunities present in the environment, and, in order to learn of these, it must acquire this information through some feedback process.

A complete analysis of this political system would have to include a determination of the interactions of the system elements, the range of sensitivity of changes in these elements, and a determination of the processes needed to take advantage of the opportunities as well as of the processes needed to deal with the threats to the system.

Furthermore, one can readily see even from this rudimentary illustration that the number of systems which affect the political system are indeed many. In fact, the political system is but a subsystem of yet a larger system. Every politician knows that the political system is linked inextricably to the economic system. Every four years the intricate ramifications of the political system with economics, with labor, with

[16] David Easton, *A Systems Analysis of Politcal Life* (New York: John Wiley & Sons, Inc., 1965), pp. 17–35.

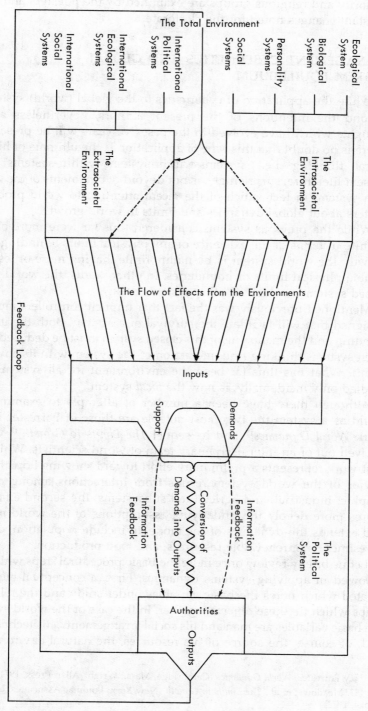

FIGURE 12-11
A Dynamic Response Model of a Political System

The Total Environment

The Intrasocietal Environment
- Ecological System
- Biological System
- Personality Systems
- Social Systems

The Extrasocietal Environment
- International Political Systems
- International Ecological Systems
- International Social Systems

The Flow of Effects from the Environments

Inputs
- Support
- Demands

The Political System

Conversion of Demands into Outputs

Information Feedback

Information Feedback

Authorities

Outputs

Feedback Loop

Source: Adapted from David Easton, *A Systems Analysis of Political Life* (New York: John Wiley & Sons, Inc., 1965), by permission of the publisher.

minority and religious groups are examined by the pollsters and their constant changes noted for the electorate.

MANAGEMENT CYBERNETICS AND GLOBAL SYSTEM EQUILIBRIUM

While the application of cybernetics to the global (world) system is beyond the intentions of the present authors, nevertheless a few thoughts which have evolved in the past five years will be presented. There is no doubt that this type of application is the ultimate of hierarchical thinking—i.e., systems-within-systems-within-systems. The present literature, surprisingly, is not devoid of treatments of the world as a system, in fact, much of the recent attention to world problems centers about global dynamics and limits to world growth.

While the previous systems considered thus far were more or less infinite in terms of the existence of any physical limits bounding their growth, the world system is by nature finite, having more or less definable physical limits or boundaries. In other words, the world is *the* closed system in which all other systems exist.

Mentioned previously was the fact that implicit controllers must be designed that will facilitate the survival of a system. Both the understanding and the managing of this closed system must be deduced from open systems thinking (the interaction of the system with its environment). What has thus far been the environment for all systems and studied only incidentally, is now the *focal* system.

Although there have been a number of attempts to examine the world as a system the two most notable are those of Forrester in his work *World Dynamics*[17] and its sequel *The Limits to Growth*,[18] which evolved out of an interdisciplinary team of world scientists. While the first work represents a preliminary effort toward showing how the behavior of the world system results from interactions among demographic, industrial, and agricultural subsystems, the second effort explores more deeply the underlying assumptions of the world model, and extends the dynamics of the model to include population, capital investment, nonrenewable resources, and food production.

Let us briefly review once more the basic procedural steps which are followed in applying systems dynamics: First, a conceptual model is created which notes the basic variables under study and the relationships which tie these objects together. In the case of the world system, the basic variables are man and his social arrangements, his technology, and, of course, the source of his resources, the natural environment.

---

[17] Jay Forrester, *World Dynamics* (Cambridge, Mass.: Wright-Allen Press, 1971).

[18] D. Meadows, et al., *The Limits to Growth* (New York: Potomac Associates, Universe Books, 1972).

Second, a mathematical model of the basic interactions is constructed, showing the systems behavior at a certain point of time. Through this model, the systems dynamics researcher gains knowledge about the consequences and the theoretical inconsistencies of the assumptions underlying the general model.

Third, the mathematical model is experimented with and changes in the basic parameters and/or relationships of the model are induced either through the computer or through man.

Finally, interpretations of the dynamic behavior of the model lead to certain recommendations for the improvement of both the theoretical conception of the real phenomenon and the level of abstraction of the modeling process.

The viewpoint advocated by world dynamics is that of a long-term, global perspective. Within this framework the world system is conceptualized as consisting of the interaction of five basic parameters/processes:

I. Population
II. Natural Resources
III. Industrial Production
IV. Agricultural Production
V. Pollution

Based upon the conceptualization of these five subsystems, the analysis followed two interrelated steps:

Step 1. Identification of the variable (subsystem) whose behavior limits the growth of the remaining variables (subsystem) and consequently of the whole system; and,

Step 2. Assessment of possible outcomes resulting from the elimination of the limiting variable.

In this manner, all combinations of the five basic variables are examined. A few of the results follow:

The global system will collapse because of limited resources.

Given "unlimited resources," pollution will limit population and economic growth.

Given "unlimited resources" and "pollution control," land becomes the limiting factor.

In short, given the finite ability of the global system to support pollution, and exploitation of resources, the conclusion is reached that exponential growth is an impossibility.

Before offering our own views of the implications of this kind of thinking for the manager of an enterprise, we briefly present the world dynamics group's recommendations or prescriptions for survival of the spaceship earth. It must be emphasized at the very outset that these prescriptions are unorthodox, and alarming.

The world model advocated by the systems dynamics group must satisfy two basic conditions: ultrastability and effectiveness. In the group's own words: "We are searching for a model output that represents the world systems that is:

"1. sustainable without sudden and uncontrollable collapse; and

"2. capable of satisfying the basic material requirements of all of its people."[19]

For the world system to be able to perform these tasks, it must necessarily be in a state of global equilibrium. The group has also set forth the necessary conditions for a global equilibrium.

1. *The capital plant and the population are constant in size.* The birth rate equals the death rate, and the capital investment rate equals the depreciation rate.

2. *All input and output rates—births, deaths, investments, and depreciation—are kept to a minimum.*

3. *The levels of capital and population and the ratio of the two are set in accordance with the values of the society.* They may be deliberately revised and slowly adjusted as the advance of technology creates new options.

An equilibrium defined in this way does not mean stagnation. Within the first two guidelines above, corporations could expand or fail, local populations could increase or decrease, income could become more or less evenly distributed. Technological advance would permit the services provided by a constant stock of capital to increase slowly. Within the third guideline, any country could change its average standard of living by altering the balance between its population and its capital. Furthermore, a society could adjust to changing internal or external factors by raising or lowering the population or capital stocks, or both, slowly and in a controlled fashion, with a predetermined goal in mind. The three points above define a *dynamic* equilibrium, which need not and probably would not "freeze" the world into the population-capital configuration that happens to exist at the present time. The object in accepting the above three statements is to create freedom for society, not to impose a straitjacket.[20]

While the results of this study have come under criticism from many who regard the authors as alarmists and doomsday advocates, it raises many serious questions for students of organizational practices. For example, it has been stressed that an organization must grow if it is to survive, and, yet the MIT group study can easily be labeled as an anti-

[19] Forrester, *World Dynamics*, pp. 151–52.
[20] Meadows, *The Limits to Growth*, pp. 173–75.

growth model. This is intuitively obvious if one accepts the first condition of global equilibrium, which is that capital investment have the same parity as depreciation and that a zero population growth prevail.

Again, the interested student is encouraged to read the original sources for the full implications of this study. The study of world dynamics was introduced here again to show the applicability of the systems approach to problems of extreme complexity.

SUMMARY AND CONCLUSIONS

This chapter has been concerned with the review of numerous attempts aimed at applying systems thinking to real-life problems. The authors have attempted to report on several ingenious frameworks devised by serious and inquisitive men whose main objective has been to obtain a better grasp of the world. This exposition obviously reflects our own biases. Furthermore, it is highly selective and condensed. Henry Kissinger's aphorism, "An apple grower would tell Newton he did not tell everything there is to be known about the apple," applies here as well. Certainly Wiener, Beer, Deutsch, Forrester, the Meadows group and others, would argue that the present authors did not tell everything there is to be known about cybernetics and systems dynamics.

The basic ideas in this chapter can best be summarized in the sentence: *The behavior of organic systems must, by necessity, be homeostatic.* Homeostatic control is a continuous automatic corrective action resulting from a continuous and automatic sensing and comparing of the system's output. As such, homeostatic control is an integral part of system behavior. That is to say, it is an implicit control.

In a natural system, this implicit control is designed by nature into the system's basic structure and function. In artificial or man-made mechanical and social systems, these controls must be rationally designed by either the creators or the managers of them, or by both.

The complexity and sophistication of the design of implicit controllers for man-created systems vary in relation to the complexity and sophistication of the operating system whose behavior is to be controlled. This is, of course, true for natural systems as well. The law of requisite variety dictates that variety can only be dealt with through variety.

In natural systems, no matter what their complexity, all three functions of implicit control (i.e., sensing, comparing, and correcting) are performed in a semi-autonomous fashion which parallels the goal-directed function of the operating system. To that extent, no extra devices are required. For example, man's physiology, the most complex natural system of all, requires no extra extensions of its sensory or motor control devices. All these devices are sufficiently complex to handle the variety inherent in the main functions of the human body.

Unfortunately, the same statement cannot be made regarding man's social nature. Here the magnitude and rate of change of the variety generated by man-to-man interaction is considerably more than man's sensory and motor devices can handle. This is, indeed, the basic thesis of such popular clichés as, for example, the "overloaded, overstimulated individual," the hero or victim of Alvin Toffler's *Future Shock*.[21] Here man's sensory and motor skills must be supplemented by man-made (artificial) devices which act as an extension of man's insufficient social-variety—handling mechanisms.

Man's role in organizations or enterprises falls under the latter category. This role is essentially sociological in nature. The organizational design of implicit controllers must supplement man's inborn or innate control devices.

These last three decades of the 20th century are destined to be the Age of Cybernation, characterized by the conscious effort toward designing complex social enterprises equipped with implicit governors which guarantee long-range survival. Management cybernetics with its emphasis on systems thinking and information technology is charged with this ultimate task of global human existence.

REVIEW FOR CHAPTER TWELVE

Class Project

The popular book by Meadows and associates, *The Limits to Growth* (New York: Potomac Associates, Universe Books, 1972), treats the world as a global, closed system subject to many constraints (limits). The book has been received with mixed enthusiasm. On the one hand, it is applauded as the most important single work within the decade and has had a profound impact on scientists as well as nontechnical personnel. On the other hand, it has been criticized as sheer nonsense and derided for creating an alarmist's doomsday-hysteria.

In treating the world from a systems approach, the study team attempts to identify the dominant elements and the interactions that influence the long-term behavior of world systems. The book explicitly and implicitly treats many of the concepts advanced in this text.

Required. Read *The Limits to Growth,* set up discussion groups, and discuss the application of both the systems concept and cybernetic principles to the model of the world system there presented.

---

[21] A. Toffler, *Future Shock* (New York: Bantam Books, 1970).

Glossary

Act A system event for the occurrence of which a change in the system or the environment is neither necessary nor sufficient.

Activity A work effort, of the project that must be done, which consumes time, money, resources, or facilities. An activity connects two events in a project.

Adaptability of Systems The ability of systems to respond appropriately to their environmental states.

ALGOL A programming language developed in Europe, primarily designed for solving scientific and mathematical problems.

APT A programming language for describing functions of numerically controlled machinery.

Arithmetic-Logic Unit Unit where the actual processing of data is done.

Assembler A computer program that translates symbolic language into machine language.

Attributes Properties of objects or relationships that manifest the way something is known, observed, or introduced in a process.

Auxiliary Devices Offline storage and input/output devices which can be used to transmit data from the user to the machine, to convey the results of data processing from the machine to the user, and, finally, to supplement internal core storage.

BASIC A simple programming language developed specifically for use in instruction, allowing the user direct interaction with the computer.

Black Box Technique The study of the relations between the experimenter and the object as well as the study of what information comes from the object, and how it is obtained.

Block Diagram A basic schematic tool for illustrating functionally the components of a control system. Four basic symbols are employed: the arrow, the block, the transfer function, and the circle.

Boundaries of a System The lines forming closed circles around selected vari-

347

ables, where there is less interchange of energy and/or information across the lines of the circles than within the delimiting circles.

Classical Organization Theory   This early theory was based on two major themes: the use of men as adjuncts to machines in the performance of routine productive tasks, and the formal structure of the organization.

COBOL   A programming language commonly used in accounting and business applications.

Compiler   An intermediate program that translates a problem-oriented language into machine language.

Computer System   A system made up of three basic subsystems:
1.   The input/output subsystem,
2.   The central processing subsystem, and
3.   The auxiliary storage subsystem.

Control Unit   It contains the circuitry requirements to direct and integrate all units of the total computer system; it performs the same function as the feedback element of a system.

Critical Path   One path composed of activities, within the network of a project, which, if delayed, would affect the expected completion date of the project. This path is also the most time-consuming (longest) path of activities within the network.

Cybernetic System   A system characterized by extreme complexity, probabilism, and self-regulation.

Cybernetics   The science of control and communication in the animal and in the machine.

Decision-Making Process   A process comprising three principal phases:
1.   Finding occasion for making a decision,
2   Finding possible courses of action, and
3.   Choosing among courses of action.

Decision System   A subsystem of a goal-seeking system, serving as a channel through which the goal-seeking system acts on the environment.

Dummy Activity   An activity that requires neither time nor resources, used to preserve the uniqueness of the activity-event numbering system or to preserve a unique predecessor-successor relationship among activities, which cannot otherwise be achieved.

Earliest Expected Date   The earliest calendar date on which an event can be expected to take place. It is the summation of all activity times through the most time-consuming chain of events, from the beginning event to the given event.

Enterprise-Environment Interaction System   An open system which functions by importing the necessary resources and/or information from the environment and by exporting the product of the combinations of these resources into the environment.

Environment of a System   That which not only lies outside the system's control but which also determines in some way how the system performs.

Environmental Factors   A set of measurable properties of the environment perceived directly or indirectly by the organization operating in the environment.

Environment-Organization Interactive System In systems terminology, it is the superordinate system—it is the whole.

Event A specific definable accomplishment in a program plan, recognizable at a particular instant in time. It may represent the beginning or ending of an activity.

Feedback Control System A system which tends to maintain a prescribed relationship of one system variable to another by comparing functions of these variables and using the difference as a means of control.

Feedback Input A portion of the system's output that is reintroduced as an input into the same system for ascertaining the difference between the goal (desired state of affairs) and the actual performance.

First Order Feedback System A single goal maintenance system.

Flow Diagram A graphical representation of the sequence of data transformations needed to produce an output data structure from an input data structure.

FORTRAN A programming language designed for solving scientific problems by mathematical formulation.

Gantt Chart A scheduling system developed in the early 1900s which preceded and influenced the PERT method. The chart lists scheduled start and finish dates, and time durations for each of the tasks or activities to be completed. It is plotted to show corresponding progress.

General Systems Theory The theory of open-organic systems which possess certain characteristics such as organization, dynamic equilibrium, self-regulation, and teleology. Its main domain is growth and evolution of general systems. It evolved from Bertalanffy's concept of "organismic revolution."

Goal-Seeking Systems Systems that can so modify their output through a feedback mechanism that they tend toward a pre-set state or goal.

GPSS A programming language designed especially for simulation.

Hardware A symbol-receiving, -processing, and -communication device. Mechanical devices that augment man's sensory and intelligence capacities.

Hierarchic System A system composed of interrelated subsystems, all of which are ranked or ordered such that each is subordinate to the one above it, until the lowest elementary subsystem level is reached.

Ideal-Seeking System A system which, upon attainment of any of its goals or objectives, seeks another goal or objective which more closely approximates its ideal.

Input A start-up force or signal that provides the system with its operating necessities. Or, it can be a stimulus or excitation applied to a system from an external source eliciting a response from the system. Or the importation of data and instructions which translate the external form into a set of symbols that can be read and interpreted by the computer's electronic circuitry.

Internal Organizing System   A subsystem of a goal-seeking system, which acts as an evaluator of the scanning system.

Internal Storage   Memory unit within the computer system which receives program instructions and the data to be processed.

Isomorphic Systems   Two systems whose elements exist in a one-to-one correspondence with each other. There is also a correspondence between the systems' operational characteristics.

Latest Allowable Date   The latest calendar date on which an event can take place without delaying the completion of the project. It is calculated by subtracting the sum of the activity times from the activities on the path between the given event and the end event of the program.

Law of Requisite Variety   There must be as many actions available to the system's controller as there are states in the system.

Management Information System   An information network designed to provide the right information to the right person at the right time.

Matrix Organization   The combination of project organization and functional organization where a pool of specialists is assigned to particular projects for the duration of the project. These personnel are subject to the horizontal authority of the project manager as well as the vertical, functional authority.

Metatechnology   The conceptualization, design, and implementation of ways of organizing man and machine into systems for the collection, storage, processing, dissemination, and use of information.

Milestone Chart   A system quite similar to the Gantt chart, except that the tasks are broken down into subtasks termed milestones. The major objective is to more closely relate task accomplishment with the time schedule.

Mismatch Signal   Conveys to the organizing system feedback information about changes in the external environment of a system with one known intended state and one known actual state.

Modern Organization Theory   Theory characterized by three parallel developments:
1. The continuation of earlier classical and neoclassical theory;
2. The emergence of behavioral science research; and
3. The emergence of operations research.

Modern Systems Approach   The view that organizations are systems which are in constant interaction with their environment.

Neoclassical Organization Theory   This theory accepts the basic postulates of the classical theory but modifies them by superimposing changes in operating methods and structure evoked by individual behavior and the influence of the informal group.

Nesting of Systems   The division and subdivision of a system into subsystems and subsubsystems depending on the particular resolution level desired.

Objects   The inputs, processes, outputs, and feedback control of a system.

Open Loop System   A system in which the output of the system is not coupled to the input for measurement.

Organization   A relative concept, depending upon the relation between the real thing (the organized complexity), the environment, and the observer.

Organized Complexity   Interactions within subsystems and interactions among subsystems, which, viewed as a hierarchy, can be described as a series of feedback loops arranged in an ascending order of complexity.

Output   The result of the process, or alternatively, the purpose for which the system exists.

Parameters   Elements outside a designated system which have an effect on one or more variables of the designated system.

PERT   A variation of network analysis used in planning, controlling, and evaluating the progress of a project. First developed and utilized for managing the Fleet Ballistic Missile program in 1958. The PERT approach is time- versus task-oriented.

PERT Network   A flow diagram consisting of activities which must be accomplished to reach the project objective. It separates the planning phase from the scheduling phase, thereby allowing consideration of alternative sources of action.

PL/1   A general-purpose programming language, combining most of the features of the special languages developed for scientific, business, simulation, and command-control applications.

Process   The manner of combining the inputs so that the system will achieve a certain result.

Programming   Analysis in exhaustive detail of data transactions, by developing outlines of the execution procedure in the form of flow charts and symbolic language.

Programming Language   Language that is used in writing the series of instructions that direct the computer to perform a specific series of operations.

Project Manager Concept   A systems approach whereby one individual, the project manager, has full authority and responsibility over all people and activities within the project. This horizontal project organization implies extension of the project manager's authority over all, within and outside the organization, involved in the project.

Random Input   Outputs from previous systems which are potential inputs to the focal system.

Reaction   A change in a system's structural properties caused by another change in a more or less deterministic manner.

Redundant Relationships   Those relationships that duplicate other relationships. They increase the probability that a system will operate all of the time and not just some of the time.

Relationships   The bonds that link the objects together.

Response   A system event for which another event that occurs in the same system or in the environment is necessary but not sufficient.

Scanning   The process whereby the organization acquires information for decision making.

Scanning System   A subsystem of a goal-seeking system, serving as a channel through which the goal-seeking system receives information about the environment.

Search   A mode of scanning which aims at finding a particular piece of information for solving a specific problem.

Second-Order Feedback System   A system which contains a memory unit and can initiate alternative courses of action in response to changed external conditions and can choose the best alternative for the particular set of conditions.

Serial Input   The result of a previous system with which the focal system is serially or directly related.

Slack   The difference between the latest allowable date and the earlist expected date of an event's completion.

SNOBOL   A programming language that can be used to manipulate information in the form of natural language.

Software   A set of computer programs, procedures, and possibly associated documentation concerned with the operation of a data processing system, e.g., compilers, library routines, manuals, circuit diagrams. Contrasted with hardware.

State of a System   It is the set of relevant processes in a system at a given moment of time.

Surveillance   A mode of scanning which aims at finding some general knowledge for the information seeker.

Symbiotic Relationship   One without which the connected systems cannot continue to function.

Synergistic Relationship   That in which the cooperative action of semi-independent subsystems taken together produces a total output greater than the sums of their outputs taken independently.

System   A set of objects, together with relationships between the objects and between their attributes, connected or related to each other and to their environment in such a manner as to form an entirety or whole.

Systems Analysis   The organized step-by-step study of the detailed procedures for the collection, manipulation, and evaluation of data about an organization for the purpose of determining not only what must be done but also to ascertain the best way to improve the functioning of the system.

System Approach   A philosophy that conceives of an enterprise as a set of objects with a given set of relationships between the objects and their attributes, connected or related to each other and to their environment in such a way as to form a whole or entirety.

System's Behavior   A series of changes in one or more structural properties of the system or of its environment.

Systems Chart   Chart that focuses on the inputs and the outputs of the system; it identifies programs, procedures and data structures by name.

Systems Thinking   Its objective is to reverse the subdivision of the sciences into smaller and more highly specialized disciplines through an inter-disciplinary synthesis of existing scientific knowledge.

Third-Order Feedback System A system similar to the second-order system with the added ability of being able to reflect upon its past decision making.

Time Estimate The most likely time, expressed in months, weeks, fractions of weeks, or days, that a task will consume.

Variables Elements within a designated system.

Bibliography
on Systems

Ackoff, R. L. "Toward a System of Systems Concepts." In *Management Science,* vol. 17, no. 11, July 1971.
Attempts to rigorously define basic concepts and terms commonly used to describe systems.

_____, and **Emery, F. E.** *On Purposeful Systems* (Chicago: Aldine Publishing Co., 1972).

Angyal, A. *Foundations for a Science of Personality* (Cambridge, Mass.: Harvard University Press, 1941).
One of the earliest attempts to develop a theory of personality based upon concepts of systems theory. Chapter 8, "A Logic of Systems," contains a good comparison of analytic and systems thinking.

Ashby, W. Ross. *Design for a Brain.* 2d ed. (London: Chapman and Hall, 1960).
Develops a logic of mechanism and applies it to the behavior of organisms. Interesting for its treatment of dynamic systems and such concepts as stability and homeostatis.

_____. *An Introduction to Cybernetics* (London: Chapman and Hall, 1956).
Step-by-step guide to basic principles, with practical exercises.

Baker, Frank. *Organizational Systems: General Systems Approaches to Complex Organizations* (Homewood, Ill. Richard D. Irwin, Inc., 1973).
An excellent collection of readings on the subject of the application of general systems theory to the study and management of complex enterprises with particular emphasis on hospitals. Includes the classic writings of Ackoff, Hagen, Emery and Trist, James Miller, J. D. Thompson, A. K. Rice, and others.

Beer, Stafford. *Cybernetics and Management.* Science Edition, vol. 5 (New York: John Wiley & Sons, Inc., 1959).

_____. *Management Science* (New York: Doubleday and Co., Inc., 1968).

_____. *Decision and Control* (New York: John Wiley & Sons, Inc., 1967).

Beer, Stafford. "Managing Modern Complexity." In *The Management of Information and Knowledge.* Committee on Science and Astronautics, U.S. House of Representatives, 1970.

———. "Below the Twilight Arch." in *Yearbook of General Systems,* vol. 5, 1960.
All Beer's works are imaginative attempts to relate or translate cybernetics to management. His *Management Science* book is the most elementary, while his *Cybernetics and Management* is the most advanced. Both books are "musts" for every student of systems.

Berrien, F. Kenneth. *General and Social Systems* (New Brunswick, N.J.: Rutgers University Press, 1968).
Traces development of general systems thinking and provides a new viewpoint for behavioral scientists interested in group phenomena. Introductory chapters provide a reliable account of the development of major general systems theory concepts.

Bertalanffy, Ludwig von. *General Systems Theory: Foundations, Development, Applications* (New York: Braziller, 1968).
A collection of previous writings on the subject, revised and brought up to date with some additional material, by one of the original proponents of a general systems theory.

———. *Problems of Life* (New York: John Wiley & Sons, Inc., 1952).

———. *Robots, Men and Minds* (New York: Braziller, 1967).
A short treatise on psychology for the modern world using general systems concepts.

Boulding, Kenneth E. "General Systems Theory: A Skeleton of Science." In P. P. Schoderbek (ed.), *Management Systems* 2d ed. (John Wiley & Sons, Inc., 1971).

———. *The Image* (Ann Arbor, Mich.: University of Michigan Press, 1956).

———. *The Meaning of the 20th Century* (New York: Harper & Row, 1964).
K. Boulding, one of the original cofounders of general system theory—along with Bertalanffy, Rapoport, and Gerard—is the first economist who attempted to relate GST to economics, management, and sociology. His classic article, "GST—A Skeleton of Science," is a must.

Buckley, Walter, ed. *Modern Systems Research for the Behavioral Scientist* (Chicago: Aldine Publishing Company, 1968).
Collection of articles on general systems research, cybernetics, and information and communication theory, with a focus on recent systems research in behavioral science. Minimum of mathematics.

———. *Sociology and Modern Systems Theory* (Englewood Cliffs, N.J.: Prentice-Hall, 1967).
Principles, ideas, and insights of general systems theory applied to modern social science.

Cannon, Walter B. *The Wisdom of the Body* (New York: W. W. Norton and Company, 1939).
Contains author's original exposition of the concept of *homeostatis.*

Churchman, C. West. *The Systems Approach* (New York: Delacorte Press, 1968).

_____. *Challenge to Reason* (New York: McGraw-Hill Book Co., 1968).

_____. "On Whole Systems: The Anatomy of Teleology" (Space Science Laboratory, University of California at Berkeley, August 1968).

_____. "Operations Research as a Profession," *Management Science,* 17, no. 2, October 1970.

C. West Churchman, a pioneer in OR, is a leading scholar in systems. Ph.D. students in management and quantitative methods will find his *Challenge to Reason* and *OR as a Profession* extremely interesting. DPA students will enjoy his Systems Approach.

Crosson, Frederick J., and **Sayre, Kenneth M.,** eds. *Philosophy and Cybernetics* (New York: Simon and Schuster, 1967).

Analyzes current state and future implications of cybernetics; identifies the philosophic issues.

Dechert, Charles R., ed. *The Social Impact of Cybernetics* (New York: Simon and Schuster, 1966).

An analysis of the social implications of the Second Industrial Revolution — the effects of culture and future social development.

Dewan, Edmond M. *Cybernetics and the Management of Large Systems* (New York: Spartan Books, 1969).

Proceedings of the Second Annual Symposium of the American Society for Cybernetics. The papers presented at this meeting fall into three categories: (1) papers of a somewhat philosophical nature that indicate general directions to solutions of certain problems and point out some problems that have been overlooked in the past; (2) papers dealing with the future of large-scale computer systems; (3) papers dealing with the direct application of cybernetics or systems techniques to such important real problems as: government, transportation, power distribution, and control of weather and pollution.

Eckman, Donald P., ed. *Systems: Research and Design* (New York: John Wiley & Sons, Inc., 1961).

Proceedings of the first systems symposium at Case Institute of Technology. Oriented to operations research and systems engineering viewpoints.

Ellis, David O., and **Ludwig, Fred J.** *Systems Philosophy* (Englewood Cliffs, N.J.: Prentice-Hall, 1962).

Discussion of major key points and trends in systems technology and systems engineering, with examples from airborne control systems.

Emery, F. E. *Systems Thinking* (Baltimore, Md.: Penguin Modern Management Readings, 1970).

A collection of classic readings about the origins of systems thinking and its relation to organizational management.

Foerster, Heinz von, ed. *Cybernetics* (New York: Macy, 1953).

_____, et al. *Purposive Systems* (New York: Spartan Books, 1968). Proceedings of the 1967 Conference of the American Society for Cybernetics (first meeting). Looks at the potential role of cybernetics in solving technical and social problems.

Foerster, Heinz von, ed., and **Zopf, George W., Jr.,** eds. *Principles of Self-Organization* (New York: Pergamon, 1962).

Forrester, J. W. *Industrial Dynamics* (Cambridge, Mass.: MIT Press, 1961).

_____. *Urban Dynamics* (Cambridge, Mass.: MIT Press, 1968).

_____. *World Dynamics* (Cambridge, Mass.: Wright-Allen, 1971).

_____. *Principles of Systems* (Cambridge, Mass.: Wright-Allen, 1970).
Forrester's applications of servomechanics concepts and principles to industrial, urban and world equilibrium problems are skillfully and elegantly simulated via the dynamo simulation technique. Familiarity with industrial dynamics is expected of every Ph.D. student. (See also Meadows, D.)

Greniewski, Henry. *Cybernetics Without Mathematics* (New York: Pergamon, 1960).

Grinker, Roy R., Sr., ed. *Toward a Unified Theory of Human Behavior: An Introduction to General Systems Theory,* 2d ed. (New York: Basic Books, 1967). Edited proceedings of four multi-disciplinary conferences. Concentrates on a general theoretical approach towards bringing the diversity of living systems under the umbrella of one unified theory.

Guidbaud, Georges T. *What Is Cybernetics?* (New York: Criterion Books, 1959). Nonmathematical explanation for the layman. Good introduction. Covers control systems, measurement of information, and communication.

Handel, S. *The Electronic Revolution* (Baltimore, Md.: Penguin Books, 1967). An account of the historical developments of electronics and information technology.

Johnson, R. A.; Kast, F. E., and **Rosenzweig, J. E.** *The Theory and Management of Systems* (New York: McGraw-Hill Book Company, 1963, 1967, 1973). A good introductory book which employs some theoretical aspects of the systems approach with traditional quantitative techniques and illustrative case studies.

Katz, D., and **Kahn, R. L.** *The Social Psychology of Organizations* (New York: John Wiley & Sons, Inc., 1966). One of the earliest and well-accepted attempts to relate general systems theory to the study of organizational behavior. A widely used text in ORG theory graduate courses.

Klir, George J. *An Approach to General Systems Theory* (Princeton: Van Nostrand, 1969). A good introductory text to fundamental concepts, problems and methodological principles concerning systems. Highly mathematical.

_____. *Trends in General Systems Theory* (New York: John Wiley & Sons, Inc., 1972). Attempts to integrate the writings on general systems theory by showing the "state of the art" and by predicting future developments. An excellent collection of the writings of the most important originators of various aspects of general systems theory such as Ludwig von Bertalanffy, Anatol Rapoport, W. Ross Asby, Walter Buckley, Mihajlo Mesarovic, and C. West Churchman. A useful source book for students with advanced knowledge of systems theory.

Klir, Jiri and **Valachi, Miraslav.** *Cybernetic Modeling* (Princeton: Van Nostrand, 1967).
Begins with a fundamental discussion of systems, cybernetic systems in particular; continues with a discussion of the application of cybernetic methods to some problems in biology, psychology, linguistics, and man-machine communication.

Koestler, Arthur. *The Ghost in the Machine* (New York: Macmillan, 1967).
Interesting for his discussion of the "holon," and other systems concepts, and for his presentation of a series of propositions concerning the general properties of "open hierarchical systems."

_____, and **Smythies, J. R.** *Beyond Reductionism* (Boston: Beacon Press, 1969).
The end product of the Alpbach Symposium: a conference attended by 15 eminent scientists gathered in the summer of 1968 to refute the mechanistic world view of reductionism—a world view evident not only in the life sciences, but in the political climate as well—and to find a new scientific synthesis that would provide a place for human values in human behavior. One of the best explanations of the origin of systems thinking. Includes papers by Bertalanffy, Bruner, Hayek, Weiss, and others.

Lange, Oskar. *Wholes and Parts: A General Theory of System Behavior* (Oxford: Pergamon Press, 1965).
Highly mathematical. Good short exposition of systems theory.

Langley, L. L. *Homeostatis* (New York: Reinhold Publishing Corp., 1965).
A short popular treatment of homeostatis, emphasizing the biological principles involved.

Laszlo, Ervin. *The Systems View of the World* (New York: George Braziller, 1972).
If one wants to change the world, or at least to make sure that it doesn't head blindly to its own destruction, one must understand the nature of this world. This book attempts to do just that by examining the so-called atomistic view and the systems view. Laszlo subsequently presents the systems view of nature and the systems view of man. Elegantly written by a well-known philosopher.

McKay, D. *Information Mechanism and Meaning* (Cambridge, Mass.: The MIT Press, 1969).
A collection of most of the papers of McKay—a British leading authority on information theory—dealing with the quantitative and semantic aspects of information.

Maurer, John G. *Readings in Organization Theory: Open-Systems Approaches* (New York: Random House, 1971).
A good collection of writings on the subject of systems-oriented theories of organizations. Includes the writings of J. D. Thompson, Emery and Trist, William Evan, and others. A good sourcebook for students interested in open-systems approaches to organization theory.

Meadows, D., et al. *The Limits to Growth* (New York: Universe Books, 1972).
A highly controversial report to the Club of Rome on the predicament of mankind. Application of Forrester's dynamics to universal equilibrium problem. A must for every student.

Mesarovic, Mihajlo D. *Views on General Systems Theory* (New York: John Wiley & Sons, Inc., 1964).
Proceedings of the second systems symposium at Case Institute of Technology.

Optner, Stanford L. *Systems Analysis for Business and Industrial Problem Solving* (Englewood Cliffs, N.J.: Prentice-Hall, Inc., 1965).
A pioneering step toward a system science; provides the principles and framework for relating systems concepts to the analysis and solution of problems.

Rapoport, Anatol. *Operational Philosophy: Integrating Knowledge and Action,* Science Editions (New York: John Wiley & Sons., Inc., 1965).
A sophisticated attempt to provide operational definitions for concepts usually regarded as purely philosphical.

Rose, J. *Automation: Its Anatomy and Physiology* (London: Oliver and Boyd, Ltd., 1967).
An elegant comparison of mechanization (First Industrial Revolution) and automation/cybernation (Second Industrial Revolution). Brief but clever explanation of the principles of feedback/automatic control.

Schoderbek, P. P. *Management System,* 2d ed. (New York: John Wiley & Sons, Inc., 1971).
A sourcebook of excellent articles on the systems concept, management information systems, and system applications. Sections on general systems theory, cybernetics, models and simulation, information technology, computers, total management systems, industrial dynamics, PERT, and real-time systems. Overall tight organization with detailed connective summaries for all major topics.

Scientific American. *Information* (San Francisco: W. H. Freeman and Company, 1966).
A compilation of articles on hardware and software computer technology.

_____. *Automatic Control* (New York: Simon and Schuster, 1955).
A survey of what is currently being done to make it possible for machines to regulate themselves by feedback.

Simon, H. A. *The Sciences of the Artificial* (Cambridge, Mass.: MIT Press, 1970).
Three out of the four chapters in this book are based on Simon's Karl Compton lectures, delivered at MIT in the Spring of 1968. In these lectures, together with his "The Architecture of Complexity," Simon attempts to construct a "Science of Design" to guide today's complex organizations.

Singh, Jaglit. *Great Ideas in Information Theory, Language and Cybernetics* (New York: Dover Publications, Inc., 1966).
Elementary but yet very methodical treatment of the subject of codification of information in both human and machines. Meaningful summaries of the work of McCulloch and Pitts, von Newmann, Turing, Wiener, Uttley, and other pioneers in information technology.

Stanley-Jones, D. and **K.** *The Cybernetics of Natural Systems: A Study in Patterns of Control* (New York: Pergamon, 1960).

A nonmathematical treatment of cybernetics, information theory, and general systems concepts.

Stulman, Julius. *Fields Within Fields Within Fields* (New York: World Institute Council).
Since 1968 The World Institute Council—a nonprofit research and educational institution—has been publishing papers on the systems viewpoint under the direction of J. Stulman. Each issue of the publication is devoted to in-depth application of the World Institute's methodology to social problems. Very original and creative effort. Recommended especially for graduate students.

Thayer, Lee. *Communication and Communication Systems* (Homewood, Ill.: Richard D. Irwin, Inc., 1968).
A systems-oriented treatise on the subject of human communication in contrived organization. Relevant to all students of human organizations.

U.S. House of Representatives. *The Management of Information and Knowledge* (Committee on Science and Astronautics, 1970).
A compilation of papers prepared for the eleventh meeting of the panel on science and technology (Bell, Beer, Kahn, Kozmetsky).

Vickers, Geoffrey. *Value Systems and Social Process* (New York: Basic Books, Inc., 1968).

_____. *Toward a Sociology of Management* (New York: Basic Books, Inc., 1967).

_____. *The Art of Judgment* (New York: Basic Books, Inc., 1965).

_____. *Freedom in a Rocking Boat: Changing Values in an Unstable Society* (London: Penguin Press, 1970).
For the last two decades Sir Geoffrey Vickers has been writing on the subjects of policy and management in terms of general systems theory and cybernetics.

Vonnegut, Kurt. *Player Piano* (New York: Avon Books, 1952).
A clever satire about managers, engineers, and automation.

Weiss, P. A. *Hierarchically Organized Systems* (New York: Hafner Publishing Co., Inc., 1971).
A demonstration of the logical cogency of a systems concept of the world, the rational superiority of its flexibility over the rigidity of the conventional, scientifically outdated, notion that the universe, including man, operates as a microprecise "cause-effect" machine (contributing authors: U. Lorenz, Forrester, and others).

Wiener, Norbert. *Cybernetics: Or Control and Communication in the Animal and the Machine*, 2d ed. (Cambridge, Mass.: MIT Press, 1961).
A study of human control functions and the mechanical and electrical systems designed to replace the human being. Requires some degree of sophistication in mathematics.

_____. *The Human Use of Human Beings: Cybernetics and Society* (Boston: Houghton Mifflin, 1950; New York: Avon, 1967).
An edited and revised version of the classic, *Cybernetics*, addressed to a larger and less specialized audience. An excellent explanation of the logic and basic principles of cybernetics.

Yovits, Marshall G., and **Cameron, Scott.**   *Self-Organizing Systems* (New York: Pergamon, 1960).
Proceedings of a 1959 conference on the evolution and proposed future of artificial self-organizing systems.

_____; **Jacobi, George T.;** and **Goldstein, Gordon D.,** eds.   *Self-Organizing Systems* (New York: Spartan Books, 1962).
Proceedings of a 1962 conference in Chicago on the evolution and proposed future of artificial self-organizing systems.

Indexes

Name Index

365

Ruesch, Jurgen, 146

Subject Index

This book has been set in 10 and 9 point Palatino, leaded 2 points. Part and chapter numbers are 30 point Baskerville italic. Part and chapter titles are 24 point Baskerville. The size of the type page is 27 × 45½ picas.